CAMBRIDGE LIBRARY COLLECTION

Books of enduring scholarly value

Mathematical Sciences

From its pre-historic roots in simple counting to the algorithms powering modern desktop computers, from the genius of Archimedes to the genius of Einstein, advances in mathematical understanding and numerical techniques have been directly responsible for creating the modern world as we know it. This series will provide a library of the most influential publications and writers on mathematics in its broadest sense. As such, it will show not only the deep roots from which modern science and technology have grown, but also the astonishing breadth of application of mathematical techniques in the humanities and social sciences, and in everyday life.

Mathematical and Physical Papers

Sir George Stokes (1819-1903) established the science of hydrodynamics with his law of viscosity describing the velocity of a small sphere through a viscous fluid. He published no books, but was a prolific lecturer and writer of papers for the Royal Society, the British Association for the Advancement of Science, the Victoria Institute and other mathematical and scientific institutions. These collected papers (issued between 1880 and 1905) are therefore the only readily available record of the work of an outstanding and influential mathematician, who was Lucasian Professor of Mathematics in Cambridge for over fifty years, Master of Pembroke College, President of the Royal Society (1885-90), Associate Secretary of the Royal Commission on the University of Cambridge and a Member of Parliament for the University.

Cambridge University Press has long been a pioneer in the reissuing of out-of-print titles from its own backlist, producing digital reprints of books that are still sought after by scholars and students but could not be reprinted economically using traditional technology. The Cambridge Library Collection extends this activity to a wider range of books which are still of importance to researchers and professionals, either for the source material they contain, or as landmarks in the history of their academic discipline.

Drawing from the world-renowned collections in the Cambridge University Library, and guided by the advice of experts in each subject area, Cambridge University Press is using state-of-the-art scanning machines in its own Printing House to capture the content of each book selected for inclusion. The files are processed to give a consistently clear, crisp image, and the books finished to the high quality standard for which the Press is recognised around the world. The latest print-on-demand technology ensures that the books will remain available indefinitely, and that orders for single or multiple copies can quickly be supplied.

The Cambridge Library Collection will bring back to life books of enduring scholarly value (including out-of-copyright works originally issued by other publishers) across a wide range of disciplines in the humanities and social sciences and in science and technology.

Mathematical and Physical Papers

VOLUME 5

GEORGE GABRIEL STOKES

CAMBRIDGE
UNIVERSITY PRESS

CAMBRIDGE UNIVERSITY PRESS

Cambridge, New York, Melbourne, Madrid, Cape Town, Singapore,
São Paolo, Delhi, Dubai, Tokyo

Published in the United States of America by Cambridge University Press, New York

www.cambridge.org
Information on this title: www.cambridge.org/9781108002677

This edition first published 1905
This digitally printed version 2009

ISBN 978-1-108-00267-7 Paperback

MATHEMATICAL

AND

PHYSICAL PAPERS.

CAMBRIDGE UNIVERSITY PRESS WAREHOUSE,
C. F. CLAY, Manager.

London: AVE MARIA LANE, E.C.
Glasgow: 50, WELLINGTON STREET.

Leipzig: F. A. BROCKHAUS.
New York: THE MACMILLAN COMPANY.
Bombay and Calcutta: MACMILLAN AND CO., Ltd.

G. G. Stokes

MATHEMATICAL

AND

PHYSICAL PAPERS

BY THE LATE

Sir GEORGE GABRIEL STOKES, Bart.,

Sc.D., LL.D., D.C.L., Past Pres. R.S.,

KT PRUSSIAN ORDER *POUR LE MÉRITE*, FOR. ASSOC. INSTITUTE OF FRANCE, *ETC.*
MASTER OF PEMBROKE COLLEGE AND LUCASIAN PROFESSOR OF MATHEMATICS
IN THE UNIVERSITY OF CAMBRIDGE.

Reprinted from the Original Journals and Transactions,
with brief Historical Notes and References.

VOL. V.

CAMBRIDGE:

AT THE UNIVERSITY PRESS.

1905

Cambridge:

PRINTED BY JOHN CLAY, M.A.
AT THE UNIVERSITY PRESS.

PREFACE.

THE present volume concludes the reprint of the Scientific Papers of the late Professor Sir George Gabriel Stokes. As in the preceding volume, the brief editorial additions take the form of undated notes enclosed in square brackets.

Various lectures and addresses on scientific subjects remain over, which were considered to be not sufficiently original for inclusion in this collection, in those cases where they are not already represented in other ways. Among them may be mentioned the following :—

Solar Physics, the introductory lecture of a course, delivered in the Theatre of the South Kensington Museum on April 6, 1881, on behalf of the Committee on Solar Research. Reprinted in *Nature*, Vol. XXIV. pp. 595—598 and 613—618.

The Absorption of Light and the Colours of Natural Bodies, and *Fluorescence*; two lectures to science teachers at South Kensington. Pp. 43, crown octavo, Macmillan and Co., 1876.

The Presidential Address to the British Association for the Advancement of Science, Exeter Meeting 1869,—mainly recounting recent progress in Science.

Presidential Addresses at the Anniversary Meetings of the Royal Society, 1886 to 1891 inclusive—mainly concerned with the business of the Royal Society.

Annual addresses on scientific topics delivered as President of the Victoria Institute :—*The Luminiferous Ether*, 1893; *The Perception of Light*, 1895; *The Röntgen Rays*, 1896; *The Perception of Colour*, 1898.

Three volumes of Burnett Lectures delivered at Aberdeen have been published by Macmillan and Co.; *On the Nature of Light*, Nov. 1883; *On Light as a Means of Investigation*, Dec. 1884; and *On the Beneficial Effects of Light*, Nov. 1885. These lectures have also appeared in a German Translation.

The courses of Gifford Lectures on Natural Theology in the University of Edinburgh for the years 1891 and 1893 were delivered by Sir George Stokes, and have been published in two volumes by A. and C. Black; they are in part concerned with the theological aspects of scientific knowledge.

Some preparations have been made for a volume of biographical character, to be occupied in part by a selection from Sir George Stokes' voluminous scientific correspondence, including some unpublished manuscript material. Stress of other engagements has hitherto delayed this undertaking, but it is hoped that substantial progress will soon be made.

It will be observed that the present volume represents the period in which Sir George Stokes' scientific activities were mainly expended in the work of the Royal Society and of public Scientific Committees, and in giving assistance to the investigations of others. The volume thus consists largely of additions and notes originally appended to memoirs by other authors. Considerable care has been taken in making its contents as complete as possible, and in the elucidation of points of ambiguity or difficulty. It is hoped that no important printed contribution has been overlooked. In one case, pp. 146—159, it has been thought right to insert an unpublished manuscript of considerable length, on the forms of water-waves of finite height, in what appeared to be its natural sequence.

Acting on the advice of high authorities, the Smith's Prize Examination Papers set by Sir George Stokes in the University of Cambridge for a long series of years, and also some early papers of Problems set in the Mathematical Tripos, have been reprinted in an Appendix.

The biographical sketch prefixed to this volume was contributed by Lord Rayleigh to the Obituary of the Royal Society. The photographic portrait which illustrated it, taken about the year 1892, has again been made available by kind permission of Mrs F. W. H. Myers.

If this collection serves to bring into due prominence a series of memoirs which have at no time been sufficiently widely known, although some of the greatest of British physicists have traced to them much of their inspiration, any trouble that has been incurred in collecting and editing the final volumes will have been amply rewarded.

                                            J. LARMOR.

St John's College, Cambridge.
     March 20, 1905.

CONTENTS.

OBITUARY NOTICE.

[By Lord Rayleigh. From the *Proceedings of the Royal Society*, 1903.]

Sir GEORGE GABRIEL STOKES, Bart. 1819—1903.

In common with so many distinguished men Sir George Stokes was the son of a clergyman. His father, Gabriel Stokes, who was Rector of Skreen, County Sligo, married Elizabeth Haughton, and by her had eight children, of whom George was the youngest. The family can be traced back to Gabriel Stokes, born 1680, a well-known engineer in Dublin and Deputy Surveyor General of Ireland, who wrote a treatise on Hydrostatics and designed the Pigeon House Wall in Dublin Harbour. This Gabriel Stokes married Elizabeth King in 1711 and among his descendants in collateral branches there are several mathematicians, a Regius Professor of Greek, two Regius Professors of Medicine, and a large sprinkling of Scholars of Trinity College, Dublin. In more recent times Margaret Stokes, the Irish antiquary, and the Celtic scholar Whitley Stokes, children of the eminent physician Dr William Stokes, have, among others, shed lustre on the name.

The home at Skreen was a very happy one. In the excellent sea air the children grew up with strong bodies and active minds. Of course great economy had to be practised to meet the educational needs of the family; but in the Arcadian simplicity of a place where chickens cost sixpence and eggs were five or six a penny, it was easy to feed them. They were all deeply attached to their mother, a beautiful and severe woman who made herself feared as well as loved.

Stokes was taught at home; he learnt reading and arithmetic from the Parish Clerk, and Latin from his father who had been a scholar of Trinity College, Dublin. The former used to tell with great delight that Master George had made out for himself new ways of doing sums, better than the book. In 1832, at 13 years of age, he was sent to Dr Wall's school in Dublin; and in 1835 for two years to Bristol College, of which Dr Jerrard was Principal. There is a tradition that he did many of the propositions of Euclid, as problems, without looking at the book. He considered that he owed much to the teaching of Francis Newman, brother of the Cardinal, then mathematical master at Bristol College, and a man of great charm of character as well as of unusual attainments.

On the first crossing to Bristol the ship nearly foundered; and his brother, who was escorting him, was much impressed by his coolness in face of danger. His habit, often remarked in after life, of answering with a plain "yes" or "no," when something more elaborate was expected, is supposed to date from this time, when his brothers chaffed him and warned him that if he gave "long Irish answers" he would be laughed at by his school-fellows.

It is surprising to learn that as a little boy he was passionate, and liable to violent, if transitory, fits of rage. So completely was this tendency overcome that in after life his temper was remarkably calm and even. He was fond of botany, and when about sixteen or seventeen, collected butterflies and caterpillars. It is narrated that one day while on a walk with a friend he failed to return the salutation of some ladies of his acquaintance, afterwards explaining his conduct by remarking that his hat was full of beetles!

In 1837, the year of Queen Victoria's accession, he commenced residence at Cambridge, where he was to find his home, almost without intermission, for sixty-six years. In those days sports were not the fashion for reading men, but he was a good walker, and astonished his contemporaries by the strength of his swimming. Even at a much later date he enjoyed encounters with wind and waves in his summer holidays on the north coast of Ireland. At Pembroke College his mathematical abilities soon attracted attention, and in 1841 he graduated as Senior Wrangler and first Smith's Prizeman. In the same year he was elected Fellow of his College.

After his degree, Stokes lost little time in applying his mathematical powers to original investigation. During the next three or four years there appeared papers dealing with hydrodynamics, wherein are contained many standard theorems. As an example of these novelties, the use of a stream-function in three dimensions may be cited. It had already been shown by Lagrange and Earnshaw that in the motion of an incompressible fluid in *two dimensions* the component velocities at any point may be expressed by means of a function known as the stream-function, from the property that it remains constant along any line of motion. It was further shown by Stokes that there is a similar function in three dimensions when the motion is symmetrical with respect to an axis. For many years the papers, now under consideration, were very little known abroad, and some of the results are still attributed by Continental writers to other authors.

A memoir of great importance on the "Friction of Fluids in Motion, etc.," followed a little later (1845). The most general motion of a medium in the neighbourhood of any point is analysed into three constituents—a motion of pure translation, one of pure rotation, and one of pure strain. These results are now very familiar; it may assist us to appreciate their novelty at the time,

if we recall that when similar conclusions were put forward by
Helmholtz twenty-three years later, their validity was disputed by
so acute a critic as Bertrand. The splendid edifice, concerning
the theory of inviscid fluids, which Helmholtz raised upon these
foundations, is the admiration of all students of Hydrodynamics.

In applying the above purely kinematical analysis to viscous
fluids, Stokes lays down the following principle:—" That the difference between the pressure on a plane passing through any point
P of a fluid in motion and the pressure which would exist in all
directions about P if the fluid in its neighbourhood were in a state
of relative equilibrium depends only on the relative motion of the
fluid immediately about P; and that the relative motion due to
any motion of rotation may be eliminated without affecting the
differences of the pressures above mentioned." This leads him to
general dynamical equations, such as had already been obtained
by Navier and Poisson, starting from more special hypotheses as to
the constitution of matter.

Among the varied examples of the application of the general
equations two may be noted. In one of these, relating to the
motion of fluid between two coaxial revolving cylinders, an error
of Newton's is corrected. In the other, the propagation of sound,
as influenced by viscosity, is examined. It is shown that the
action of viscosity (μ) is to make the intensity of the sound
diminish as the time increases, and to render the velocity of propagation less than it would otherwise be. Both effects are greater
for high than for low notes; but the former depends on the first
power of μ, while the latter depends only on μ^2, and may usually
be neglected.

In the same paragraph there occur two lines in which a
question, which has recently been discussed on both sides, and
treated as a novelty, is disposed of. The words are—" we may
represent an arbitrary disturbance of the medium as the aggregate
of series of plane waves propagated in all directions."

In the third section of the memoir under consideration, Stokes
applies the same principles to find the equations for an elastic
solid. In his view the two elastic constants are independent and
not reducible to one, as in Poisson's theory of the constitution of
matter. He refers to indiarubber as hopelessly violating Poisson's
condition. Stokes' position, powerfully supported by Lord Kelvin,
seems now to be generally accepted. Otherwise, many familiar
materials must be excluded from the category of elastic solids.

In 1846 he communicated to the British Association a Report
on Recent Researches in Hydrodynamics. This is a model of what
such a survey should be, and the suggestions contained in it have
inspired many subsequent investigations. He greatly admired the
work of Green, and his comparison of opposite styles may often
recur to the reader of mathematical lucubrations. Speaking of the
Reflection and Refraction of Sound, he remarks that "this problem

had been previously considered by Poisson in an elaborate memoir. Poisson treats the subject with extreme generality, and his analysis is consequently very complicated. Mr Green, on the contrary, restricts himself to the case of plane waves, a case evidently comprising nearly all the phenomena connected with this subject which are of interest in a physical point of view, and thus is enabled to obtain his results by a very simple analysis. Indeed Mr Green's memoirs are very remarkable, both for the elegance and rigour of the analysis, and for the ease with which he arrives at most important results. This arises in a great measure from his divesting the problems he considers of all unnecessary generality; where generality is really of importance he does not shrink from it. In the present instance there is one important respect in which Mr Green's investigation is more general than Poisson's, which is, that Mr Green has taken the case of any two fluids, whereas Poisson considered the case of two elastic fluids, in which equal condensations produce equal increments of pressure. It is curious that Poisson, forgetting this restriction, applied his formulae to the case of air and water. Of course his numerical result is quite erroneous. Mr Green easily arrives at the ordinary laws of reflection and refraction. He obtains also a very simple expression for the intensity of its reflected sound...." As regards Poisson's work in general there was no lack of appreciation. Indeed, both Green and Stokes may be regarded as followers of the French school of mathematicians.

The most cursory notice of Stokes' hydrodynamical researches cannot close without allusion to two important memoirs of somewhat later date. In 1847 he investigated anew the theory of oscillatory waves, as on the surface of the sea, pursuing the approximation so as to cover the case where the height is not very small in comparison with the wave-length. To the reprint in *Math. and Phys. Papers* are added valuable appendices pushing the approximation further by a new method, and showing that the slopes which meet at the crest of the highest possible wave (capable of propagation without change of type) enclose an angle of 120°.

The other is the great treatise on the Effect of Internal Friction of Fluids on the Motion of Pendulums. Here are given the solutions of difficult mathematical problems relating to the motion of fluid about vibrating solid masses of spherical or cylindrical form; also, as a limiting case, the motion of a viscous fluid in the neighbourhood of a uniformly advancing solid sphere, and a calculation of the *resistance* experienced by the latter. In the application of the results to actual pendulum observations, Stokes very naturally assumed that the viscosity of air was proportional to density. After Maxwell's great discovery that viscosity is independent of density within wide limits, the question assumed a different aspect; and in the reprint of the memoir Stokes explains how it happened that the comparison with theory was not more prejudiced by the use of an erroneous law.

In 1849 appeared another great memoir, on the Dynamical Theory of Diffraction, in which the luminiferous aether is treated as an elastic solid so constituted as to behave as if it were nearly or quite incompressible. Many fundamental propositions respecting the vibration of an elastic solid medium are given here for the first time. For example, there is an investigation of the disturbance due to the operation at one point of the medium of a periodic force. The waves emitted are of course symmetrical with respect to the direction of the force as axis. At a distance, the displacement is transverse to the ray and in the plane which includes the axis, while along the axis itself there is no disturbance. Incidentally a general theorem is formulated connecting the disturbances due to initial displacements and velocities. "If any material system in which the forces acting depend only on the positions of the particles be slightly disturbed from a position of equilibrium, and then left to itself, the part of the subsequent motion which depends on the initial displacements may be obtained from the part which depends upon the initial velocities by replacing the arbitrary functions, or arbitrary constants, which express the initial velocities by those which express the corresponding initial displacements, and differentiating with respect to the time."

One of the principal objects of the memoir was to determine the law of vibration of the secondary waves into which, in accordance with Huygens' principle, a primary wave may be resolved, and thence, by a comparison with phenomena observed with gratings, to answer a question then much agitated but now (unless restated) almost destitute of meaning, viz., whether the vibrations of light are parallel or perpendicular to the plane of polarisation. As to the law of the secondary wave Stokes' conclusion is expressed in the following theorem: "Let $\xi = 0$, $\eta = 0$, $\zeta = f(bt - x)$ be the displacements corresponding to the incident light; let O_1 be any point in the plane P, dS an element of that plane adjacent to O_1; and consider the disturbance due to that portion only of the incident disturbance which passes continually across dS. Let O be any point in the medium situated at a distance from the point O_1 which is large in comparison with the length of a wave; let $OO_1 = r$, and let this line make angles θ with the direction of propagation of the incident light, or the axis of x, and ϕ with the direction of vibration, or the axis of z. Then the displacement at O will take place in a direction perpendicular to OO_1, and lying in the plane ZO_1O; and if ζ' be the displacement at O, reckoned positive in the direction nearest to that in which the incident vibrations are reckoned positive,

$$\zeta' = \frac{dS}{4\pi r}(1 + \cos\theta)\sin\phi \cdot f'(bt - r).$$

In particular, if

$$f(bt - x) = c \sin \frac{2\pi}{\lambda} (bt - x),$$

we shall have

$$\zeta' = \frac{cdS}{2\lambda r} (1 + \cos\theta) \sin\phi . \cos\frac{2\pi}{\lambda}(bt - r)."$$

Stokes' own experiments on the polarisation of light diffracted by a grating led him to the conclusion that the vibrations of light are perpendicular to the plane of polarisation.

The law of the secondary wave here deduced is doubtless a possible one, but it seems questionable whether the problem is really so definite as Stokes regarded it. A merely mathematical resolution may be effected in an infinite number of ways; and if the problem is regarded as a physical one, it then becomes a question of the character of the obstruction offered by an actual screen.

As regards the application of the phenomena of diffraction to the question of the direction of vibration, Stokes' criterion finds a better subject in the case of diffraction by very small *particles* disturbing an otherwise uniform medium, as when a fine precipitate of sulphur falls from an aqueous solution.

The work already referred to, as well as his general reputation, naturally marked out Stokes for the Lucasian Professorship, which fell vacant at this time (1849). It is characterised throughout by accuracy of thought and lucidity of statement. Analytical results are fully interpreted, and are applied to questions of physical interest. Arithmetic is never shirked.

Among the papers which at this time flowed plentifully from his pen, one "On Attractions, and on Clairaut's Theorem," deserves special mention. In the writings of earlier authors the law of gravity at the various points of the earth's surface had been deduced from more or less doubtful hypotheses as to the distribution of matter in the interior. It was reserved for Stokes to point out that, in virtue of a simple theorem relating to the potential, the law of gravity follows immediately from the form of the surface, assumed to be one of equilibrium, and that no conclusion can be drawn concerning the internal distribution of attracting matter.

From an early date he had interested himself in Optics, and especially in the Wave Theory. Although, not long before, Herschel had written ambiguously, and Brewster, the greatest living authority, was distinctly hostile, the magnificent achievements of Fresnel had converted the younger generation; and, in his own University, Airy had made important applications of the theory, *e.g.*, to the explanation of the rainbow, and to the diffraction of object-glasses. There is no sign of any reserve in the attitude of Stokes. He threw himself without misgiving into the discussion of outstanding difficulties, such as those connected with the aberration of light, and by further investigations succeeded in bringing new groups of phenomena within the scope of the theory.

An early example of the latter is the paper " On the Theory of certain Bands seen in the Spectrum." These bands, now known after the name of Talbot, are seen when a spectrum is viewed through an aperture half covered by a thin plate of mica or glass. In Talbot's view the bands are produced by the interference of the two beams which traverse the two halves of the aperture, darkness resulting whenever the relative retardation amounts to an odd number of half wave-lengths. This explanation cannot be accepted as it stands, being open to the same objection as Arago's theory of stellar scintillation. A body emitting homogeneous light would not become invisible on merely covering half the aperture of vision with a half wave-plate. That Talbot's view is insufficient is proved by the remarkable observation of Brewster— that the bands are seen only when the retarding plate is held towards the blue side of the spectrum. By Stokes' theory this polarity is fully explained, and the formation of the bands is shown to be connected with the limitation of the aperture, viz., to be akin to the phenomena of diffraction.

A little later we have an application of the general principle of reversion to explain the perfect blackness of the central spot in Newton's rings, which requires that when light passes from a second medium to a first the coefficient of reflection shall be numerically the same as when the propagation is in the opposite sense, but be affected with the reverse sign—the celebrated " loss of half an undulation." The result is obtained by expressing the conditions that the refracted and reflected rays, due to a given incident ray, shall on reversal reproduce that ray and no other.

It may be remarked that on any mechanical theory the reflection from an infinitely thin plate must tend to vanish, and therefore that a contrary conclusion can only mean that the theory has been applied incorrectly.

A not uncommon defect of the eye, known as astigmatism, was first noticed by Airy. It is due to the eye refracting the light with different power in different planes, so that the eye, regarded as an optical instrument, is not symmetrical about its axis. As a consequence, lines drawn upon a plane perpendicular to the line of vision are differently focussed according to their direction in that plane. It may happen, for example, that vertical lines are well seen under conditions where horizontal lines are wholly confused, and vice versâ. Airy had shown that the defect could be cured by cylindrical lenses, such as are now common; but no convenient method of testing had been proposed. For this purpose Stokes introduced a pair of plano-cylindrical lenses of equal cylindrical curvatures, one convex and the other concave, and so mounted as to admit of relative rotation. However the components may be situated, the combination is upon the whole neither convex nor concave. If the cylindrical axes are parallel, the one lens is entirely compensated by the other, but as the axes diverge the combination forms an astigmatic lens of gradually increasing power, reaching a maximum when the axes are perpendicular. With the aid of this instrument, an eye, already focussed as well as possible by means (if necessary) of a suitable spherical lens, convex or concave, may be corrected for any degree or direction of astigmatism; and from the positions of the axes of the cylindrical lenses may be calculated, by a simple rule, the curvatures of a single lens which will produce the same result. It is now known that there are comparatively few eyes whose vision may not be more or less improved by an astigmatic lens.

Passing over other investigations of considerable importance in themselves, especially that on the composition and resolution of streams of polarised light from different sources, we come to the great memoir on what is now called Fluorescence, the most far-reaching of Stokes' experimental discoveries. He "was led into the researches detailed in this paper by considering a very singular phenomenon which Sir J. Herschel had discovered in the case of a weak solution of sulphate of quinine and various other salts of the same alkaloid. This fluid appears colourless and transparent, like water, when viewed by transmitted light, but exhibits in certain aspects a peculiar blue colour. Sir J. Herschel found that when the fluid was illuminated by a beam of ordinary daylight, the blue light was produced only throughout a very thin stratum of fluid adjacent to the surface by which the light entered. It was unpolarised. It passed freely through many inches of the fluid. The incident beam after having passed through the stratum from which the blue light came, was not sensibly enfeebled or coloured, but yet it had lost the power of producing the usual blue colour when admitted into a solution of sulphate of quinine. A beam of light modified in this mysterious manner was called by Sir J. Herschel *epipolised*.

Several years before, Sir D. Brewster had discovered in the case of an alcoholic solution of the green colouring matter of leaves a very remarkable phenomenon, which he has designated as *internal dispersion*. On admitting into this fluid a beam of sunlight condensed by a lens, he was surprised by finding the path of the rays within the fluid marked by a bright light of a blood-red colour, strangely contrasting with the beautiful green of the fluid itself when seen in moderate thickness. Sir David afterwards observed the same phenomenon in various vegetable solutions and essential oils, and in some solids. He conceived it to be due to coloured particles held in suspension. But there was one circumstance attending the phenomenon which seemed very difficult of explanation on such a supposition, namely, that the whole or a great part of the dispersed beam was unpolarised, whereas a beam reflected from suspended particles might be expected to be polarised

by reflection. And such was, in fact, the case with those beams
which were plainly due to nothing but particles held in suspension.
From the general identity of the circumstances attending the two
phenomena, Sir D. Brewster was led to conclude that epipolic
was merely a particular case of internal dispersion, peculiar only in
this respect, that the rays capable of dispersion were dispersed
with unusual rapidity. But what rays they were which were capable
of affecting a solution of sulphate of quinine, why the active rays
were so quickly used up, while the dispersed rays which they
produced passed freely through the fluid, why the transmitted
light when subjected to prismatic analysis showed no deficiencies in
those regions to which, with respect to refrangibility, the dispersed
rays chiefly belonged, were questions to which the answers appeared
to be involved in as much mystery as ever."

Such a situation was well calculated to arouse the curiosity
and enthusiasm of a young investigator. A little consideration
showed that it was hardly possible to explain the facts without
admitting that in undergoing dispersion the light *changed its
refrangibility,* but that if this rather startling supposition was
allowed, there was no further difficulty; and experiment soon
placed the fact of a change of refrangibility beyond doubt. "A
pure spectrum from sunlight having been formed in air in the
usual manner, a glass vessel containing a weak solution of sulphate
of quinine was placed in it. The rays belonging to the greater
part of the visible spectrum passed freely through the fluid, just
as if it had been water, being merely reflected here and there from
motes. But from a point about halfway between the fixed lines
G and *H* to far beyond the extreme violet, the incident rays gave
rise to a light of a sky-blue colour, which emanated in all directions
from the portion of the fluid which was under the influence of the
incident rays. The anterior surface of the blue space coincided, of
course, with the inner surface of the vessel in which the fluid was
contained. The posterior surface marked the distance to which
the incident rays were able to penetrate before they were absorbed.
This distance was at first considerable, greater than the diameter
of the vessel, but it decreased with great rapidity as the refrangibility of the incident rays increased, so that from a little beyond
the extreme violet to the end, the blue space was reduced to an
excessively thin stratum adjacent to the surface by which the
incident rays entered. It appears, therefore, that this fluid, which
is so transparent with respect to nearly the whole of the visible
rays, is of an inky blackness with respect to the invisible rays,
more refrangible than the extreme violet. The fixed lines
belonging to the violet and the invisible region beyond were
beautifully represented by dark planes interrupting the blue space.
When the eye was properly placed these planes were, of course,
projected into lines."

At a time when photography was of much less convenient

application than at present—even wet collodion was then a novelty—the method of investigating the ultra-violet region of the spectrum by means of fluorescence was of great value. The obstacle presented by the imperfect transparency of *glass* soon made itself apparent, and this material was replaced by *quartz* in the lenses and prisms, and in the mirror of the heliostat. When the electric arc was substituted for sunlight a great extension of the spectrum in the direction of shorter waves became manifest.

Among the substances found "active" were the salts of uranium—an observation destined after nearly half a century to become in the hands of Becquerel the starting point of a most interesting scientific advance, of which we can hardly yet foresee the development.

In a great variety of cases the refrangibility of the dispersed light was found to be *less* than that of the incident. That light is always degraded by fluorescence is sometimes referred to as Stokes' law. Its universality has been called in question, and the doubt is perhaps still unresolved. The point is of considerable interest in connection with theories of radiation and the second law of Thermodynamics.

Associated with fluorescence there is frequently seen a "false dispersion," due to suspended particles, sometimes of extreme minuteness. When a horizontal beam of falsely dispersed light was viewed from above in a vertical direction, and analysed, it was found to consist chiefly of light polarised in the plane of reflection. On this fact Stokes founded an important argument as to the direction of vibration of polarised light. For "if the diameters of the (suspended) particles be small compared with the length of a wave of light, it seems plain that the vibrations in a reflected ray cannot be perpendicular to the vibrations in the incident ray." From this it follows that the direction of vibration must be perpendicular to the plane of polarisation, as Fresnel supposed, and the test seems to be simpler and more direct than the analogous test with light diffracted from a grating. It should not be overlooked that the argument involves the supposition that the effect of a particle is to *load* the aether.

It was about this time that Lord Kelvin learned from Stokes "Solar and Stellar Chemistry." "I used always to show [in lectures at Glasgow] a spirit lamp flame with salt on it, behind a slit prolonging the dark line *D* by bright continuation. I always gave your dynamical explanation, always asserted that certainly there was sodium vapour in the sun's atmosphere and in the atmospheres of stars which show presence of the *D*'s, and always pointed out that the way to find other substances besides sodium in the sun and stars was to compare bright lines produced by them in artificial flames with dark lines of the spectra of the lights of the distant bodies*."

* Letter to Stokes, published in Edinburgh address, 1871.

Stokes always deprecated the ascription to him of much credit in this matter; but what is certain is, that had the scientific world been acquainted with the correspondence of 1854, it could not have greeted the early memoir of Kirchhoff (1859) as a new revelation. This correspondence will appear in Vol. IV. of Stokes' Collected Papers, now being prepared under the editorship of Prof. Larmor. The following is from a letter of Kelvin, dated March 9, 1854: " It was Miller's experiment (which you told me about a long time ago) which first convinced me that there must be a physical connection between agency going on in and near the sun, and in the flame of a spirit lamp with salt on it. I never doubted, after I learned Miller's experiment, that there *must* be such a connection, nor can I conceive of any one knowing Miller's experiment and doubting....If it could only be made out that the bright line D never occurs without soda, I should consider it perfectly certain that there is soda or sodium in some state in or about the sun. If bright lines in any other flames can be traced, as perfectly as Miller did in his case, to agreement with dark lines in the solar spectrum, the connection would be equally certain, to my mind. I quite expect a qualitative analysis of the sun's atmosphere by experiments like Miller's on other flames."

By temperament, Stokes was over-cautious. " We must not go too fast," he wrote. He felt doubts whether the effects might not be due to some constituent of sodium, supposed to be broken up in the electric arc or flame, rather than to sodium itself. But his facts and theories, if insufficient to satisfy himself, were abundantly enough for Kelvin, and would doubtless have convinced others. If Stokes hung back, his correspondent was ready enough to push the application and to formulate the conclusions.

It is difficult to restrain a feeling of regret that these important advances were no further published than in Lord Kelvin's Glasgow lectures. Possibly want of time prevented Stokes from giving his attention to the question. Prof. Larmor significantly remarks that he became Secretary of the Royal Society in 1854. And the reader of the Collected Papers can hardly fail to notice a marked falling off in the speed of production after this time. The reflection suggests itself that scientific men should be kept to scientific work, and should not be tempted to assume heavy administrative duties, at any rate until such time as they have delivered their more important messages to the world.

But if there was less original work, science benefited by the assistance which, in his position as Secretary of the Royal Society, he was ever willing to give to his fellow-workers. The pages of the " Proceedings" and "Transactions" abound with grateful recognitions of help thus rendered, and in many cases his suggestions or comments form not the least valuable part of memoirs which appear under the names of others. It is not in human nature for an author to be equally grateful when his mistakes are indicated, but from the point of view of the Society and of science in general, the service may be very great. It is known that in not a few cases the criticism of Stokes was instrumental in suppressing the publication of serious errors.

No one could be more free than he was from anything like an unworthy jealousy of his comrades. Perhaps he would have been the better for a little more wholesome desire for reputation. As happened in the case of Cavendish, too great an indifference in this respect, especially if combined with a morbid dread of mistakes, may easily lead to the withholding of valuable ideas and even to the suppression of elaborate experimental work, which it is often a labour to prepare for publication.

In 1857 he married Miss Robinson, daughter of Dr Romney Robinson, F.R.S., astronomer of Armagh. Their first residence was in rooms over a nursery gardener's in the Trumpington Road, where they received visits from Whewell and Sedgwick. Afterwards they took Lensfield Cottage, where they resided until her death in 1899. Though of an unusually quiet and silent disposition, he did not like being alone. He was often to be seen at parties and public functions, and, indeed, rarely declined invitations. In later life, after he had become President of the Royal Society, the hardihood and impunity with which he attended public dinners were matters of general admiration. The nonsense of fools, or rash statements by men of higher calibre, rarely provoked him to speech; but if directly appealed to, he would often explain his view at length with characteristic moderation and lucidity.

His experimental work was executed with the most modest appliances. Many of his discoveries were made in a narrow passage behind the pantry of his house, into the window of which he had a shutter fixed with a slit in it and a bracket on which to place crystals and prisms. It was much the same in lecture. For many years he gave an annual course on Physical Optics, which was pretty generally attended by candidates for mathematical honours. To some of these, at any rate, it was a delight to be taught by a master of his subject, who was able to introduce into his lectures matter fresh from the anvil. The present writer well remembers the experiments on the spectra of blood, communicated in the same year (1864) to the Royal Society. There was no elaborate apparatus of tanks and "spectroscopes." A test-tube contained the liquid and was held at arm's length behind a slit. The prism was a small one of 60°, and was held to the eye without the intervention of lenses. The blood in a fresh condition showed the characteristic double band in the green. On reduction by ferrous salt, the double band gave place to a single one, to re-assert itself after agitation with air. By such simple means was a fundamental reaction established. The impression left upon the hearer was that Stokes felt himself as much at home in chemical and botanical questions as in Mathematics and Physics.

At this time the scientific world expected from him a systematic treatise on Light, and indeed a book was actually advertised as in preparation. Pressure of work, and perhaps a growing habit of procrastination, interfered. Many years later (1884—1887) the Burnett Lectures were published. Simple and accurate, these lectures are a model of what such lectures should be, but they hardly take the place of the treatise hoped for in the sixties. There was, however, a valuable report on Double Refraction, communicated to the British Association in 1862, in which are correlated the work of Cauchy, MacCullagh and Green. To the theory of MacCullagh, Stokes, imbued with the ideas of the elastic solid theory, did less than justice. Following Green, he took too much for granted that the elasticity of aether must have its origin in *deformation*, and was led to pronounce the incompatibility of MacCullagh's theory with the laws of Mechanics. It has recently been shown at length by Prof. Larmor that MacCullagh's equations may be interpreted on the supposition that what is resisted is not deformation, but *rotation*. It is interesting to note that Stokes here expressed his belief that the true dynamical theory of double refraction was yet to be found.

In 1885 he communicated to the Society his observations upon one of the most curious phenomena in the whole range of Optics— a peculiar internal coloured reflection from certain crystals of chlorate of potash. The seat of the colour was found to be a narrow layer, perhaps one-thousandth of an inch in thickness, apparently constituting a twin stratum. Some of the leading features were described as follows:—

(1) If one of the crystalline plates be turned round in its own plane, without alteration of the angle of incidence, the peculiar reflection vanishes twice in a revolution, viz., when the plane of incidence coincides with the plane of symmetry of the crystal.

(2) As the angle of incidence is increased, the reflected light becomes brighter, and rises in refrangibility.

(3) The colours are not due to absorption, the transmitted light being strictly complementary to the reflected.

(4) The coloured light is not polarised.

(5) The spectrum of the reflected light is frequently found to consist almost entirely of a comparatively narrow band. In many cases the reflection appears to be almost total.

Some of these peculiarities, such, for example, as the evanescence of the reflection at perpendicular incidence, could easily be connected with the properties of a twin plane, but the copiousness of the reflection at moderate angles, as well as the high degree of selection, were highly mysterious. There is reason to think that they depend upon a regular, or nearly regular, alternation of twinning many times repeated.

It is impossible here to give anything more than a rough sketch of Stokes' optical work, and many minor papers must be passed over without even mention. But there are two or three contributions to other subjects as to which a word must be said.

Dating as far back as 1857 there is a short but important discussion on the effect of wind upon the intensity of sound. That sound is usually ill heard up wind is a common observation, but the explanation is less simple than is often supposed. The velocity of moderate winds in comparison with that of sound is too small to be of direct importance. The effect is attributed by Stokes to the fact that winds usually increase overhead, so that the front of a wave proceeding up wind is more retarded above than below. The front is thus tilted; and since a wave is propagated normally to its front, sound proceeding up wind tends to rise, and so to pass over the heads of observers situated at the level of the source, who find themselves, in fact, in a sound shadow.

In a more elaborate memoir (1868) he discusses the important subject of the communication of vibration from a vibrating body to a surrounding gas. In most cases a solid body vibrates without much change of volume, so that the effect is represented by a distribution of sources over the surface, of which the components are as much negative as positive. The resultant is thus largely a question of *interference*, and it would vanish altogether were it not for the different situations and distances of the positive and negative elements. In any case it depends greatly upon the *wave-length* (in the gas) of the vibration in progress. Stokes calculates in detail the theory for vibrating spheres and cylinders, showing that when the wave-length is large relatively to the dimensions of the vibrating segments, the resultant effect is enormously diminished by interference. Thus the vibrations of a piano-string are communicated to the air scarcely at all directly, but only through the intervention of the sounding board*.

On the foundation of these principles he easily explains a curious observation by Leslie, which had much mystified earlier writers. When a bell is sounded in hydrogen, the intensity is greatly reduced. Not only so, but reduction accompanies the actual addition of hydrogen to rarefied air. The fact is that the hydrogen increases the wave-length, and so renders more complete the interference between the sounds originating in the positively and negatively vibrating segments.

The determination of the laws of viscosity in gases was much advanced by him. Largely through his assistance and advice, the first decisive determinations at ordinary temperatures and pressures were effected by Tomlinson in 1886. In 1881 he brilliantly took advantage of Crookes' observations on the decrement of oscillation of a vibrator in a partially exhausted space, to prove that Maxwell's law holds up to very high exhaustion and to trace the mode of subsequent departure from it. Throughout the course of Crookes' investigations on the electric discharge in vacuum tubes, in which he was keenly interested and closely concerned, he upheld the British

---

* It may be worth notice that similar conclusions are more simply reached by considering the particular case of a *plane* vibrating surface.

view that the cathode stream consists of projected particles which excite phosphorescence in obstacles by impact: and accordingly, after the discovery of the Röntgen rays, he came forward with the view that they consisted of very concentrated spherical pulses travelling through the aether, but distributed quite fortuitously because excited by the random collisions of the cathode particles.

A complete estimate of Stokes' position in scientific history would need a consideration of his more purely mathematical writings, especially of those on Fourier series and the discontinuity of arbitrary constants in semi-convergent expansions over a plane, but this would demand much space and another pen. The present inadequate survey may close with an allusion to another of those "notes," suggested by the work of others, where Stokes in a few pages illuminated a subject hitherto obscure. By an adaptation of Maxwell's colour diagram he showed (1891) how to represent the results of experiments upon ternary mixtures, with reference to the work of Alder Wright. If three points in the plane represent the pure substances, all associations of them are quantitatively represented by points lying within the triangle so defined. For example, if two points represent water and aether, all points on the intermediate line represent associations of these substances, but only small parts of the line near the two ends correspond to *mixture*. If the proportions be more nearly equal, the association separates into two parts. If a third point (off the line) represents alcohol, which is a solvent for both, the triangle may be divided into two regions, one of which corresponds to simple mixtures of the three components, and the other to proportions for which a simple mixture is not possible.

A consideration of Stokes' work, even though limited to what has here been touched upon, can lead to no other conclusion than that in many subjects, and especially in Hydrodynamics and Optics, the advances which we owe to him are fundamental. Instinct, amounting to genius, and accuracy of workmanship are everywhere apparent; and in scarcely a single instance can it be said that he has failed to lead in the right direction. But, much as he did, one can hardly repress a feeling that he might have done still more. If the activity in original research of the first fifteen years had been maintained for twenty years longer, much additional harvest might have been gathered in. No doubt distractions of all kinds multiplied, and he was very punctilious in the performance of duties more or less formal. During the sitting of the last Cambridge Commission he interrupted his holiday in Ireland to attend a single meeting, at which however, as was remarked, he scarcely opened his mouth. His many friends and admirers usually took a different view from his of the relative urgency of competing claims. Anything for which a date was not fixed by the nature of the case, stood a poor chance. For example, owing to projected improvements and additions, the third volume of his Collected Works was delayed until eighteen years after the second, and fifty years after the first appearance of any paper it included. Even this measure of promptitude was only achieved under much pressure, private and official.

But his interest in matters scientific never failed. The intelligence of new advances made by others gave him the greatest joy. Notably was this the case in late years with regard to the Röntgen rays. He was delighted at seeing a picture of the arm which he had broken sixty years before, and finding that it showed clearly the united fracture.

Although this is not the place to dilate upon it, no sketch of Stokes can omit to allude to the earnestness of his religious life. In early years he seems to have been oppressed by certain theological difficulties, and was not exactly what was then considered orthodox. Afterwards he saw his way more clearly. In later life he took part in the work of the Victoria Institute : the spirit which actuated him may be judged from the concluding words of an Address on Science and Revelation. " But whether we agree or cannot agree with the conclusions at which a scientific investigator may have arrived, let us, above all things, beware of imputing evil motives to him, of charging him with adopting his conclusions for the purpose of opposing what is revealed. Scientific investigation is eminently truthful. The investigator may be wrong, but it does not follow he is other than truth-loving. If on some subjects which we deem of the highest importance he does not agree with us—and yet he may agree with us more nearly than we suppose—let us, remembering our own imperfections, both of understanding and of practice, bear in mind that caution of the Apostle : ' Who art thou that judgest another man's servant ? To his own master he standeth or falleth.' "

Scientific honours were showered upon him. He was Foreign Associate of the French Institute, and Knight of the Prussian Order *Pour le Mérite*. He was awarded the Gauss Medal in 1877, the Arago on the occasion of the Jubilee Celebration in 1899, and the Helmholtz in 1901. In 1889 he was made a Baronet on the recommendation of Lord Salisbury. From 1887 to 1891 he represented the University of Cambridge in Parliament, in this, as in the Presidency of the Society, following the example of his illustrious predecessor in the Lucasian Chair. He was Secretary of the Society from 1854 to 1885, President from 1885 to 1890, received the Rumford Medal in 1852, and the Copley in 1893.

But the most remarkable testimony by far to the estimation in which he was held by his scientific contemporaries was the gathering at Cambridge in 1899, in celebration of the Jubilee of his Professorship. Men of renown flocked from all parts of the world to do him homage, and were as much struck by the modesty and simplicity of his demeanour as they had previously been by the brilliancy of his scientific achievements. The beautiful lines by his colleague, Sir R. Jebb, cited below, were written upon this occasion.

There is little more to tell. In 1902 he was chosen Master of Pembroke. But he did not long survive. At the annual dinner of the Cambridge Philosophical Society, held in the College about a month before his death, he managed to attend though very ill, and made an admirable speech, recalling with charming simplicity and courtesy his lifelong intimate connection with the College, to the Mastership of which he had recently been called, and with the Society through which he had published much of his scientific work. Near the end, while conscious that he had not long to live, he retained his faculties unimpaired; only during the last few hours he wandered slightly, and imagined that he was addressing the undergraduates of his College, exhorting them to purity of life. He died on the first of February, 1903.

> Clear mind, strong heart, true servant of the light,
> True to that light within the soul, whose ray
> Pure and serene, hath brightened on thy way,
> Honour and praise now crown thee on the height
> Of tranquil years. Forgetfulness and night
> Shall spare thy fame, when in some larger day
> Of knowledge yet undream'd, Time makes a prey
> Of many a deed and name that once were bright.
>
> Thou, without haste or pause, from youth to age,
> Hast moved with sure steps to thy goal. And thine
> That sure renown which sage confirms to sage,
> Borne from afar. Yet wisdom shows a sign
> Greater, through all thy life, than glory's wage;
> Thy strength hath rested on the Love Divine.

R.

MATHEMATICAL AND PHYSICAL PAPERS.

NOTE ON CERTAIN FORMULÆ IN THE CALCULUS OF OPERATIONS. (In a letter to Prof. TAIT.)

[From the *Proceedings of the Royal Society of Edinburgh*, XI, pp. 101–2.]

"*January* 14*th*, 1876.

"FORMULÆ like those you sent me[*] are readily *suggested* by supposing the function operated on to be of the form $\Sigma A x^a$, or say, for shortness, x^a, with the understanding that no transformations are to be made which are not equally valid for $\Sigma A x^a$.

Thus

$$\left(\frac{d}{dx} x \frac{d}{dx}\right)^n x^a = a^2 (a-1)^2 \ldots (a-n+1)^2 x^{a-n}$$

$$= a (a-1) \ldots (a-n+1) \left(\frac{d}{dx}\right)^n x^a$$

$$= \left(\frac{d}{dx}\right)^n x^n \left(\frac{d}{dx}\right)^n x^a \ ;$$

and

$$\left(x \frac{d}{dx} x\right)^n x^a = (a+1)(a+2) \ldots (a+n) x^{a+n}$$

$$= (a+n)(a+n-1) \ldots (a+1) x^{a+n}$$

$$= x^n \left(\frac{d}{dx}\right)^n x^n x^a \ .$$

The direct transformation may readily be effected by noticing, in the first instance, that any two operations of the form

$$x^{-m+1} \frac{d}{dx} x^m$$

[*] See *ante* [*Proc. R. S. Edin.*], p. 95.

are convertible.   We find, in fact,

$$x^{-m+1}\frac{d}{dx}x^{m}\cdot x^{-n+1}\frac{d}{dx}x^{n} = x^{2}\left(\frac{d}{dx}\right)^{2} + (m+n+1)x\frac{d}{dx} + mn,$$

into which m and n enter symmetrically.

Replacing the operations in the left hand member of the first formula by convertible operations, which will be separated by points, we find

$$\frac{d}{dx}x\frac{d}{dx} = x^{-1}\times x\frac{d}{dx}\cdot x\frac{d}{dx},$$

$$\frac{d}{dx}x\frac{d}{dx}x^{-1} = x^{-2}\times x^{2}\frac{d}{dx}x^{-1}\cdot x^{2}\frac{d}{dx}x^{-1},$$

and so on.   Hence

$$\left(\frac{d}{dx}x\frac{d}{dx}\right)^{n} = x^{-n}\left(x^{n}\frac{d}{dx}x^{-n+1}\right)^{2}\left(x^{n-1}\frac{d}{dx}x^{-n+2}\right)^{2}\dots\left(x\frac{d}{dx}\right)^{2},$$

$$= x^{-n}\left\{x^{n}\frac{d}{dx}x^{-n+1}\cdot x^{n-1}\frac{d}{dx}x^{-n+2}\dots x\frac{d}{dx}\right\}^{2},$$

$$= x^{-n}\left\{x^{n}\left(\frac{d}{dx}\right)^{n}\right\}^{2} = \left(\frac{d}{dx}\right)^{n}x^{n}\left(\frac{d}{dx}\right)^{n}.$$

Again,

$$x\frac{d}{dx}x = x\times\frac{d}{dx}x,$$

$$x\frac{d}{dx}x^{2} = x^{2}\times x^{-1}\frac{d}{dx}x^{2},$$

and so on.   Hence

$$\left(x\frac{d}{dx}x\right)^{n} = x^{n}\times x^{-n+1}\frac{d}{dx}x^{n}\cdot x^{-n+2}\frac{d}{dx}x^{n-1}\dots\frac{d}{dx}x,$$

$$= x^{n}\times\frac{d}{dx}x\cdot x^{-1}\frac{d}{dx}x^{2}\dots x^{-n+1}\frac{d}{dx}x^{n},$$

$$= x^{n}\left(\frac{d}{dx}\right)^{n}x^{n}.\text{''}$$

An Experiment on Electro-Magnetic Rotation.
By W. Spottiswoode, Treas.R.S. [Extract.]

[From the *Proceedings of the Royal Society*, xxiv, *Feb.* 24, 1876, pp. 403—407.]

[The experiment is on the rotation and spirality assumed in a magnetic field by the luminous electric discharge through a gas.]

The following explanation of the phenomenon is due to Prof. Stokes, from whose correspondence it is substantially taken. The mathematical solution, although only roughly approximate, is perhaps still quite sufficient to give the general character of the experimental results *.

The magnetic field will be supposed uniform, and the lines of force parallel straight lines from pole to pole. The path of the current when undisturbed is also a straight line from pole to pole. In such a condition of things, everything being symmetrical, no rotation would take place. But if through any local circumstance, as in the experiment in air, or through heating of the chamber as in the exhausted tube, or otherwise, the path of the current be distorted and displaced, then each element will be subject to the action of two forces. To estimate these, let ds be an element of the path, with rectangular components dx, dy, dz, C the strength of the current, and R the magnetic force with components X, Y, Z, which in the first instance will be treated generally. Then one force will be that tending to impel the current in the direction of the axes respectively, and may be expressed by

$$C(Ydz - Zdy)/ds, \quad C(Zdx - Xdz)/ds, \quad C(Xdy - Ydx)/ds.$$

Besides this, there will be the tendency of the current to follow

[* The electric discharge is considered as represented by a current in a flexible inextensible conducting thread; see p. 243 *supra*. On the catenary, cf. Larmor, *Proc. Lond. Math. Soc.* 1884, p. 170; and on the physical problem, cf. J. J. Thomson, *Conduction of Electricity through Gases*, Ch. iv.]

1—2

the shortest path so as to diminish the resistance. Representing this as a tension τ, the components at one end of ds will be

$$- \tau dx/ds, \ - \tau dy/ds, \ - \tau dz/ds,$$

and those at the other

$$(\tau dx/ds) + d\,(\tau dx/ds), \ldots,$$

the algebraical sums of which are

$$d\,(\tau dx/ds), \ d\,(\tau dy/ds), \ d\,(\tau dz/ds),$$

and the equations of equilibrium then become

$$C\,(Ydz - Zdy) + d\,(\tau dx/ds) = 0 \ldots\ldots\ldots\ldots(1),$$
$$C\,(Zdx - Xdz) + d\,(\tau dy/ds) = 0 \ldots\ldots\ldots\ldots(2),$$
$$C\,(Xdy - Ydx) + d\,(\tau dz/ds) = 0 \ldots\ldots\ldots\ldots(3);$$

taking s as the independent variable and multiplying by dx/ds dy/ds, dz/ds respectively, and adding, we obtain $d\tau = 0$, or $\tau = $ constant. Again, multiplying by X, Y, Z and adding we obtain

$$X d^2x/ds^2 + Y d^2y/ds^2 + Z d^2z/ds^2 = 0 \ldots\ldots\ldots(4),$$

which expresses that the absolute normal (or normal in the osculating plane) is perpendicular to the resultant magnetic force.

In the case of a uniform tint, X, Y, Z will be constant. Integrating (4) and putting i for the angle between the tangent and the lines of magnetic force, we find

$$X dx + Y dy + Z dz = R ds \cos i,$$

so that the tangent line is inclined at a constant angle to the line joining the poles.

Again, the following combinations,

$$(2)\, dz - (3)\, dy = 0, \ \ (3)\, dx - (1)\, dz = 0, \ \ (1)\, dy - (2)\, dx = 0$$

give

$$Cdx\,(Xdx + \ldots) - CX ds^2 + \tau \left(\frac{dz}{ds}\frac{d^2y}{ds^2} - \frac{dy}{ds}\frac{d^2z}{ds^2}\right) ds^2 = 0, \ \&c.,$$

or $\qquad C\,(R \cos i\, dx - X ds) + \tau \left(\frac{dz}{ds}\frac{d^2y}{ds^2} - \frac{dy}{ds}\frac{d^2z}{ds^2}\right) ds = 0, \ \&c.$

Transposing, squaring, and adding, and putting ρ for the radius of curvature, we obtain

$$C^2 R^2 \sin^2 i = \tau^2/\rho^2, \ \text{ or } \ \rho = \tau/CR \sin i,$$

which is constant. The curve is therefore a helix. Also the radius of curvature of the projection of the curve on a plane perpendicular to the axis will be $\rho \sin^2 i$, viz. $= \tau \sin i/CR$.

" The value of τ depends doubtless on the nature and pressure of the gas, and perhaps also on the current; but it must be the same for equal values of C of opposite signs. Hence the handedness of the helix will be reversed by reversing either the current or the magnetic polarity. If the left-hand magnetic pole be north (*i.e.* austral, or north-pointing), and the left-hand terminal positive, the helix will be right-handed."

The general nature of the phenomenon may therefore now be described as follows:—" First, we have the bright spark of no sensible duration which strikes nearly in a straight line between the terminals. This opens a path for a continuous discharge, which being nearly in a condition of equilibrium, though an unstable one, remains a short time without much change of place. Then it moves rapidly to its position of equilibrium, the surface which is its locus forming the sheet. Then it remains in its position of equilibrium during the greater part of the discharge, approaching the axis again as the discharge falls, so that its equilibrium position is not so far from the axis. Thus we see two bright curves corresponding to the two positions of approximate rest united by a less bright sheet, the first curve being nearly a straight line, and the second nearly a helix traced on a cylinder of which the former line is a generating line.

" It was noticed that the sheet projected a little beyond the helix. This may be explained by considering that at first the discharge is more powerful than can be maintained, so that the curve reaches a little beyond the distance that can be maintained."

The appearance of the discharge when viewed in a revolving mirror (except the projection beyond the sheet, the illumination of which was too feeble to be observed) confirmed the above remarks.

ON THE FOCI OF LINES SEEN THROUGH
A CRYSTALLINE PLATE.

[From the *Proceedings of the Royal Society*, XXVI, pp. 386—401, *June* 21, 1877.]

AT the Soirée of the Royal Society on the 25th of April Mr Sorby showed me the method he had recently devised for discriminating between minerals by focusing a microscope over a delicate image of cross lines, which image was viewed, first directly, and then through a crystalline plate, having previously been adjusted to be at the distance of the lower surface of the plate. With glass and singly refracting substances the alteration of the focus produced by the interposition of the plate affords a measure of its refractive index. But with a plate cut from a doubly refracting crystal, not only is there more than one focal distance, but for one at least of the pencils there is (except in special cases) no true focus, but the foci of the two systems of cross lines are found at two different depths, or else there is no sharply defined image at all, according to the orientation of the lines relatively to lines fixed in the crystalline plate. Moreover the result obtained on applying the formula which, for a singly refracting plate, gives the refractive index from the measured displacement of the focus is often widely different from what is known to be the refractive index of the crystal, for the pencil under examination, in a direction perpendicular to the plate.

The phenomena will be described in detail by Mr Sorby in his own paper. My object is to show how they flow from the known laws of double refraction, as consequences of which they will necessarily come under review*.

[* It seemed pretty certain that some of the phenomena must have been noticed before, though I am not aware that they have been described, or their theory worked out in any detail. I find that Prof. Clifton has been in the habit of using an instrument somewhat similar to Mr Sorby's, which was procured several years ago for the Museum of the University of Oxford, and that he was familiar with

The simplest case is that of a uniaxal crystal, such as Iceland spar, cut perpendicular to its axis. As regards the ordinary ray, a plate cut from a uniaxal crystal, in whatever direction, behaves, of course, like a plate of glass, so far as focusing is concerned, and the index obtained is the true ordinary index. To find what takes place as regards the extraordinary ray, we must have recourse to Huyghens's construction.

Let O be any point in the further surface of the crystalline plate, OA perpendicular to the surface the direction of the axis, OP the direction of any extraordinary ray. Let the plane of the

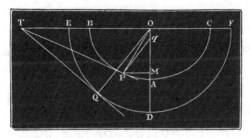

Fig. 1.

paper be the plane of incidence, AOP; take OA to represent the velocity of propagation (a) within the crystal in the direction of the axis, and OD in OA produced to represent the velocity of propagation (unity) in air. With O as centre, construct the half-spheroid, BAC, which is the extraordinary sheet of the wave-surface, and the hemisphere EDF representing the wave into which a disturbance emanating from O would have spread in air

such things as the low apparent index of calcite for the extraordinary pencil nearly in the direction of the axis, and the astigmatism in general of a pencil refracted across a plate otherwise than by ordinary refraction; and, further, that he utilized these phenomena for the instruction of students as to the general form of the wave-surface. No one, however, so far as I know, before Mr Sorby, had applied the phenomena to the practical discrimination of minerals, or had worked them out quantitatively and in detail; and it is my desire to complete the subject, by supplying the mathematical theory, that must be my excuse for offering to the Society an investigation which in itself consists merely in easy deductions from well-known principles.

It is perhaps hardly necessary to refer to a paper by Dr Quincke in *Poggendorff's Annalen* (1862, Vol. xxvii, p. 563), containing some elaborate observations on the focal lines formed within a doubly refracting plate, as his experiments were made in a very different manner from those of Mr Sorby, and with a totally different object in view. October, 1877.]

in a unit of time, and let OB or OC be denoted by c. Let OP cut the half-spheroid in P. At P draw a tangent plane to the spheroid, the trace of which on the surface of the crystal is projected in T; and through the trace T draw a tangent plane to the hemisphere, touching it in Q, and join OQ. Then if an extra-ordinary ray travel within the crystal in the direction OP, the refracted ray to which it will give rise will travel in a direction parallel to OQ. Hence if we now take OP to denote the whole path of the ray within the plate, and draw Pq parallel to QO, cutting OA in q, the ray OP, after refraction at P, will proceed as if it came from q. Hence the limiting position of q, as P moves up to A, will be the geometrical focus, after refraction, of a small pencil proceeding from O, and having OA for its axis.

Draw PM perpendicular to OA, and let m represent the ratio of the sine of refraction to the sine of incidence. Then

$$m = OP : Pq;$$

and by similar triangles

$$Pq : PM = OT : OQ = OB^2 : PM . OQ,$$

since

$$OB^2 = OT . PM;$$

also

$$OP = OA$$

ultimately. Hence as

$$OA : OB : OQ = a : c : 1,$$

we have ultimately

$$m = \frac{a}{c^2} = \frac{\mu'^2}{\mu} \quad \dots\dots\dots\dots\dots\dots\dots(1),$$

where μ, μ' denote the ordinary and the principal extraordinary indices of refraction, which are the reciprocals of a, c.

In this particular case the ordinary and extraordinary images cannot be distinguished directly by their polarization, since each consists of rays polarized in all azimuths. But if the objective of the microscope be limited by a narrow aperture, so as to give a predominance to rays lying in one plane, there will in the ordinary image be a predominance of polarization in a plane parallel to the length of the aperture, and in the extraordinary image of polariza-tion in the perpendicular plane.

Next take the case of a uniaxal crystal cut parallel to the axis. In this case, as regards the extraordinary pencil, the divergence after refraction will be different in the axial and equatorial planes, so that a small pencil diverging from a point at the under surface of the crystal will, after refraction, diverge from two focal lines; and in order that a line may be seen distinctly, it must lie in one of the planes of symmetry, in which case, at a certain focal adjustment of the microscope, each element of the line would be seen as a short line in the direction of the actual line, and therefore the line as a whole will be seen sharply defined.

In the equatorial plane the extraordinary ray obeys the ordinary law of refraction; and as regards divergence, therefore, in this plane, on which depends clear vision of a line parallel to the axis, the apparent index will be the same as the real index, μ'. In the axial plane everything will be the same in respect of divergence as in the first case, except that the principal axes of the ellipse which is the section of the extraordinary wave-surface will be interchanged. Accordingly a line in the equatorial plane will be seen distinctly at a focal adjustment which will give an apparent refractive index $\mu^2 : \mu'$.

There will therefore, on the whole, be three focal adjustments of the microscope at which one or other of the systems of cross lines, or both together, will be seen distinctly, namely, one for the extraordinary pencil, which is polarized in the equatorial plane, at .which the lines in the axial plane are seen distinctly; another at which the lines in the equatorial plane are seen distinctly; and, intermediate between these, a third for the ordinary pencil, which is polarized in the axial plane, at which both systems at once will be seen distinctly. And the ordinary index, which will be given by the ordinary image, will be a geometric mean between the two apparent extraordinary indices, of which one, namely, that got from the lines in the axial plane, will be the real extraordinary index.

There are two uniaxal crystals, calcite and quartz, for which we know accurately the principal refractive indices for the principal lines of the spectrum from the measures of Rudberg. The principal indices for these two minerals and the apparent indices in the two directions mentioned above are given in the following Table. The indices are given to four places of decimals, and the fixed lines C, D, E are chosen, whence the results applicable to the kinds of light most likely to be employed may be obtained, directly or by interpolation.

Lines	Calcite				Quartz			
	μ	μ'	$\dfrac{\mu'^2}{\mu}$	$\dfrac{\mu^2}{\mu'}$	μ	μ'	$\dfrac{\mu'^2}{\mu}$	$\dfrac{\mu^2}{\mu'}$
C	1·6545	1·4846	1·3321	1·8438	1·5418	1·5509	1·5601	1·5328
D	1·6585	1·4864	1·3322	1·8505	1·5442	1·5533	1·5624	1·5352
E	1·6636	1·4887	1·3322	1·8590	1·5471	1·5563	1·5656	1·5380

It is well known that the double refraction of quartz differs from that of the generality of uniaxal crystals. Its wave-surface for any colour, instead of being the sphere and spheroid of Huyghens, is a surface of two distinct sheets, which, instead of touching, only make a very close approach along the axis. The polar diameters of the outer, or ordinary, and of the inner, or extraordinary, sheet differ by minute and practically equal quantities from the equatorial diameter of the ordinary sheet. The effect of this, however, on the indices, real or apparent, determined by Mr Sorby's method on a plate cut perpendicular to the axis, would not be sensible. The peculiarity would show itself by giving the two images at different depths circularly polarized, one right-handedly and the other left-handedly.

It may be noticed that the refractive index is given by the reciprocal of the radius of curvature of a section of the wave-surface by a plane perpendicular to the lines seen in focus, and that in order that the lines may be seen distinctly, they must be perpendicular to one of the planes of principal curvature. This rule, as I proceed to show, is general; and it will much simplify the calculation in more complicated cases, by enabling us to dispense with the direct application of Huyghens's construction.

Let O be a point in the first surface of the plate, and consider a small pencil emanating from O in such a direction that its axis, after refraction, is perpendicular to the plate. With centre O describe half a wave-surface, of which only one sheet, DEF, is represented in the figure to avoid confusion. In a direction parallel to the surfaces of the plate draw a tangent plane to the wave-surface, touching it in E. Join OE, and draw EG a normal

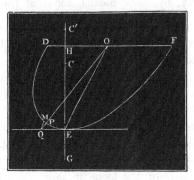

Fig. 2.

to the plate, and produce it to cut DF in H. Then OE is the course of a ray within the crystal which, after refraction, proceeds in a direction perpendicular to the plate, and is therefore the axis of the pencil. Let OPQ be an adjacent ray, cutting the wave-surface in P, and the tangent plane, which we may suppose to coincide with the second surface of the plate, in Q. Then the retardation of the wave on arriving at Q, relatively to the wave at E, will be the time the ray of light takes to travel from P to Q. The form of the wave after refraction will depend only on the value of this retardation, regarded as a function of the two co-ordinates which determine the position of Q on the plate. This follows at once from Huyghens's principle. If we regard QE as a small quantity of the first order, the retardation will be a small quantity of the second order; and in determining the foci of the refracted pencil we only want to know the retardation true to this order, and we may substitute for the actual retardation any quantity which bears to it a ratio that is ultimately one of equality. Hence, as the wave progresses within the crystal beyond DEF, we may feign it to be travelling in an ordinary medium, with a velocity of propagation equal to the actual wave-velocity in the direction HE normal to the plate. For if from Q we conceive a normal QM drawn to the wave-surface, the actual wave-velocity along MQ will differ from that in the direction HE by a small quantity of the first order; and since the whole distance MQ is a small quantity of the second order, we may neglect the variation of wave-velocity, and treat the medium as if it were a singly refracting one, in which a wave was travelling which had already, by some means, acquired the form DEF.

Through the normal EH draw the two rectangular planes of principal curvature of the surface at E, and let C, C' be the centres of curvature, and ρ, ρ' the radii of curvature, on the same scale in which HE represents the wave-velocity v in the direction HE. Then the rays in that plane of principal curvature, the normals in which intersect in C, may be thought of as diverging from C in an ordinary medium, of which the refractive index is v^{-1}. If τ be the thickness of the plate, the distance of C from the second surface will be $\rho\tau/v$, and the product of this by v, or $\rho\tau$, will give the distance of the focus in that plane after refraction into air; and therefore the apparent index of refraction will be ρ^{-1}. Similarly ρ'^{-1} will be the apparent index in the other plane. And in order that one or other of the two rectangular systems of lines may be seen distinctly at the proper focus, the lines must be placed perpendicular respectively to the two planes of principal curvature.

A similar construction applies to the other pencil which the plate is capable of transmitting independently, to which corresponds the other sheet of the wave-surface, and which is polarized in a plane perpendicular to the plane of polarization of the former pencil. In a biaxal crystal, in which neither sheet of the wave-surface is a sphere, there will in general be four focal distances at which lines in proper directions can be seen distinctly. For either pencil the two required directions are perpendicular to each other; and if the plate be perpendicular to one of the principal planes, or planes of optical symmetry of the crystal, the required directions of the cross lines are the same for both pencils, namely, parallel and perpendicular to the plane of symmetry.

The case next in order of simplicity to that of a uniaxal crystal cut parallel to the axis is that of a biaxal crystal cut in a direction perpendicular to one of the principal axes; but before proceeding to this it may be well to complete the investigation for a uniaxal crystal, by considering a plate cut in any manner.

Let θ be the inclination of the axis of the crystal to the normal to the plate; a, c, as before, the polar and equatorial semiaxes of the spheroid. We need only consider the extraordinary ray and the spheroid corresponding to it. Let ρ be the radius of curvature of the elliptic section made by the principal plane, ρ' the radius of curvature of the perpendicular section, which will be the length of the normal drawn as far as the axis of revolution. ρ and ρ' are to be expressed in terms of θ. We have

$$\rho^{-1} = a^{-2} c^{-2} (a^2 \cos^2 \theta + c \sin^2 \theta)^{\frac{3}{2}} \quad \dots\dots\dots\dots(2),$$

$$\rho'^{-1} = c^{-2} (a^2 \cos^2 \theta + c^2 \sin^2 \theta)^{\frac{1}{2}} \quad \dots\dots\dots\dots(3).$$

(2) gives the apparent index as obtained by focusing on a line perpendicular to the principal plane, and (3) as obtained by focusing on a line in the principal plane.

We see from (2) that as θ changes from $0°$ to $90°$, ρ changes from $a^{-1} c^2$ to $c^{-1} a^2$, of which one is greater than a and the other less than a. Hence for an intermediate value of θ $\rho = a$. For this value we have from (2)

$$\tan^2 \theta = a^{\frac{2}{3}} c^{-\frac{2}{3}}(a^{\frac{2}{3}} + c^{\frac{2}{3}}) = \mu^{-\frac{2}{3}} \mu'^{\frac{2}{3}}(\mu^{\frac{2}{3}} + \mu'^{\frac{2}{3}})\dots\dots\dots(4).$$

In this case, as in that of a uniaxal crystal cut perpendicular to the axis, there are only two focal distances at which a distinct image is seen. But the two cases are easily distinguished; for in the present case the ordinary and extraordinary images are both polarized in definite planes; also at one of the focal distances only one of the systems of cross lines, namely, those parallel to the principal plane, are seen distinctly; and, further, either extraordinary image becomes confused when the plate is rotated in its own plane.

For this particular inclination we have, in the case of Iceland spar, according to the indices above quoted for the line D, $\theta = 53° 34'$.

In this mineral the normal to the plane of easy cleavage is inclined to the axis at the angle $44° 37'$. Substituting this value in (2) and (3), writing μ^{-1}, μ'^{-1} for a, c, we have, for the apparent indices of the extraordinary pencil:—

		For lines perpendicular to the principal plane		For lines parallel to the principal plane
For the fixed line	C	1·5777	......	1·4094
	D	1·5809	......	1·4104
	E	1·5849	......	1·4116

In biaxal crystals the simplest case, and at the same time the most important, is that of a plate cut perpendicular to one of the principal axes, or so-called *axes of elasticity*. As the calculation for both pencils is precisely the same as for the extraordinary pencil in a plate of a uniaxal crystal cut parallel to the axis, it will be sufficient to give the result.

Let the principal axes be designated as those of *x, y, z*, to which relate the parameters or principal velocities of propagation, *a, b, c*, and their reciprocals, the principal indices, μ, μ', μ''. I will suppose *a, b, c*, taken in descending, and consequently μ, μ', μ'' in ascending order of magnitude*. For the arrangement of the Table, it will be convenient to specify the direction of the line seen in focus, and that of the normal to the plane of polarization of the image observed. When these directions are different, the plane of the plate is defined, being that containing them both. When they are the same, the plane of the plate may be either of the principal planes containing that common direction; and, indeed, it might be any plane containing it, only that the case of a plate cut obliquely is not at present under consideration. The apparent indices obtained by focusing will accordingly be arranged as follows:—

		Direction of line brought to focus		
		x	*y*	*z*
Direction of normal to plane of polarization	*x*	μ	$\dfrac{\mu''^2}{\mu}$	$\dfrac{\mu'^2}{\mu}$
	y	$\dfrac{\mu''^2}{\mu'}$	μ'	$\dfrac{\mu^2}{\mu'}$
	z	$\dfrac{\mu'^2}{\mu''}$	$\dfrac{\mu^2}{\mu''}$	μ''

It may be well to give the numerical results for aragonite

---

* In the previous investigation for Iceland spar *a* was taken for the ordinary wave-velocity, in order to conform to the notation in Airy's tract, so that *c* was greater than *a*.

and topaz, as calculated from Rudberg's indices. I have chosen the same fixed lines as before.

	Fixed lines	Aragonite			Topaz		
		x	y	z	x	y	z
x	C	1·5282	1·8513	1·8420	1·6093	1·6284	1·6135
	D	1·5301	1·8576	1·8481	1·6116	1·6296	1·6148
	E	1·5326	1·8653	1·8554	1·6145	1·6338	1·6189
y	C	1·6862	1·6778	1·3919	1·6262	1·6114	1·6072
	D	1·6902	1·6816	1·3922	1·6285	1·6137	1·6115
	E	1·6953	1·6863	1·3929	1·6315	1·6167	1·6123
z	C	1·6736	1·3885	1·6820	1·6040	1·5999	1·6188
	D	1·6773	1·3887	1·6859	1·6063	1·6041	1·6211
	E	1·6818	1·3892	1·6908	1·6093	1·6056	1·6241

The numbers in the squares xx, yy, zz are, of course, the real principal indices.

I proceed now to the case of a plate cut in any direction perpendicular to one of the principal planes, to which I propose to limit myself, merely observing that the leading features of the most general case have already been noticed.

The principal plane perpendicular to the plate being a plane of optical symmetry, the proper directions for the cross lines are parallel and perpendicular to that plane. Let the plane of symmetry be the plane of xz, without necessarily implying thereby that the axis of y is that of mean parameter, and let θ be the inclination of the normal to the plate to the axis of z. The section of the wave-surface, which I assume to be that of Fresnel, by the principal plane being a circle and an ellipse, the formulæ for the foci of a line perpendicular to the principal plane will be the same as for a uniaxal crystal, inasmuch as the relation which, in the case of a uniaxal crystal, subsists between the radius of the circle and one of the semiaxes of the ellipse is not involved in the formulæ. For light polarized perpendicularly to the principal plane, then, the apparent index for a line parallel to y is given by (2), while for the other pencil it is simply b^{-1} or μ'.

To find the foci for a line lying in the plane of symmetry, we must have recourse to the wave-surface itself, and not merely to its principal section. We have to find the radius of curvature at any point in the principal section for a normal section perpendicular to the principal plane.

Let P be a point in the principal section, PN the normal at P, M a point in PN near P, and through M draw MQ parallel to y, cutting in Q the sheet of the wave-surface to which P belongs. Then the limit of $MQ^2 \div 2PM$, as M moves up to P, will be the radius of curvature required.

Taking the equation of the wave-surface under the form

$$
\left.
\begin{aligned}
(x^2 + y^2 + z^2)(a^2 x^2 + b^2 y^2 + c^2 z^2) - a^2(b^2 + c^2) x^2 \\
- b^2(c^2 + a^2) y^2 - c^2(a^2 + b^2) z^2 + a^2 b^2 c^2 = 0
\end{aligned}
\right\} \quad \ldots\ldots(5),
$$

let x, 0, z be the coordinates of P, and $x + \delta x$, y, $z + \delta z$ those of Q. Substituting in (5), which, by hypothesis, is satisfied by the coordinates x, 0, z, observing that δx, δz are small quantities of the order y^2, and omitting small quantities of the order y^4, we find

$$
\left.
\begin{aligned}
\{a^2 x^2 + c^2 z^2 + a^2(x^2 + z^2 - b^2 - c^2)\}\, 2x\delta x \\
+ \{a^2 x^2 + c^2 z^2 + c^2(x^2 + z^2 - a^2 - b^2)\}\, 2z\delta z \\
+ \{a^2 x^2 + c^2 z^2 + b^2(x^2 + z^2 - a^2 - c^2)\}\, y^2 = 0
\end{aligned}
\right\} \quad \ldots\ldots(6).
$$

Let $PM = p$, and first suppose P to lie in the circular section. Then $x^2 + z^2 = b^2$, $x = b \sin\theta$, $z = b \cos\theta$. Also, as the normal coincides with the radius vector,

$$
\frac{\delta x}{x} = \frac{\delta z}{z} = -\frac{p}{b}, \quad x\delta x + z\delta z = -bp.
$$

Substituting in (6) and putting $y^2 = 2p\rho'$, we find for the curvature

$$
\rho'^{-1} = \frac{b(a^2 \sin^2\theta + c^2 \cos^2\theta + b^2 - a^2 - c^2)}{b^2(a^2 \sin^2\theta + c^2 \cos^2\theta) - a^2 c^2},
$$

or

$$
\rho'^{-1} = \frac{b\{(b^2 - a^2)\cos^2\theta + (b^2 - c^2)\sin^2\theta\}}{c^2(b^2 - a^2)\cos^2\theta + a^2(b^2 - c^2)\sin^2\theta} \quad \ldots\ldots\ldots(7),
$$

which gives the apparent index for light polarized in the principal plane when a line in that plane is brought into focus.

Next let P lie in the elliptic section. Then

$$a^2x^2 + c^2z^2 = a^2c^2,$$

which reduces (6) to

$$2(x^2 + z^2 - b^2)(a^2x\delta x + c^2z\delta z) + \{b^2(x^2 + z^2 - a^2 - c^2) + a^2c^2\}\,y^2 = 0$$
$$\dots\dots(8),$$

and
$$\frac{\delta x}{a^2x} = \frac{\delta z}{c^2z} = \frac{-p}{\sqrt{(a^4x^2 + c^4z^2)}} = \frac{a^2x\delta x + c^2z\delta z}{a^4x^2 + c^4z^2},$$

which, on putting ρ' for $y \div 2p$, reduces (8) to

$$(x^2 + z^2 - b^2)(a^4x^2 + c^4z^2)^{\frac{1}{2}} - \{b^2(x^2 + z^2 - a^2 - c^2) + a^2c^2\}\,\rho' = 0 \dots(9).$$

We have also

$$\tan\theta = \frac{a^2x}{c^2z},$$

which, combined with the equation to the ellipse, gives

$$\frac{a^4x^2}{\sin^2\theta} = \frac{c^4z^2}{\cos^2\theta} = \frac{a^2c^2}{a^{-2}\sin^2\theta + c^{-2}\cos^2\theta} = a^4x^2 + c^4z^2$$
$$= \frac{x^2 + z^2}{a^{-4}\sin^2\theta + c^{-4}\cos^2\theta}.$$

On substituting in (9), and reducing, we find

$$\rho'^{-1} = \frac{\{(b^2 - a^2)\cos^2\theta + (b^2 - c^2)\sin^2\theta\}(a^2\cos^2\theta + c^2\sin^2\theta)^{\frac{1}{2}}}{a^2(b^2 - a^2)\cos^2\theta + c^2(b^2 - c^2)\sin^2\theta} \dots(10),$$

which gives the apparent index for light polarized perpendicularly to the principal plane, when a line in that plane is brought into focus.

To sum up. For the pencil which is polarized in the principal plane the apparent index for a line perpendicular to that plane is the real index b^{-1} or μ', while for a line in the principal plane it is given by (7). For the pencil which is polarized perpendicularly to the principal plane the apparent index for a line perpendicular to that plane is given by (2), and for a line in the plane by (10).

On examining the expressions (7) and (10) and for radii of curvature of normal sections perpendicular to the principal plane, we see that if b be the greatest or least parameter they remain constantly positive. But if b be the mean parameter, both expressions change sign twice, once in passing through zero, and once through infinity, as θ changes from $0°$ to $90°$.

The radii of curvature become infinite together when

$$\tan^2 \theta = \frac{a^2 - b^2}{b^2 - c^2} \quad \dots\dots\dots\dots\dots\dots\dots(11) ;$$

that is, when the plate is perpendicular to the optic axis. For a point in the circular section the radius vanishes when

$$\tan^2 \theta = \frac{c^2 (a^2 - b^2)}{a^2 (b^2 - c^2)} \quad \dots\dots\dots\dots\dots\dots(12) ;$$

that is, when the plate is perpendicular to the ray-axis. For a point in the elliptic section the radius vanishes when

$$\tan^2 \theta = \frac{a^2 (a^2 - b^2)}{c^2 (b^2 - c^2)} \quad \dots\dots\dots\dots\dots(13) ;$$

that is, when the plate is perpendicular to the normal to the elliptic section at the point where the two sections intersect.

A figure may make these changes clearer. Let xOz be a quadrant of the plane perpendicular to the axis of mean parameter. Let BB' be the circular, and AC the elliptic section,

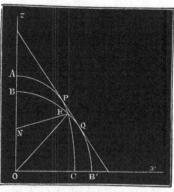

Fig. 3.

intersecting in R, PQ the common tangent, RN a normal at R to the elliptic section. Conceive a plate cut perpendicular to the plane of xz, its normal being inclined at the angle θ to Oz; and imagine θ to change continuously from $0°$ to $90°$: and let ρ', ρ_1' represent the radii of curvature in the secondary plane (xOz being deemed the primary plane) for points in the sections AC, BB' respectively. As θ starts from zero, ρ' starts from a and increases, and ρ_1' starts from $b^{-1}c^2$ and decreases. When θ becomes BOR, ρ_1' vanishes, and beyond that becomes negative, while ρ' continues to increase. As θ increases to BOQ, ρ' increases positively, and ρ_1' negatively, to infinity, and beyond that both change sign, ρ' becoming negative and ρ_1' positive. As θ increases to ANR, ρ' decreases negatively to zero, while ρ_1' decreases positively from infinity. On passing ANR, ρ' becomes positive, and increases to its final value, c, which it reaches when $\theta = 90°$, while ρ_1' decreases to its final value, $b^{-1}a^2$. Thus though $a > c$ we may say that as θ increases from 0° to 90°, ρ' *increases* from a to c by passing through ∞ and 0, and ρ_1' *decreases* from $b^{-1}c^2$ to $b^{-1}a^2$ by passing through 0 and ∞.

The extravagant changes of apparent index in the immediate neighbourhood of the wave- and ray-axes could probably not well be followed by the microscope, on account of the necessity of working with pencils of finite angular aperture, which would make the phenomena of focusing blend themselves with those of conical refraction. But there can be little doubt that a large increase or diminution of apparent index on approaching the critical region would be readily discernible. That these changes are not confined to the principal plane is evident, inasmuch as one principal radius of curvature of the wave-surface becomes infinite at any point of the circle of contact of the surface with the tangent plane perpendicular to the optic axis, and one principal radius of curvature vanishes at the conical point, to whatever normal section it be thought of as belonging.

Let us now resume the equations (2), (10), which give the principal curvatures for the elliptic section, without deciding beforehand anything as to the relative magnitude of the parameters.

As θ changes from 0° to 90°, the radius of curvature in the primary plane changes from $a^{-1}c^2$ to $c^{-1}a^2$, and that in the secondary plane from a to c; and the ratio of the radii, therefore, changes from $a^{-2}c^2$ to $c^{-2}a^2$, of which one is greater than 1 and the other less than 1. If, then, both radii remain positive, as is the case in the two principal planes passing through the mean axis, the two radii must be equal for some intermediate value of θ.

Hence there must be four umbilici in each of these planes. To find the umbilici we must equate the values of ρ, ρ' given by (2), (10), whence

$$a^2 c^2 \{(b^2 - a^2) \cos^2 \theta + (b^2 - c^2) \sin^2 \theta\}(\cos^2 \theta + \sin^2 \theta)$$
$$= \{a^2 (b^2 - a^2) \cos^2 \theta + c^2 (b^2 - c^2) \sin^2 \theta\} (a^2 \cos^2 \theta + c^2 \sin^2 \theta),$$

which gives, after reduction,

$$\tan^4 \theta = \frac{a^2(b^2 - a^2)}{c^2(b^2 - c^2)} \quad \ldots\ldots\ldots\ldots\ldots\ldots(14).$$

This expression shows that the umbilici in the elliptic section made by the principal plane of greatest and least parameters are imaginary. If we take v_{xy}, v_{zy} to denote the inclinations to Oy of the normals to the umbilici in the planes of xy, zy, we have, by the requisite interchanges of letters,

$$\tan^4 v_{xy} = \frac{a^2(a^2 - c^2)}{b^2(b^2 - c^2)}; \quad \tan^4 v_{zy} = \frac{c^2(a^2 - c^2)}{b^2(a^2 - b^2)} \quad \ldots\ldots(15).$$

If a plate be cut perpendicular to the normal at one of these umbilici, one of the polarized pencils which it transmits will give the images of both systems of cross lines distinct together; and the distinctness will not be affected by rotating the plate in its own plane while the cross lines are fixed. In this respect it agrees with a plate of a uniaxal crystal cut in an arbitrary direction, with which it might easily be confounded. But if the double refraction be strong enough to give a sensible *lateral* separation of the two oppositely polarized images, the two cases may be distinguished thereby in either of two ways:—First, if the images be compared with a mark fixed to the focus of the eyepiece, and the crystal be rotated in its own plane, while the object viewed through it is fixed, in the case of a uniaxal crystal the image free from astigmatism will remain fixed, while any point of the other describes a small circle round its mean position, whereas in a plate of a biaxal crystal cut perpendicular to the normal at one of the umbilici above considered it is the reverse; the image affected by astigmatism remains fixed, though its distinctness alters, while any point in the other describes a small circle about its mean position. Secondly, if the plane of separation of the two oppositely polarized images be noticed, in a uniaxal crystal the plane of polarization of the image which is free from astigmatism will be parallel to the plane of separation, while in a biaxal crystal cut as above supposed it will be perpendicular to the plane of separation*.

There are no umbilici in the circular sections of the wave-surface made by the principal planes. If we equate ρ' given by (7) to be the radius of curvature in the primary plane, we get, in fact, $\cos^2 \theta + \sin^2 \theta = 0$, which cannot be satisfied.

The formulæ (15) give for v_{xy}, v_{zy} (ray D) in aragonite 69° 26' and 45° 8' ; in topaz 46° 49' and 55° 27'.

In the employment of his method Mr Sorby has chiefly had in view the discrimination of minerals, but it admits of one or two interesting applications to optical theory.

At the time when Fresnel invented his theory of double refraction it had been supposed, from the observations of those who had specially examined the question, that in biaxal crystals one of the rays obeyed the ordinary law of refraction; and Fresnel proved by two methods, both requiring skill on the part of the optician who cut the crystals, that the anticipation that his theory led him to entertain that that would not prove to be the case was verified. It is interesting to find that the extraordinary character of the refraction of both rays in a biaxal crystal admits of being established by such a comparatively simple mode of observation as that of Mr Sorby.

The theory of Fresnel is confessedly wanting in rigour; and though the observations of Huyghens, of Wollaston, and of Malus proved that in Iceland spar Huyghens's construction, if not rigorously true, was at least a very close approximation to the truth, it seemed desirable to put it to a sharper observational test, more especially as different theories might lead to Huyghens's construction *as a near approximation*. For instance, in a paper read before the Cambridge Philosophical Society in 1849, I obtained a formula† which led me to perceive that double refraction would be simply accounted for by attributing it to a difference of inertia in different directions, such as would be produced if a

[* The first test supposes the surfaces of the plate to be pretty truly parallel, as otherwise it would produce displacement in consequence of its slightly wedge-shaped form ; and Mr Sorby thinks this requirement would prevent this test from being of much use. As to the second test, it is needless to observe that in any case in which we know independently which is the principal plane we need not attempt to observe the lateral separation of the images. October, 1877.]

† *Cambridge Philosophical Transactions*, Vol. VIII, p. 111. [*Ante*, Vol. I, p. 28.]

fluid had to make its way among a number of bodies on the average regularly arranged, that arrangement being different in different directions, and that the wave-velocity on this theory would be related to the direction of the wave normal just as in the theory of Fresnel, with the exception that the reciprocals of wave-velocities would take the place of the velocities themselves. I refrained, however, from putting forward that theory either in the memoir referred to or elsewhere (though I have incidentally alluded to it in my report on double refraction*), because, on calculating the difference of refraction of the extraordinary ray on this theory and according to Huyghens's construction, at about 45° from the axis, where the difference would be greatest, I found it barely small enough, as seemed to me, to have escaped detection. Still this theory, which has occurred independently, in the same or a similar shape, to others†, led me to wish for a more exact verification : and in the report referred to I have proposed a method which seemed to me well calculated to lead to the desired result. This method I carried out some years later in the case of Iceland spar, though I did not publish the results; and I found that, to the limit of error of my observations (about 0·0001 in the index), Huyghens's construction was fully confirmed, while the error of the other was nearly a hundred times as great as the limit of error of the observations‡. The accuracy of the Huyghenian law has also been confirmed by the elaborate observations of M. Abria§.

In the method of prismatic refraction employed by M. Abria and myself, the difference between Huyghens's construction and the result of the theory just referred to is greatest about 45° from the axis, while extremely close to the axis, or to the equator, it would hardly be sensible. Mr Sorby's method is remarkable for this, that it brings out into prominence variations of refraction with change of direction, though the absolute refractions which are involved may be nearly the same. Thus Mr Sorby informs me

---

* *Report of the British Association* for 1862, Part I, p. 269 [*ante*, Vol. IV, p. 182].

† See papers by the late Professor Rankine in the *Philosophical Magazine*, Vol. I (1851), p. 441, and by Lord Rayleigh in the same, Vol. XLI (1871), p. 519 [*Scientific Papers*, Vol. I, p. 118.]

‡ This result is briefly mentioned in the *Proceedings of the Royal Society*, Vol. XX, p. 443 [*ante*, Vol. IV, p. 336].

§ *Annales de Chimie*, tom. I (1874), p. 289. [Also R. T. Glazebrook, *Phil. Trans.* 1880, pp. 421—449; and C. S. Hastings, *Amer. Jour. Sci.* 1888. Cf. *ante*, Vol. IV, p. 336.]

that his method shows with perfect distinctness the two widely different foci for a plate of Iceland spar cut perpendicular to the axis, even though the inclination of the rays concerned to the axis is so small that, when polarized light is used, which is extinguished by an analyzer, the field remains dark after the interposition of the crystal.

In the theory referred to above the extraordinary sheet of the wave-surface is generated by the revolution of the curve which is the envelope of straight lines whose distance, v, from the origin is connected with their inclination, $90° - \theta$, to the axis by the relation

$$v^{-2} = a^{-2} \cos^2 \theta + c^{-2} \sin^2 \theta.$$

The radius of curvature of this envelope at the axis is $c^{-2}(2ac^2 - a^3)$, and accordingly the apparent index, m, is given by

$$m = \frac{\mu^3}{2\mu^2 - \mu'^2} \quad \dots\dots\dots\dots\dots\dots(16),$$

which exceeds the apparent index, $\mu^{-1}\mu'^2$, given by the spheroid of Huyghens, by

$$\frac{(\mu^2 - \mu'^2)^2}{\mu(2\mu^2 - \mu'^2)} \quad \dots\dots\dots\dots\dots(17).$$

The same formula will apply to a point in the equator if we interchange μ and μ'.

Putting these excesses into numbers, according to Rudberg's indices for the line D in Iceland spar, we find 0·0536 and 0·1180, giving for apparent indices 1·3858 instead of 1·3322 for a plate perpendicular to the axis, and 1·9685 instead of 1·8505 for a plate parallel to the axis when the microscope is focused on a line in the equatorial plane. These differences are much too large to escape detection.

Postscript.

Being anxious to complete the theory of the images free from astigmatism by determining in what cases, if in any, they could be formed by transmission in a perpendicular direction across a crystalline plate cut otherwise than perpendicular to a principal plane, I have since worked out the differential equation (between two parameters) of the lines of curvature of the wave-surface, the discussion of which shows that there are no umbilici out of the principal planes. Hence the four directions determined by equations (15) are the only ones perpendicular to which if a plate be cut one of the images is free from astigmatism.—*October* 1877.

On Certain Movements of Radiometers.

[From the *Proceedings of the Royal Society*, XXVI, pp. 546—555, 1877.]

NEARLY two years ago Mr Crookes was so good as to present me with two of his beautiful radiometers of different constructions, the disks of one being made of pith, and those of the other of roasted mica, in each case blackened with lampblack on one face. With these I was enabled to make some experiments, having relation to their apparently anomalous movements under certain circumstances, which were very interesting to myself, although the facts are only such as have already presented themselves to Mr Crookes, either in the actual form in which I witnessed them, or in one closely analogous, and have mostly been described by him. Although it will be necessary for me to describe the actual experiments, which have all been repeated over and over again so as to make sure of the results, I do not bring forward the facts as new. My object is rather to endeavour to coordinate them, and point to the conclusions to which they appear to lead.

I do not pretend that these conclusions are established; I am well aware that they need to be further confronted with observation; but as I have not leisure to engage in a series of experiments which would demand the expenditure of a good deal of time, and have lately been urged by a friend to publish my views, I venture to lay them before the Royal Society, in hopes that they may be of some use, even if only in the way of stimulating inquiry.

In describing my experiments I will designate that direction of rotation in which the white face precedes as positive, and the reverse as negative. It will be remembered that, under ordinary circumstances, radiation towards either radiometer produces positive rotation.

1. If a glass tumbler be heated to the temperature of boiling water, and inverted over the mica radiometer, there is little or

no *immediate* motion of the fly; but quickly a *negative* rotation sets in, feeble at first, but rapidly becoming lively, and presently dying away.

2. If after the fly has come to rest the hot tumbler be removed, a *positive* rotation soon sets in, which becomes pretty lively, and then gradually dies away as the apparatus cools.

3. If the tumbler be heated to a somewhat higher temperature, on first inverting it over the radiometer there is a *slight positive* rotation, commencing with the promptitude usual in the case of a feeble luminous radiation, but quickly succeeded by the negative rotation already described. If the tumbler be heated still more highly, the initial positive rotation is stronger and lasts longer, and the subsequent negative rotation is tardy and feeble.

4. If the pith radiometer be treated as in § 1, the result is the same, with the remarkable difference that the rotation is positive instead of negative; it is also less lively.

5. But if the tumbler be removed when the fly has come to rest, it remains at rest, or nearly so.

6. If the tumbler be more strongly heated, positive rotation begins as promptly as with light. In this case the tumbler must not be left long over the radiometer, for fear the vacuum should be spoiled by the evolution of gas from the pith.

7. If the tumbler be heated by holding it over the spout of a kettle from which steam is issuing, and held there till the condensation of water has approximately ceased, and be then inverted over the pith radiometer, the bulb is immediately bedewed, and a *negative* rotation is almost immediately set up, though sometimes, just at the very first moment, there is a trace of positive rotation. The negative rotation is lively, but not lasting; and, after 15 seconds or so, is exchanged for a positive rotation, which is not lively, but lasts longer.

8. If the tumbler be lifted when the negative rotation has ceased, and the dewed surface be strongly blown upon, a lively, but brief, positive rotation is set up.

9. To produce positive rotation by blowing, it is not *essential* that the bulb be wet. If it be merely warm, and the circum-

stances are such that the fly is at rest for the moment, or nearly so, blowing produces positive rotation, though much less strongly than when the bulb is wet.

10. If the tumbler be heated as in § 7, and inverted over the mica radiometer, the rotation is positive, as when the tumbler is dry.

11. If the tumbler or a cup be smoked inside (to facilitate radiation), heated to a little beyond the temperature of boiling water, and inverted over the pith radiometer, a positive rotation is produced; and if, when this has ceased, which takes place in a couple of minutes or so, the heated vessel be removed, a negative, though not lively, rotation is produced as the apparatus cools.

12. These results do not seem difficult to coordinate so far as to reduce them to their proximate cause.

As regards the small quantity, if any, of heat radiated directly across the glass of the bulb, the action of which was experimentally distinguishable by its promptitude, both radiometers behaved in the ordinary way.

13. As regards the mica radiometer, when the bulb gets heated, and radiates towards the fly, the fly is impelled in the negative direction, *as if* the white, pearly mica were black and the lampblack were white. And there is nothing opposed to what we know in supposing that such is *really* their relative order of darkness as regards the heat of low refrangibility absorbed and radiated by the glass; for the researches of Melloni and others have shown that lampblack is, if not absolutely white, at any rate very far from black as regards heat of low refrangibility. On the other hand, glass and mica are both silicates, not so very dissimilar in chemical composition; and it would not therefore be very wonderful, but rather the reverse, if there were a general similarity in their mode of absorption of radiant heat, so that the heat most freely radiated by glass, and accordingly abounding in the radiation from *thin* glass such as that of the bulb, were greedily absorbed by mica. The explanation of the reversal of the action when heat and cold were interchanged is too well known to require mention.

14. With the pith radiometer, when the bulb as a whole is heated, and radiates towards the fly, the impulse is positive,

though less strong than in the case of the mica (§ 4); and when the bulb as a whole is cooler than the fly the impulse is negative (§ 11).

But to explain all the phenomena we must dissect the total radiation from or towards the bulb. When I first noticed the negative rotation produced by a heated wet tumbler, I was disposed to attribute it to radiation from the water, which possibly the glass of the bulb might be thin enough to let pass; but when I found that hot water in a glass vessel outside, even though the glass of it were thin, produced no sensible effect, and that blowing on the heated bulb when it was dry produced a similar effect to blowing on it when dewed, though of much less amount, I perceived that the moisture acted, not by direct radiation from it, and in consequence of a difference of quality between the radiations from glass and water, but by causing a rapid *superficial* heating of the bulb; and, similarly, the blowing on the dewed surface acted by causing a rapid superficial cooling. When the dry tumbler radiates to the bulb, the radiation is absorbed at various depths; the absorption is most copious, it is true, at the outer strata; but still the change of temperature is not by any means so much confined to the immediate surface as when we have to deal with the latent heat of vapour condensed on it, or obtained from it by rapid evaporation.

Hence, thin as is the glass of the bulb (about 0·02 in. thick), we must still, in imagination, divide it into an outer and inner stratum, and examine the effects of these separately. The heat radiated by either stratum depends only on its temperature; but the radiation from the outer, on its way to the fly, is sifted by passing through the inner, and the portion for which glass is most excessively opaque is in great part stopped. It appears from the observed results that the residue acts decidedly negatively, while when the bulb is pretty uniformly heated there is positive action. We may infer that if it were possible to heat the inner stratum alone it would manifest a very decided positive action.

15. In the struggle between the opposing actions of the outer and inner strata we see the explanation of the strange behaviour of the pith radiometer. In the experiment of § 7 the outer stratum at first shows its negative action; but quickly

the inner also gets heated, partly by conduction from the outer, partly by direct radiation from the tumbler, and then the inner prevails.   In the experiment of § 5 the whole bulb cools, partly by radiation, partly by convection, while the fly remains warmer; and the slightly greater coolness of the outer than of the inner stratum makes up for the superiority of the inner when the two are equally cool, so that the antagonistic actions nearly balance, and slight causes, such, as greater or less agitation of the air, suffice to make the balance incline one way or other.   That the inner stratum *would* prevail if the two were about equally cooled may be inferred from the behaviour of the radiometer when the bulb is pretty uniformly heated (§§ 4, 11), or shown more directly by cooling the bulb with snow, when a negative rotation may be obtained.

16.   The complete definition of a radiation would involve the expression of the intensity of each component of it as a function of some quantity serving to define the quality of the component, such as its refractive index in a standard medium, or its wave-length, or the squared reciprocal of the wave-length *.   The experimental determination of the character, as thus defined, of a radiation consisting of invisible heat-rays is beset with diffi-culties, at least in the case of heat of extremely low refrangibility; and in general we can do little more than speak in a rough way of the radiation as being of such or such a kind.   It is obvious that the behaviour of radiometers by itself alone affords no in-dication of the refrangibilities of the kinds of heat with which we have to deal; nevertheless, by combining what we know of the behaviour of bodies in respect to radiations in general (especially luminous radiations, which are the most easily studied) with what we observe as to the motions of radiometers, we may arrive at some probable conclusions.

17.   We may evidently *conceive* a series of ethereal vibrations of any periodic time, however great, to be incident on a homo-geneous medium such as glass, and inquire in what manner the

* A map of the spectrum, constructed with the squared reciprocals of the wave-lengths for abscissæ, would be referred to a natural standard, no less than that of Ångström, which is constructed according to wave-lengths; while it would have the great advantage of admitting of ready comparison with refraction spectra, the kind almost always used.

rate of absorption would change with the period; though whether we can actually *produce* ethereal vibrations of a very long period is another question, seeing that we can only act on the ether by the intervention of matter, and are limited to such periods of vibration as matter can assume when vibrating molecularly, in a manner communicable to the ether, and not as a continuous mass, in the manner of the vibrations which produce sound. We may inquire whether, on continually increasing the period of vibration, the glass (or other medium) would ultimately become and remain very opaque, or whether, after passing through a range of opacity, it would become transparent again, on still further increasing the period of the incident vibrations.

18. This is a question the experimental answer to which, as it seems to me, could only be given, in so far as it could be given at all, as a result of a long series of experiments, of a kind that Melloni has barely touched on. A variety of considerations, which I could not explain in short compass, lead me to regard the second alternative as the more probable, namely, that, on increasing the periodic time, homogeneous substances in general (perhaps even metals, though this is doubtful) become at last transparent, or at least comparatively so. The limit of opacity, in all probability, varies from one substance to another; and the lower it is, the lower would be the lowest refrangibility of the radiation which the same substance is capable of emitting.

19. In what immediately follows I shall suppose accordingly that glass is strongly absorbing through a certain range of low refrangibility, on *both* sides of which it gradually becomes transparent again*. Imagine a spectrum containing radiations of all refrangibilities with which we have to deal; let portions of this spectrum on the two sides of the region of powerful absorption for glass be called *wings* of that region, and let left to right be the order of increasing refrangibility. Then the spectrum of the radiation from a thin plate of glass, if it could be observed, would be seen to occupy the region of chief absorbing (and therefore emitting) power and its wings. The spectrum of the radiation from the outer stratum of the bulb of the pith radiometer, after transmission through the inner, would consist of two wings, with

---

* It may be noticed that this supposition, which, as appearing the more probable, is adopted for clearness of conception, is not essentially involved in the explanation that follows, which would hardly be changed if the "left wing" were not terminated on the left.

a blank, or nearly blank, space between; it would resemble, in fact, a widened bright spectral line, with a dark band of reversal in its middle, save that, instead of being confined to extremely narrow limits of refrangibility, the central space and its wings would be of wide extent. It follows from the experiments that, in the complete radiation from glass, the portions of the spectrum called the wings together act negatively, the portion between positively.   It does not, of course, follow that each wing acts negatively, but only that the balance of the two is negative. When the tumbler is heated a little over 212° there is a slight positive action from radiation which passes directly through the bulb. The circumstances lead us to regard this as an extension of the right wing; for it comes from a depth, measured from the inner surface of the bulb in glass, *i.e.* not counting the intervening air, somewhat greater than the thickness of the wall of the bulb; and we know that the more a solid body is heated, the higher, as a rule, does the refrangibility of the radiation which it emits extend, and the greater the proportion of rays of high to those of low refrangibility. It is simplest, therefore, to suppose that the action of the right wing, like that of the space between the wings, is positive, and that the observed negative action in the experiment of § 7 is due to the excess of negative action of the left wing over positive action of the right. In the mica radiometer the experiments indicate no such difference of action in the different layers of the bulb as in the case of the pith radiometer. Hence taking, in accordance with what now appears to be made out to be the theory of the motion of the radiometer, the direction in which the fly is impelled as an indication which is the warmer of the two faces of the disks, and that again as an indication which is the darker with respect to the radiation to which it is exposed, we arrive at the following results as regards the order of darkness of the substances for the three regions into which the spectrum of the incident radiation has been supposed to be divided, the name of the lighter substance being in each case placed above that of the darker :—

	Left wing.	Region of intense absorption by glass.	Right wing.
From pith radiometer	Lampblack. Pith.	Pith. Lampblack.	Pith. Lampblack.
From mica radiometer	Lampblack. Mica.	Lampblack. Mica.	Mica. Lampblack.

Hence, on descending in refrangibility, the order of darkness of the two substances of either pair is at first the same as for the visible spectrum, and at last the opposite; and the reversal of the order takes place sooner with mica and lampblack than with pith and lampblack. The order falls in very well with that of the chemical complexity of the three substances.

20. The whole subject of the behaviour of bodies with respect to radiant heat of the lowest degrees of refrangibility seems to me to need a thorough experimental investigation. The investigation, however, is one involving considerable difficulty. We can do little towards classifying the rays with which we are working unless we can form a pure spectrum. A refraction-spectrum is the most convenient; but the only substance known which would be approximately suitable for forming the prism, lens, &c. required for such a spectrum, and for confining liquids, is rock-salt, of which it is extremely difficult to procure perfectly limpid specimens of any size; and even rock-salt itself, as Professor Balfour Stewart has shown, is defective in transparency for certain kinds of radiant heat. Then, again, the only suitable measuring-instrument for such researches, the thermopile, demands a thorough examination with reference to the coating to be employed for absorbing the incident radiation. Hitherto lampblack has been used almost exclusively for the purpose; and it is commonly assumed, in accordance with certain of Melloni's results, that lampblack absorbs equally heat-rays of all kinds. But the experiments by which Melloni established the partial diathermancy of lampblack prove that rays exist for the absorption of which that substance is unsuitable.

On calling on Mr Crookes after the above was written, I was surprised to find that all his mica radiometers behaved towards a heated glass shade in the opposite way to that he had given me, going round positively instead of negatively. Mr Crookes showed me and gave me a specimen of the kind of mica he employs, as eminently convenient for manipulation. It is found naturally in a condition resembling artificially roasted mica. It is not, however, quite so opaque for transmitted light, nor of quite such a pearly whiteness for reflected light as that which has been artificially roasted at a high temperature. The mica radiometer that Mr Crookes first gave me, which I will call M_1, was, Mr Gimingham told me, the only one they had made with roasted mica.

Mr Crookes was so kind as to give me, for comparative experiment, a mica radiometer, which I will call M_2, made from the natural foliated mica. It revolves a good deal more quickly than M_1 under the influence of light; it also gets more quickly under way, indicating that the mica is thinner. When covered with a hot glass it revolves positively, as already remarked; there is, however, but little negative rotation when the glass is removed.

The difference in the thickness and condition of the mica sufficiently explains the difference of behaviour of M_1 and M_2. Any radiant heat incident on the white face that reaches the middle of the mica, whether it afterwards is absorbed by the mica or reaches and is absorbed by the lampblack, tends to heat the second or blackened face more than the first, and therefore conspires with the heat incident on the lampblack, and absorbed by it, to produce positive rotation; and the smaller thickness and less fine foliation of the natural mica are favourable to the transmission of radiant heat to such a depth.

P.S.—It might be supposed at the first sight that the change of rotation from negative to positive (in § 7) was due, not to a change in the conditions of absorption, but to the circumstance that the inner surface of the bulb had become warm by conduction, so as to be warmer than the surfaces of the fly instead of colder. For we now know that the " repulsion resulting from radiation," as in some way or other it undoubtedly does result, is an indirect effect, in which radiation acts only through the alterations it occasions in the superficial temperatures of the solids in contact with the rarefied gas; and it might be supposed that when the inner surface of the bulb passed from colder than the fly to warmer, the direction of rotation would, on that account alone, be reversed. This, however, is not so. If bulb and fly are at a common temperature, and the instrument is protected from radiation, the fly remains at rest whether the common temperature be high or low. If a small portion of the total surface in contact with the rarefied gas be warmed by any means, repulsion takes place, through the intervention of the rarefied gas, between the warmed surface and the opposed surfaces, if not too distant; if it be cooled, the result is attraction. It does not matter whether the surface at the exceptional temperature belong to the fly or the bulb. The former takes place in the ordinary case of a radiometer exposed to radiation, the latter in that of a radiometer at a uniform temperature and protected from radiation when a small portion of the bulb is warmed or cooled, in which case the part at the exceptional temperature repels or attracts the disk irrespectively of its colour or the nature of its coating*. Suppose now that the fly is being warmed by radiation from without, the bulb being cool, at least at its inner surface. Let A, B be the two kinds of faces of the disks, and suppose A to be the better absorber of the total radiation. Then A will be the warmer, and therefore will be more strongly repelled than B. Suppose now that the bulb is heated till its inner surface becomes warmer than the fly. Then the fly will still be receiving heat by radiation, to some extent also by communication from the gas; but this will be the same for both faces. Hence if A be still the better absorber of the two (A, B), A will be the warmer, and being less below the temperature of the interior surface of the fly will be less attracted, or, which is the same, more repelled. Hence, whether the inner surface of the bulb be cooler or hotter than the fly, a reversal in the direction of rotation, while the fly is being heated, indicates a reversal in the order of absorbing power of the two faces, and that, again, shows that the order is different for different components of the total radiation, and that the ratio of the intensity of those components has been changed.

It is perhaps hardly necessary to observe that the radiometers mentioned in this paper are of the usual form—that is to say, that their arms are symmetrical, so far as *figure* is concerned, with respect to a vertical plane passing through the point of support. Accordingly the rotation which is attained, for instance,

---

* Theoretically there would be a minute difference of temperature, produced, other circumstances being alike, by the difference in the absorbing or emitting power of the two faces of a disk, as regards the radiation which is the difference between the radiations from or towards the affected portion of the bulb and the same portion at the normal temperature. But this, and the repulsion or attraction corresponding to it, would be only a small quantity of the second order, the main effect being deemed one of the first order.

with a radiometer with concave disks of aluminium, alike as to material on both faces (of which kind, again, I owe a beautiful specimen to Mr Crookes's kindness), has not been referred to. This rotation, depending on the more favourable presentation, to the bulb, of the outer (and therefore nearer and more efficient) portions of the fly on the convex than on the concave side, has nothing to do with the one isolated subject to which the present paper relates, namely, the elucidation of the peculiar behaviour in certain cases of certain kinds of radiometers, by a consideration of the heterogeneous character of the total heat-radiation.

November 20, 1877.

2nd P.S.—This morning I received from Mr Crookes an account of the behaviour of a kind of radiometer which he was so good as to construct at my suggestion. The consideration of an experiment mentioned in a paper of his presented to the Royal Society, which will shortly be read, and which he has kindly permitted me to refer to, suggested to me the desirability of investigating the effect of mere roughness of surface, all other circumstances being alike, and the disk of the radiometer being metallic, so that the two faces may be regarded as practically at the same temperature. Mr Crookes's experiment, above referred to, led me to suspect that mere roughness might increase the efficiency of a surface; and I suggested to him some experiments with heated glass shades, or with a hot poker presented to the radiometer, the bulb being covered with a cool tumbler to defend it from being heated by the rays easily absorbed by glass. The result in every case answered my expectation; and it may be stated shortly that the law of the motion is that when the fly is hotter than the bulb the rough surface is repelled, or, say, the motion is positive; when cooler, negative.

I subjoin Mr Crookes's memorandum of the results of experiment:—

"Aluminium Radiometer (1326), one side of the vanes being ruled closely with a sharp knife.

"1. Exposed to standard candle 3 inches off. Continuous *positive* rotation (ruled side repelled) at rate of $3\frac{1}{4}$ revolutions a minute.

" 2. Exposed to non-luminous flame of a Bunsen burner 3 inches off. Continuous *positive* rotation at the rate of $7\frac{1}{2}$ turns a minute.

" 3. The Bunsen burner removed. The positive rotation gradually diminished till it stopped. No negative rotation.

" 4. The bulb heated with Bunsen burner. Good *negative* rotation; then stopped, and rotated *positively* till quite cold.

" 5. Bulb covered with a cold glass shade, and a large red-hot ring applied round equatorially. *Positive* rotation, but not very strong.

" 6. On removing the shade and ring the positive movement soon comes to rest.

" 7. Covered with a hot glass shade, *negative* rotation, with *positive* rotation on cooling (the same as 4).

" 8. Plunged into hot water. *Negative* rotation.

" 9. Removed from the hot water, and immediately plunged into cold. *Positive* rotation."

Results nearly identical were obtained with another radiometer described as " silver radiometer (1327), one side coated with finely divided silver, electro-deposited."

We must accordingly recognize three distinct conditions under which motion may be obtained in a radiometer, namely:—(1) difference of temperature of the two faces, as in a pith radiometer coated on one face with lampblack ; (2) more favourable presentation of one face than the other, as in a radiometer with curved disks; (3) roughness of surface on one face (if this be really different from 2). These three conditions may be variously combined so as to assist or oppose each other, as the case may be, in producing motion.

December 20, 1877.

ON THE QUESTION OF A THEORETICAL LIMIT TO THE APERTURES OF MICROSCOPIC OBJECTIVES.

[From the *Journal of the Royal Microscopical Society*, June 5, 1878.]

I HAVE just received from Mr Mayall, jun., a photograph of Professor R. Keith's computations relative to an immersion $\frac{1}{8}$ microscopic objective by Mr Tolles. I have not at present leisure to go through this long piece of calculation, which I am the less disposed to do as the calculation is perfectly straightforward, and has evidently been made with great care, and I can see no reason to question the result. The only reason for scepticism as to the results of such calculations seems to be a notion derived from *à priori* considerations, that it is impossible to collect into a focus a pencil of rays emanating from a radiant immersed in water or balsam of wider aperture than that which in such a medium corresponds to 180° in air, or, in other words, than 2γ, where γ is the critical angle.

I do not wish to enter into controversy on the subject, or to criticise the arguments by which this statement has been sustained; I prefer to show directly that it has no foundation.

To disprove an alleged proposition, the shortest and least invidious plan is often to show by one or more particular instances that it is untrue.

Suppose a pencil of parallel rays is incident upon a refracting medium of index μ, and let it be required that it be brought without aberration to a focus q within the medium. By a well-known proposition, the form of the surface must be that of a prolate ellipsoid of revolution generated by the revolution of an ellipse of which q is the further focus and μ^{-1} the eccentricity, about its major axis, which is parallel to the incident rays. Conversely, if q be a radiant within the medium, the emergent rays are parallel to the axis.

The limit of the incident parallel rays in any section through the axis is the pair that touch at the extremities of the minor axis. Consequently in the reversed pencil the limiting rays are those that proceed from q to the extremities of the minor axis. If we suppose the index to be 1·525, for which $\gamma = 40° 59'$, the extreme rays will be inclined to the axis at the complementary angle 49° 1'. Hence a radiant within glass may send a pencil of aperture 98° 2', which by a single refraction shall be brought accurately to a second focus at infinity. The double of the critical angle is only 81° 58', so that the aperture exceeds that supposed limit by 16° 4'.

If it were required that the pencil after the single refraction should converge to a real focus, the surface would have to be generated by the revolution of a Cartesian oval instead of an ellipse. If the distance of the point of convergence were considerable compared with the dimensions of the glass, it is evident that the oval would not differ much from the ellipse considered in the first instance, nor would the extreme aperture in glass fall much short of the limit assigned above. Or again, the rays emerging from the ellipsoid might be brought to converge to a second focus q' in air by receiving them on a prolate spheroid of which q' is the further focus and μ^{-1} the eccentricity, and allowing them to emerge from the glass by a spherical surface of which q' is the centre. Or the parallel rays might be brought to a focus without sensible aberration as is done in telescopes.

I do not, of course, propose this as a practical construction of a microscope. It is intended simply and solely to show the fallacy of the supposed limit of 2γ assigned to the aperture, within a medium, of a pencil which can be brought without sensible aberration to a focus in air. As the sphericity rather than spheroidicity of the surfaces employed does not enter in any way into the arguments by which the limit in question is attempted to be established, the spheroidal or Cartesian surfaces are quite admissible in argument.

Nevertheless, as an example of what can be done without going beyond spherical surfaces, and as bearing in a very direct way on actual practice, I will take another instance.

Let it be required to make a pencil diverging from a radiant point Q in glass diverge from a virtual focus q after a single refraction into air.

If P be a point in the required surface, $\mu QP - qP$ must be constant, which gives, according to the value we arbitrarily assign to the constant, an infinity of Cartesian ovals, any one of which, by its revolution round Qq, would generate a surface which may be taken for the bounding surface of the glass. In one particular case the oval becomes a circle, namely, when the constant $= 0$, in which case we have a circle cutting the line Qq internally and externally in the ratio of 1 to μ.

This case is represented in the figure, in which O is the centre of the circle HAL, which by revolution round the line qQA generates the sphere. Rays diverging from Q within the glass proceed after refraction at the surface of the sphere as if they came from q. To find the limit of the pencil, we have only to draw the tangents qHK, qLM, and HK, LM will be the extreme rays after refraction.

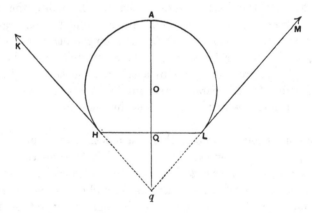

The incident rays QH, QL corresponding to these are inclined to the normals OH, OL at the critical angle. It is easy to prove that the lines QH, QL are prolongations of each other, so that the aperture *in glass* of the pencil which, after refraction into air, diverges without aberration from q, is 180°. The aperture HqL of this pencil, after refraction into air, is 2γ, which with the above value of γ, for which the figure is drawn, comes to 81° 58'. Setting aside chromatic variations, the refracted rays proceed, of course, as if they came from q, forming a pencil of aperture 81° 58′. A pencil of aperture in air no greater than 81° 58′ is one which all parties allow can practically be brought to a focus; it could be brought *exactly* to a focus by the use of surfaces other than spherical.

The spherical surface of no aberration accords with the form of the first lens to which the makers of immersion objectives have been led. By reducing somewhat the excessively large segment of a sphere represented in the figure, say reducing it to a hemisphere, the space gained in front (of thickness QO if the reduction be to a hemisphere) is available for the cover or interposed balsam, which have both nearly the same index as the crown glass of the first lens; and the aperture in glass, though reduced from the extreme of 180°, still remains very large.

On an Easy and at the same time Accurate Method of Determining the Ratio of the Dispersions of Glasses intended for Objectives.

[From the *Proceedings of the Royal Society*, xxvii, June 18, 1878.]

In examining the dispersive powers of a great variety of glasses prepared by the late Rev. W. Vernon Harcourt, I had occasion to examine several prisms which were too much striated to show clearly even the boldest dark lines of the solar spectrum. I found that I was able to get a fair measure of the dispersive powers even of these by a method depending on the achromatizing of one prism by another. If the method succeeded even with such prisms, it stands to reason that it would be still more successful with prisms of good glass.

For the construction of an objective we require but one datum as regards the dispersions, namely, the ratio of the dispersions, or rather, the ratio which on being treated as if it were the ratio of the dispersions gives the best results in practice.

If it were not for irrationality, the matter would be comparatively simple. The ratio of the dispersions would then be the same for whatever interval of the spectrum it were taken; and we should merely have to take two well-defined lines, bright or dark, situated as nearly as may be at the extremities of the spectrum, so that any errors of observation should be divided by as large a quantity as practicable, to measure the refractive indices of the two glasses for each of those two lines, and to take the ratio of the increments in passing from one line to the other. But in consequence of irrationality we get a different ratio according to the particular interval we choose; we are obliged, unless we adopt some different method altogether, to observe more than two lines in each glass; and when we have got the results, that is to say, the indices for several fixed lines in each glass, it still becomes a question what it is best to do with them so as to produce the best result.

Fraunhofer proposed an empirical rule for combining the results*, but remarked that the number thus obtained was not exactly that which answered best in practice. I have elsewhere shown reason to think that the rule may be taken to be simply that the ratio of dispersions to be chosen is that for an infinitely small portion of the spectrum at its brightest part†; or in other words, that the focal length of the combination of two glasses must be made a maximum or minimum (it is practically a minimum) at the brightest part of the spectrum.

The refractive index of one glass may be expressed in terms of that of another, or of some standard glass, by an interpolation formula with three, or at most four, terms. The most accurate observations with prisms are only just sufficient to show that a fourth term is needed, and for the practical purpose of the construction of object-glasses, where in consequence of irrationality the ends of the spectrum are sure to be a good deal out of focus, it will be amply sufficient to confine ourselves to three terms. The three coefficients in the interpolation formula may be determined by observing in each glass the indices of refraction for three selected lines, though it is well to observe more than three lines, and combine the results.

The angular extent of the spectrum being but small in practical cases, the necessity of determining three constants by observations made in it requires great delicacy of measurement. Small errors of observation would easily produce an error in the deduced ratio which would be sensible, or even material, in practice.

For the actual construction of an object-glass we require, as has been already remarked, the knowledge of only a single constant relating to the dispersion, not of two; and if we can find some test whereby to know when the required condition is satisfied, we may dispense with such extreme accuracy in the angular measurements.

* *Denkschriften der K. Akad. der Wiss. zu München*, Vol. v, p. 215.
† *Report of the British Association* for 1855, Part ii, p. 14 [*ante*, Vol. iv, p. 63].

Such a test is afforded by the secondary tints, which change with extreme rapidity when the refracting angle of one of the mutually achromatizing prisms, or the position of one of them in azimuth, is altered. If a moderately narrow object with vertical sides, black on a white ground, or white on a black ground, be viewed through opposed prisms, one of crown and one of flint glass, with a small telescope, and the prisms be set to achromatize each other as nearly as may be, it will be found that the slightest touch altering one of the prisms in azimuth alters notably the secondary tints.

The secondary tints in an objective are readily shown by directing the telescope to a vertical line separating light from dark, such as the edge of a chimney seen in the shade against the sky, and covering half the object-glass with a screen having a vertical edge. So delicate is this test that on testing different telescopes by well-known opticians, a difference in the mode of achromatism may be detected. The best results are said to be obtained when the secondary green is intermediate [in tint] between green and yellow. This corresponds to making the focal length a minimum for the brightest part of the spectrum.

To enable me to form a judgment as to the sharpness of the test furnished by the tint of the secondary green, as compared with the performance of an object-glass, I tried the following experiment.

A set of parallel lines of increasing fineness was ruled with ink on a sheet of white paper, and a broader black object was laid on it as well, parallel to the lines. The paper was placed, with the black lines vertical, at a considerable distance in a lawn, and was viewed through two opposed prisms, one of crown glass and the other of flint, of such angles as nearly to achromatize each other in the positions of minimum deviation, and then through a small telescope. The achromatism was now effected, and varied in character, by moving one of the prisms slightly in azimuth, and after each alteration the telescope was focused afresh, to get the sharpest vision that could be had. I found that the azimuth of the prism was fixed within decidedly narrower limits by the condition that the secondary green should be of such or such a tint, even though no attempt was made to determine the tint otherwise than by memory, than by the condition that the vision of the fine lines should be as sharp as possible. Now a small element of a double object-glass may be regarded, so far as chromatic compensation is concerned, as a pair of opposed prisms; and therefore we may infer that the tint of the secondary green ought to be at the very least as sharp a test of the goodness of the chromatic compensation as the actual performance of the telescope. And such Mr Thomas Grubb, to whom I mentioned the test, found to be actually the case in the progress of the construction of the 15-inch refractor for the Royal Society, the instrument at present in the hands of Mr Huggins.

It follows therefore that two opposed prisms, representing the glasses to be employed in an objective, are to be deemed to achromatize each other when one of the two secondary colours is about midway between yellow and green.

The condition of achromatism of two opposed prisms is given in the ordinary treatises on optics, but, so far as I have seen, rather as an optical curiosity than as a matter of practical utility. Sir David Brewster in his treatise on " New Philosophical Instruments," p. 292, alludes to it as furnishing a conceivable mode of determining dispersive powers, but mentions it only to condemn it. I cannot imagine why, for, at least with the modifications which I have been led to introduce in putting it in practice, I find it to be no less excellent than easy.

The experimental arrangements are a good deal simplified by making the prisms to be compared achromatize successively one and the same prism chosen arbitrarily, and retained in a fixed position, instead of making them directly achromatize each other. I will first describe the method as I used it, premising that as the prisms for which it was primarily designed were not of good glass, I was content with less perfect arrangements than would have been desirable for really good glasses. Nevertheless the apparatus when used with good prisms gave very good results. The object observed was a vertical slit, so wide that when viewed through the opposed prisms a broad white stripe was seen, merely fringed at the two edges with the secondary green and purple. The slit was fixed at one end of a horizontal plank, near the other end of which was fixed a vertical lens of about four feet focus. The slit was in the principal focus of the lens. The plank rested on three knobs, one under the slit, the other two near the other end. This mode of support prevented torsion of the plank, which would have produced a lateral derangement of the slit. The rays from the slit, rendered parallel by the lens, fell upon a prism, which I will call the primary prism. This prism is fixed during the observation. Its angle and dispersive power, and the azimuth in which it is set, are arbitrary within wide limits.

The beam emerging from the primary prism falls, at a little distance, on a second prism resting on a stand movable along with the verniers of a protractor reading to minutes (for the loan of which I am indebted to Professor Miller), which latter rests on the plank. As a matter of convenience, the prism has an independent motion in azimuth, but when once placed in a convenient position it is not afterwards moved except along with the verniers.

The beam retracted and dispersed in contrary directions by the two prisms is received on the object-glass of an achromatic telescope unconnected with the graduation, with which the image of the slit, if so wide an aperture may so be called, is viewed. In general the edges of the aperture are seen coloured blue and red, which colours, on turning the prism through its azimuth of achromatism, are changed into red and blue by passing through the secondary green and purple. That edge of the slit at which the green is seen is alone attended to, and is treated as the fiducial edge. The prisms are deemed to achromatize each other when this colour is midway between yellow and green.

Should the observer wish to aid his judgment by observing the purple as well, it would be proper to use an aperture of the form represented in the figure, so that the green and purple should be seen right and left along a common edge, and therefore at the same angle of incidence, or rather at angles of incidence having the same horizontal projection. I do not know that anything would be gained by this; I have found it sufficient to attend to the green.

The light employed had best be that of the sky, reflected horizontally by a looking-glass. It should be fairly bright, but not approaching to dazzling. Thus the light reflected from the sky near the sun would be too bright. The room need not be darkened; in fact, it is better that the eye should be kept fresh for the appreciation of colour by habitually looking about on ordinary objects. A simple collimating lens such as I have described is sufficient, though doubtless an achromatic would theoretically be an improvement. It should be of longish focus,

at least in the case of a simple lens, lest any slight displacement right or left of the middle of the incident beam should introduce a minute dispersion due to the lens acting as a prism. If the middle of the beam be not quite central, that does not signify, provided the eccentricity be constant; for then the minute dispersion is merely added to that of the primary prism, which is arbitrary. To ensure constancy of incidence on the lens, it is well to limit it by an aperture with vertical sides, and to take care that the beam employed is wide enough to fill the aperture. Similarly it is well to take care that the beam falls centrically, or pretty fairly so, on the achromatic viewing telescope. But what is of much more consequence is that the ray passing through the optical centre of the object-glass should pass centrically through the eye-piece, as otherwise the dispersion of the eye-piece for eccentrical pencils would alter the secondary tint. It is well, therefore, that the viewing telescope should be furnished with cross wires. The telescope is then moved a little till the fiducial edge is on the cross wires when the tint is observed.

The determination of the azimuth of achromatism is the capital observation on which the accuracy of the result depends; in comparison with it, the rest of the measurements required may be deemed exact. Accordingly a number of observations of this azimuth should be taken, and the mean adopted. The mean error will vary with circumstances, but it may be taken ordinarily as a few minutes.

The mean reading for achromatism by itself alone gives nothing, and must be combined with another determination in order to be available. The second determination which I found it most convenient to make was that of the angle of incidence of a particular part of the spectrum on the second prism for a known azimuth of this prism.

If the brightest part of the spectrum were marked by a definite line, we should choose that. As it is, the line D, though not exactly at the brightest part, lies sufficiently near it for our purpose, as will be better understood presently. No dark lines can be seen in the reflected light, because we are using an aperture, and not a mere slit. But if the slit be illuminated by a soda flame instead of daylight, a well-defined yellow image of the aperture will be seen, the fiducial edge of which can be pointed at with precision.

To determine the angle of incidence I have employed two methods, both of which I think it will be worth while to describe. The first is the less direct, and involves a little subsequent calculation, but has the advantage of not requiring any additional apparatus beyond what is wanted for the determination of the azimuth of achromatism. In carrying out the first, and in applying either, I suppose the angle of the prism and its index for the line D known from a determination in the usual way.

First Method.—The azimuth of minimum deviation for the line D could be determined pretty fairly by bringing the wire of the viewing telescope to the fiducial edge when the aperture is illuminated by a soda flame, and the edge is in or near its stationary position, and taking the mean of several determinations of the azimuth at which the edge is stationary. The angle of incidence for minimum deviation is virtually measured in the process of finding the index, and is therefore known; and by applying with its proper sign the difference of mean readings for achromatism with daylight and for minimum deviation with a soda flame, we get the angle of incidence for achromatism.

This method is mentioned merely as naturally leading up to that actually employed. It must be rejected as too slovenly, since the uncertainty of the determination of the azimuth of minimum deviation is liable to be greater than that of the azimuth of achromatism.

If we place the wire of the viewing telescope some way off the stationary position of the fiducial edge, there are two azimuths of the prism at which the edge will be on the wire, across which it will move with a finite velocity as the prism is moved in azimuth. Hence either azimuth could be determined with accuracy, and if they were equidistant from the azimuth of minimum deviation, the latter could be determined at once. This, however, is not the case, but nevertheless the azimuth of minimum deviation can be determined from the result.

Since the course of light may be reversed in refraction, it readily follows that in passing from one to the other of two azimuths which give equal deviations, the angles of incidence and emergence are simply interchanged. Hence when the prism is turned from one to the other of two azimuths for which the image of the fiducial edge is on the wire of the viewing telescope, the angle moved through is equal to the difference between the angles of incidence and emergence in either position. Both azimuths, and therefore their difference, can be determined with accuracy, provided the azimuths be sufficiently remote from that of minimum deviation.

The process of observation, then, is this. Set the prism a good way, such as 10° or 15°, from the azimuth of minimum deviation, and read the graduation. Turn the viewing telescope till the fiducial edge is on its wire. Taking care to keep the telescope fixed, turn the prism through the position of minimum deviation till the edge is again on the wire, and read again.

Let ϕ, ψ be the angles of incidence and emergence or emergence and incidence, of which let ϕ be the greater. Let 2α be the angle, i, of the prism, 2β the sum of 2α and the minimum deviation, 2γ the measured angle through which the prism has been turned, or $\phi - \psi$. Since the sum of the internal angles $= 2\alpha$, we may represent them by $\alpha + x$ and $\alpha - x$. Let $\phi + \psi$ be denoted by $2y$. Then we shall have

$$\frac{\sin(y+\gamma)}{\sin(\alpha+x)} = \frac{\sin(y-\gamma)}{\sin(\alpha-x)} = \frac{\sin\beta}{\sin\alpha}.$$

Eliminating x from these two equations, we find

$$\cos^2 y = \frac{\cos^2\alpha \cos(\beta-\gamma)\cos(\beta+\gamma)}{\cos(\alpha-\gamma)\cos(\alpha+\gamma)} \quad \ldots\ldots\ldots\ldots(1),$$

which gives y, and then $y \pm \gamma$, *i.e.* ϕ or ψ, is known. The angle of incidence for a known reading of the graduated circle being known, we have only to apply the difference between this reading and that for achromatism to that angle of incidence in order to get the angle of incidence for the azimuth of achromatism.

Second Method.—In this a little telescope is used, which is attached to the stand of the prism, so as to move with the verniers of the circle. The telescope need not be achromatic, but has cross wires in its principal focus.

After determining the azimuth of achromatism, the slit is illuminated by a soda flame, and the prism with the verniers turned till the fiducial edge is on the wires of the measuring telescope, when the circle is read. The prism is then removed, and the measuring telescope turned till the fiducial edge is seen directly, and the circle is read again. Half the supplement of the difference of the readings gives the angle of incidence when the reflected image was on the cross wires; and by applying the difference of readings for reflection and for achromatism, we get the angle of incidence in the position of achromatism.

This angle, ψ, having been determined by either of the above methods, we have, by the known formula

$$\frac{d\mu}{\operatorname{cosec} i \cos \phi' \cos \psi} = \frac{d\mu_{,}}{\operatorname{cosec} i_{,} \cos \phi_{,}' \cos \psi_{,}} \quad \ldots\ldots\ldots(2),$$

where the letters with the suffix $_{,}$ refers to the second prism. For the prisms would achromatize each other, as is supposed in the deduction of the above formula in treatises on optics, under the same circumstances in which they would achromatize, in succession, the same spectrum. In the application of the formula, it is to be remembered that, of the two angles ϕ, ψ, the former is that which lies on the side of the white light, and is, therefore, the angle of incidence for the first, but of emergence for the second, of two prisms which mutually achromatize each other.

In the application of the formula (2) the angles ϕ', ψ belong, strictly speaking, to the brightest part of the spectrum, which for shortness I will call M, for which the value of the differential coefficient $d\mu_{,} : d\mu$ is supposed to be sought. But the distance of M from D is so small that it will hardly make any sensible error if we use the values of the angles belonging to D, for not only is the correction to the product $\cos \phi' \cos \psi$ for either prism very small, but the two corrections are in the same direction, and therefore tend to neutralise each other in the ratio of the products, with which alone we are concerned. If, however, we care to introduce the correction, it can be done at the expense of a little additional calculation. In a crown glass the index for M may be taken as greater by 0·001 than that for D, and in a flint glass as greater by 0·001 multiplied by a rough value of $d\mu_{,} : d\mu$. The angle of emergence ϕ may be taken to be the same for M as for D. For the deviation, regarded as the function of the index, is a maximum or minimum for M; and D being so near M, the difference of deviations for D and M will be quite insensible compared with the errors of observation of the azimuth of achromatism, with which it is associated. Let the letters $'\mu$, $'\psi$, &c., refer to M, while μ, ψ, &c., refer to D. ψ is obtained by observation, and ψ', ϕ' must, in any case, be calculated from thence by the known values of μ and i. We have now merely to calculate the $'\phi'$ and $'\psi$ for each glass from the formulæ

$$\sin '\phi' = \frac{'\mu}{\mu} \sin \phi', \quad '\psi' = i - '\phi', \quad \sin '\psi = '\mu \sin '\psi',$$

and substitute these values in the equation (2) instead of those belonging to the line D.

The primary prism had best be made of very low flint glass (or else be a compound prism formed of two prisms of crown and flint glass respectively, with their angles turned the same way), so as to fall about midway between the glasses to be compared, and thereby divide between them the irrationality which has to be encountered in the observation. The observation of the azimuth of achromatism is most accurate when there is little or no irrationality ; and, if preferred, the crown glasses might be compared with a standard crown, and the flint glasses with a standard flint, the primary prism being in the one case any crown glass prism that happens to be at hand, and in the other case a flint glass prism. The crown glass to be measured being compared with the standard crown by using them in succession to achromatize the same primary crown in the same position, and similarly for the flint glass to be measured and the standard flint, we can deduce the ratio of the dispersions of the crown and flint glasses to be measured if we know, once for all, the ratio of the dispersions of the standard crown and flint glasses. This may be determined by a specially careful series of observations of the kind above described, made once for all, or, if preferred, by the method of indices.

The direct comparison of a crown and flint glass is, however, so accurate, especially if a glass of intermediate quality be used for the primary prism, that I feel satisfied it will suffice for practical purposes. It is hardly necessary to observe that, if the primary prism be of intermediate quality to the two compared, the right hand edge of the aperture will be the fiducial edge in the one case, and the left hand edge in the other. In saying this, I assume that we have not extravagant inclinations or differences of angles to deal with, for there is a *quasi*-irrationality observed even when two prisms of the same glass, but of different angles, achromatize each other, which is, however, ordinarily so small that it may be neglected in comparison with the real irrationality of the media.

By turning the primary prism into a different azimuth, or substituting a different primary prism, and repeating the observation, an estimate may be formed of the degree of reliance that may be placed on the results.

To give an idea of the degree of accuracy of which the results are susceptible, I subjoin a few numbers extracted from my note-book. The prisms designated H 74, H 88, H 98, were experimental prisms of phosphatic glass of different compositions. They were more or less striated, but were good enough to show the principal fixed lines of the spectrum. In the different determinations of the ratio of dispersions, the primary prism was set at different azimuths. In the calculation of (2) the indices for D were used. The differences from the means are exhibited.

H 74 to H 98, 1·882, 1·892 ; mean, 1·887 ; differences, − 0·005, + 0·005.

H 74 to H 98, 1·755, 1·761, 1·781 ; mean, 1·766 ; differences, − 0·011, − 0·005, + 0·015.

It will be seen that, even with prisms such as these, by taking the mean of different determinations, the uncertainty can hardly be as great as one-half per cent.

Extremely small prisms are quite sufficient for the determination of the ratio of the dispersions of the glasses by the
above method. It may, however, happen that an optician cannot
afford to remove even so small a piece of glass from a disk
intended for an objective, and has not a specimen of glass on the
identity of which with the glass of his disk he can thoroughly
rely. In such a case it is necessary to determine the optical
constants of the disk by means of facets cut on the disk itself.
A heavy and costly disk cannot be treated like a small prism, and
mounted on a small graduated instrument in the manner I have
supposed a small prism treated. To compare the ratio of the
dispersions of two such disks, one of crown glass and the other of
flint, the most convenient way would seem to be to leave the
disk fixed, let the light pass through it first, and then achromatize
it by a small prism of very low flint glass, mounted on a small
graduated instrument in the manner already explained. The
dispersions of the disks would be compared with each other by
comparing them in succession with the same intermediate
prism.

This, however, demands an additional determination beyond
what was required in the other process, since the prism through
which the light first passes is not the same in the two cases. The
element which best lends itself to measurement is the angle of
incidence on the first surface. The most convenient mode
of measuring this must depend on the general disposition of the
apparatus adopted to take the measurements for the angle of the
disk and the deviation of some one line, which must be made in
any case; it is accordingly best left to the choice of the observer.

ON A METHOD OF DETECTING INEQUALITIES OF UNKNOWN PERIODS IN A SERIES OF OBSERVATIONS.

[Note appended to a paper by Prof. B. Stewart and W. Dodgson*; *Proceedings of the Royal Society*, XXIX, pp. 122–3, May 29, 1879.]

As the search for periodic inequalities of unknown period must always be more or less laborious, it seems desirable to point out another mode in which the search might be conducted, and which seems to offer great facilities for the object, assuming the possession of the required instrument.

It seems to me that the harmonic analyser of Sir William Thomson is singularly well adapted to this purpose, which, as I have ascertained from him, was one of the applications of his machine that he has had in view.

If $f(t)$ be any function of the time t given by observation, and $2\pi/n$ a period p assumed at pleasure, then by plotting if necessary the function on a scale adapted to the paper cylinder of the machine, we shall get, by a simple mechanical process, the values of the integrals

$$\int f(t) \sin nt\, dt, \qquad \int f(t) \cos nt\, dt,$$

between any limits. We may take the inferior limit for the origin of the time, and then by reading off the cylinders of the machine for as many values of the superior limit t as we please, we shall get the corresponding values, as many as we like, of the integrals.

Suppose, now, that $f(t)$ contains a small term of the form

$$c \sin (n't + \alpha),$$

where n' is not much different from n, so that the period tried approaches closely to the period p' of this inequality. The corresponding part of the integrals will be

$$\frac{c}{2(n'-n)} \sin \{(n'-n)t + \alpha\} - \frac{c}{2(n'+n)} \sin \{(n'+n)t + \alpha\},$$

[* This paper was republished with alterations and additions in *Report of the Committee on Solar Physics*, 1882, pp. 173—206 : cf. note reprinted *infra*.]

and

$$-\frac{c}{2(n'-n)} \cos\{(n'-n)t+\alpha\} - \frac{c}{2(n'+n)} \cos\{(n'+n)t+\alpha\},$$

taken between the proper limits. The terms divided by $n'+n$ having but a small coefficient in the numerator, and having a denominator which is not small, may be left to take their chance with casual fluctuations; but the terms divided by $n'-n$ rise into importance from the smallness of the denominator, and express an inequality in the integrals of comparatively large amount and long period.

We may therefore confine our attention to the terms

$$\frac{c}{2(n'-n)} \sin[(n'-n)t+\alpha], \quad \frac{c}{2(n'-n)} \cos[(n'-n)t+\alpha],$$

occurring in the indefinite sine integral and cosine integral respectively.

If, therefore, the values of these two integrals, obtained from the two cylinders respectively, be plotted, we shall obtain a periodic fluctuation of a period more or less long as we hit on a period more or less near to that of the inequality which we have supposed to exist. The zero points of the fluctuation in the sine integral will correspond to the maxima and minima in that of the cosine integral, and *vice versâ*. The reciprocal of the period of the fluctuation will give the difference of the reciprocals of p and p', and thus p' will be known from p provided only we know which of the two, p, p', is the greater. This will be shown by comparing the phases of the fluctuations of the two integrals. If $n' > n$, the fluctuation of the cosine integral will be a quarter of a period behind, if $n' < n$ before, that of the sine integral.

If $f(t)$ be subject to known periodic inequalities with approximately known coefficients, the integrals should be cleared of the terms thence arising, if of sufficient moment, by using their analytical expressions, and the residues only plotted.

Of course $f(t)$ is not necessarily a function of the time given by *direct* observation; it might be a function deduced from one so obtained. For example, $f(t)$ might be the coefficient of the principal term in the daily fluctuation of the element when each day's record is separately subjected to harmonic reduction.

DESCRIPTION OF THE CARD SUPPORTER FOR SUNSHINE RE-
CORDERS ADOPTED AT THE METEOROLOGICAL OFFICE.

[From the *Quarterly Journal of the Meteorological Society*, April, 1880.]

THE method of recording sunshine by the burning of an object
placed in the focus of a glass sphere freely exposed to the rays of
the sun, which was devised by Mr Campbell, commends itself by
its simplicity, and seems likely to come into pretty general use.
In the original form of the instrument the rays were received on a
hemispherical wooden bowl, concentric with the glass sphere, and
of such a radius that the focus should fall on its inner surface. The
instrument in this form will give total effects, but only in a very
rude manner the results for individual days, since the burnings
produced on neighbouring days run into one another, and to use a
fresh bowl for each day on which the sun shone would be out of the
question on account of the expense. Accordingly it is expedient
to adopt Mr Scott's modification of the instrument, and replace the
wood by a slip of card, which can be renewed from day to day;
and it is necessary to support the slip in such a manner that the
image of the sun shall not run off it from sunrise to sunset, and
moreover that the focus shall fall, approximately at least, on the
surface during that interval.

The most obvious way of supporting the slip would be to make
it rest against the inner surface of a hemispherical bowl formed of
metal, slate, or earthenware, and such is the plan adopted at the
Royal Observatory. But this method could hardly be intrusted to
inexperienced observers; for, in order that the slips may sufficiently
nearly fit the surface of the hemisphere, they must be narrow;
and in that case a moderate error in the placing of a slip would
suffice to make the image of the sun run off it in some part of the
day. Yet there is nothing to guide the observer as to the proper
placing but certain marks on the hemisphere, respecting which he

might easily make a mistake, especially as the slip has to be fastened by clamps. The slips have to be cut of a particular form, varying with the declination of the sun ; and though correctly cut slips could be furnished from head-quarters, so many different patterns are required in the course of the year that there is risk of confusion on the part of the observer. If a dated blank slip were returned to the Office, there would be nothing to indicate whether the absence of marks of burning really arose from cloud, or was due to a misplacement of the slip in the bowl.

Other forms of support have been devised in which the card is simply slipped into its place, so that the fastening presents no difficulty. But these mostly labour under one or other of two defects, namely, that in the course of a long day the image is liable to run off the slip, or that when the sun is a good way from the meridian its image is too much out of focus.

It seemed therefore desirable to devise a form in which these defects should be avoided, while at the same time it should be sufficiently cheap in construction, and should demand little skill on the observer's part in the placing of the cards.

It is believed that these requirements are satisfied in the form adopted by the Meteorological Office.

It is well known that a piece of paper or card, regarded as a flexible but inextensible plane surface, cannot be bent, without rumpling or tearing, into the form of a sphere, but only into that of some developable surface. The locus of the focus of the sun's image for several successive days is a zone of the focal sphere, and this zone may be replaced without sensible error by a zone of a developable surface touching the sphere along the middle of the zone, that is, along a small circle which represents the path of the image for a single day, the change of declination during the day being neglected. The developable surface which touches a sphere along a small circle is of course a right circular cone. When the sun is in the equator, and the small circle becomes accordingly a great circle, the cone passes into a cylinder.

We must make provision for receiving the image through a range from 23° 28′ north of the equator to 23° 28′ south, or say for round numbers, and to allow a little for the breadth of the image, from 24° N. to 24° S. Let C (Fig. 1) be the centre of the sphere,

AE an arc of a section of the focal sphere by a meridian plane through *C*, the arc extending from 24° N. at *A* to 24° S. at *E*, *GG* a portion of a section of the sphere. Divide the arc of 48° *AE* into three equal parts, *AB, BD, DE*, and draw tangents through the middle points of these arcs, cutting the radii through *A, B, D*,

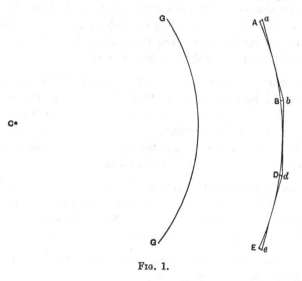

FIG. 1.

E, in *a, b, d, e*. Then the zone of the spherical surface generated by the revolution of *AE* may be replaced with very little error by zones of two conical and one cylindrical surface, these zones being generated by the revolution of *ab, de, bd*; and these being developable surfaces, bits of card may be applied so as to fit them accurately. If *R* be the radius of the focal sphere, the extreme error of focus committed will be $R(\sec 8° - 1)$. If the glass spheres be of 4 inches diameter, and the glass be free from lead, *R* will be a little under 3 inches, and $3(\sec 8° - 1) = 0.0275$ nearly, so that the greatest error of focus would be the 1-36th of an inch. If instead of the central tangents we take lines parallel to them, and passing through the middle points of *Aa, Bb, Dd*, the greatest error will be halved, or reduced to the 1-72nd of an inch; and as the spheres burn very well through a range of 0·1 inch in distance from the centre, the small error of the 1-72nd of an inch is of little consequence. If we add, say 0·020 inch for the thickness of the card, and deduct $\frac{1}{2}$ of 0·0275, or say 0·014, for the reason above mentioned, we get 0·006 to be added to the distance of the best burning focus from the centre to get the perpendicular distance from the centre on any one of the three supporting surfaces. This correction is so small that it may be neglected. In the pattern adopted this perpendicular is taken at 2·89 inches.

The fiducial supporting surface is now reduced to that generated by the revolution of *abde* about the polar axis through *C*, a line therefore parallel to *bd*. This forms the inner surface of the supporting material; and from the winter to the summer solstice the image travels from *a* to *e*. From about October 14 to February 27 the image is on some part of *ab*; from February 28 to April 10 in some part of *bd*; from April 11 to September 1 on *de*; and from September 2 to October 13 again on *bd*.

If the cards were in section no larger than the exact sizes *ab*, *bd*, *de*, here given, the image would at certain times of the year fall exactly on the edge of a card. The cards must therefore be a little larger; and in order that they may lie without any rumpling on their fiducial developable surfaces, the material of the support must be slightly cut away by prolonging a little in each direction the cuts *ab*, *bd*, *de*. The prolongations of the cut *bd* might even be made to extend a good way towards the middle points of *ab*, *de*, and similarly for the others, without removing more of the supporting surfaces than can perfectly well be spared. The prolongation of the cuts may be utilised for the support of the cards by undercutting, so as to leave flanges under which the edges of the cards may be slipped. The form thus finally assumed by a section of the supporting surface is represented in Fig. 2.

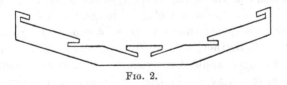

FIG. 2.

The construction of the supporting surface is not expensive. A ring is cast of the approximate form generated by the revolution of the section in Fig. 2, the middle zig-zag line being however replaced by a circle as in Fig. 1, and the shaping of the inner surface and the undercutting are then done in the lathe. When shaped, the ring is cut into two by a plane through its centre, and inclined to its axis at an angle roughly equal to the latitude of the place for which it is intended.   One ring will thus serve for two instruments, each half being mounted independently on a suitable support.

The forms of the cards are easily found.   The equinoctial slips are of course straight.   The form of the ends of these and of the other slips is not now considered for a reason which will presently appear.   If we regard the summer and winter slips as infinitely thin, and coinciding with the surfaces on which they rest, it is evident that when developed they will form portions of circular annuli, the bounding arcs having a common centre where the line ab or de in Fig. 1 (p. 56) cuts the polar axis.   The radius of the developed slip, measured to the arc corresponding to the point of contact, will accordingly be $2\cdot89 \cot 16°$.   If we allow $0\cdot02$ for the thickness of the card, it will be more accurate to take $2\cdot88$ for the coefficient; and if we suppose the depth of the cuttings at the two sides the same, so that the point of contact is in the middle of the card, and if b be the breadth of a card in inches, the outer and inner radii will be $10\cdot04 \pm \frac{1}{2}b$ inches.

The pattern of the rings, which form the only part of the apparatus involving much nicety in construction, is common to all the earth, and at whatever place the ring is to be used the circle in which the inner surface is cut by the ideal equatorial plane is divided by the plane of actual section into two equal parts; it is only the inclination of the cutting plane to the equator which changes from place to place.   If a common mean latitude were adopted for the whole of England, little error would be produced; the semi-rings in the more northerly stations would merely rise slightly above the horizontal plane through the centre of the ball on the northern side of the east and west points, and pass a little below it on the southern side, while for stations south of the mean latitude the error would be reversed.   As the sun hardly ever burns when very near the horizon, this would be practically of no moment.   In that case a common pattern might be adopted for the ends of the cards of any one of the three kinds, but as it is desirable to take in somewhat wider ranges of latitude, such as from Jersey to the Orkneys, and as it is just as easy as not to divide the rings according to the actual latitudes of the places where they have to be used, it has been decided merely to provide that the cards shall be long enough for all the stations, and to leave it to the observers to cut off the ends of the cards level with the horizontal edges of the semi-rings, where the complete ring has been divided. It is, however, only in the case of the equinoctial cards that there is any occasion to cut off the projecting ends as the ends of the summer and winter cards are not in the way of the sun's rays, even at sunrise and sunset.

Each semi-ring is marked inside down its middle, that is, along the line in which it would be cut by a bisecting plane passing through the polar axis. In mounting the stands in the first instance, once for all, this line is to be brought into the plane of the meridian; and in the daily use of the instrument the cards are to be pushed till the noon mark comes to the marked line.

The cards may be graduated beforehand by printing on the cardboard. In planning them, if a batch of cards of the same kind are drawn with the back of one in the bosom of the next, there is very little waste, and the cards can be afterwards cut out by a suitable punch.

For the equinoctial cards the hour lines are evidently a series of parallel straight lines. The interval from one hour line to the next may be taken as $2\cdot88\pi \div 12$, or $0\cdot754$ inch. For the summer and winter slips the hour lines will be straight lines converging (as they lie on the cardboard) to the common centre of curvature of the outer and inner bounding circles. The distance from one hour line to the next, measured along an arc passing through the point of contact, will be $0\cdot754 \cos 16°$, or $0\cdot725$ inch; and as it subtends at the centre of curvature an angle of only $15° \sin 16°$, or $4° 7'$, the length of the chord will be sensibly the same. If the length of the prolongations of the cuts ab, de in Fig. 1 is the same towards, as from, the equator, the arc of contact will be equidistant from the two bounding arcs, otherwise not.

Each half ring is mounted on a slab of slate, to which is fastened a brass upright ending above in a flat surface, about 1 inch or $1\frac{1}{2}$ inch square, inclined to the plane of the base by an angle representing an average latitude for the kingdom. The half ring is fastened by screws to this flat piece, being cast for the purpose in a form which is flat in the middle outside, as represented in Fig. 2 in section. The complete ring, as cast, differs from a solid

of revolution in having two such planes opposite to each other
outside, for the purpose of attachment to the slanting plane of the
upright.   A pedestal ending in a small cup carries the glass ball,
which rests there by its own weight; and even should it be blown
aside by a very violent gale, it cannot fall out, at least in the
instruments suited for our latitudes, as the horns of the half ring
are not wide enough to let it through.   It may, however, be

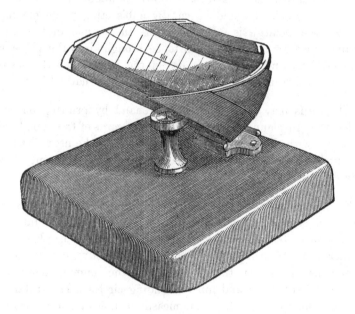

Fig. 3, showing Stand complete.

cemented for security's sake.   The instrument is mounted in its
place as a fixture, and then contains nothing movable except the
glass ball, which (if not cemented) can be lifted out and replaced
in its cup at pleasure.   The cards are introduced at the edges
of the half ring, with their upper and lower edges under their
proper flanges, and then readily slip into their places.   They are
pushed till the noon hour line is a prolongation of the line marked
inside on the brass, and the ends (in the case of the equinoctial
cards) are then cut off level with the horizontal edges of the half
ring, unless they should have been previously cut in the house,
from the pattern given by one of the cards that had been mounted
and cut in the instrument.

In mounting the instrument in place, the points to attend to are, (1) that it shall be level as regards east and west, (2) that the axis of the ring shall be inclined to the horizon at an angle equal to the latitude of the place, (3) that the plane passing through the axis of the ring and the meridian line marked on its inside shall be in the plane of the meridian. There is no occasion to change the pattern of the upright supporting the half ring, since variations of latitude may be allowed for in bedding the slate.

Directions for Adjusting and Using the Sunshine Recorder, issued by the Meteorological Council.

The instrument when in position faces the south; the glass ball rests on the pedestal, and when the sun is shining casts an image which chars a slip of card previously placed in the instrument. As the sun travels from east to west, the place of the image gradually moves along the card, which is thus scored during sunshine, and left untouched when the sun is hid.

1. *Adjustment for Concentricity.*—It is possible that the instrument may require this adjustment. To see whether it does, put the ball into its cup, and see whether in the horizontal plane passing through the ball's centre the surface of the glass stands at the same distance all round from the middle points of the belts on which the cards are destined to lie. If not, the pillar supporting the ball may be adjusted by loosening the screw underneath which fixes it, moving the pillar in the required direction, and when it is right, turning the screw home.

If the adjustment is not within the range of the hole in the slate, which for this object was designedly made a little large, the hole may be enlarged a little in the required direction by filing. Unless you are confident of being able to effect the adjustment thus, you had best not attempt the filing, but write to the Office.

2. *Choice of Position.*—It is almost needless to remark that a position should be chosen where a clear view of the sky, or at least of such portions of it as the sun is liable to occupy, is as little as may be interfered with by buildings, trees, &c. The instrument itself when roughly in position will show what portions the sun is liable to occupy.

3. *Adjustment for Level.*—The instrument is to be placed level as regards east and west, though at most stations, as will be mentioned presently, it requires to be tilted a little in the plane of the meridian. The plane of the top of the instrument, that is, the plane of section of the bowl, may perhaps not be quite parallel to the upper surface of the slate base, and in adjusting for level it is well not to trust to the surface of the slate, but to use the plane of the top of the bowl.

4. *Adjustment for Latitude.*—In most of the instruments which have been made for the United Kingdom, the brackets supporting the bowl have been made to a common pattern, suited to a mean latitude of about 53°. Except for stations very nearly in that latitude, the stand will require to be tilted a little in the plane of the meridian, through an angle equal to the difference between 53° and the latitude of the place. At stations north of 53°, the northern edge of the stand will require to be raised, at stations south of 53° the southern. For the moderate differences of latitude with which we are concerned, the elevation of edge required may be taken nearly enough at one-eighth of an inch for each degree of difference between the latitude and 53°.

In some few of the instruments the brackets have been made to suit a different latitude. In such cases the above rule will apply on substituting that latitude for 53°.

The above rule will suffice for making the adjustment for latitude very nearly right. To test and if need be correct it, the height of the image of the sun should be noted on some day when

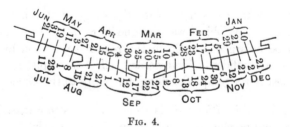

FIG. 4.

the sun is shining within an hour or so of noon, and compared with the proper height for that day. This may be obtained from the accompanying woodcut, which represents a section of the inner surface of the bowl by a plane passing through the polar axis of the ball. The figure is graduated for every 2° of the sun's declination, as well as for the maximum declination ; and the days in the spring and autumn halves of the year at which the sun has most nearly any one of these declinations are written on the woodcut. Should the day on which it is wished to test the adjustment be some intermediate day, the proper place of the image may be obtained by estimation, remembering that the declination changes very slowly about each solstice.

5. *Adjustment for the Meridian.*—This adjustment is best made by means of the time ; and as fairly correct time can now nearly everywhere be obtained, it seems needless to give methods of adjustment in which the time is supposed unknown.

Supposing then the instrument placed roughly in the plane of the meridian, it may be adjusted, provided the sun is shining about noon, by turning it a little, if necessary, in azimuth, so as to make the image of the sun cast by the ball fall on the meridian mark in the instrument at the moment of *apparent local* noon.

We are not restricted to noon for the adjustment. Any other hour may be taken, supposing the card to have been properly inserted, by taking advantage of the hour lines marked on the card. At the moment when any hour is reached according to apparent local time, the instrument is to be turned so as to cause the image of the sun to fall on the corresponding hour line. Should it be cloudy at noon, it would be well to choose for the adjustment an hour not very far from noon, as in that way defects in the other adjustments would have no appreciable effect on the adjustment for the meridian.

This supposes the correct time to be at least fairly well known. The time got from a railway clock will probably be Greenwich or Dublin, &c. time, and to get the local mean time we must first add or subtract a time proportional to the difference of longitude between the station of observation and the place the time of which is given by the clock, at the rate of 4 minutes per degree, adding or subtracting according as the station of observation lies east or west of the place for which the time is given by the clock. Having thus got the local mean time, the local apparent time will be obtained by adding or subtracting the equation of time, as given in the accompanying table.

TABLE giving for every THIRD DAY in LEAP YEAR the EQUATION OF TIME to the NEAREST HALF MINUTE, to be ADDED TO or SUBTRACTED FROM LOCAL MEAN TIME, according as the Sign is + or −, in order to get Local Apparent Time.

Day	January	February	March	April	May	June	July	August	September	October	November	December
1	− 3½	−14	−12½	−4	+3	+2½	−3½	−6	+ ½	+10½	+16½	+10½
4	− 5	−14	−12	−3	+3½	+2	−4	−6	+ 1½	+11½	+16½	+ 9½
7	− 6½	−14½	−11	−2	+3½	+1½	−4½	−5½	+ 2½	+12½	+16	+ 8
10	− 7½	−14½	−10½	−1	+4	+1	−5	−5	+ 3½	+13	+16	+ 7
13	− 9	−14½	− 9½	− ½	+4	0	−5½	−4½	+ 4½	+14	+15½	+ 5½
16	−10	−14½	− 8½	+ ½	+4	− ½	−6	−4	+ 5½	+14½	+15	+ 4
19	−11	−14	− 8	+1	+3½	−1	−6	−3½	+ 6½	+15	+14½	+ 2½
22	−11½	−14	− 7	+1½	+3½	−2	−6	−2½	+ 7½	+15½	+13½	+ 1
25	−12½	−13½	− 6	+2	+3½	−2½	−6	−2	+ 8½	+16	+12½	− ½
28	−13	−13	− 5	+2½	+3	−3	−6	−1	+ 9½	+16	+11½	− 2
31	−13½	−12	− 4	+3	+2½	−3½	−6	0	+10½	+16½	+10½	− 3½

6. *Confirmation of Adjustments.*—In order that each of the adjustments mentioned above should be sufficiently exact the other adjustments would have to be nearly right. Hence, when the adjustments are deemed to be right, they should be tested, which may be easily done when the sun shines, even though not quite continuously.

The adjustment for the meridian and the adjustment for level east and west are tested together by seeing whether at *apparent local* noon (or at 11 a.m., 1 p.m., &c.) the image falls on the noon hour line (or on the 11 a.m., 1 p.m., &c., hour line), and whether the line scored by the sun on a card runs parallel to the nearest edge of a flange confining the card. Theoretically, it should not be *quite* parallel, on account of the change of declination of the sun during the day; but even near the equinoxes this is too small to come under notice.

If reasonable care has been taken in levelling the top of the bowl in an east and west direction, no material error of level is to be feared; and a defect of parallelism of the score to the flange, though such as might be produced by an error of level, should lead the observer rather to question and to re-examine the adjustment for the meridian. It may be that when the adjustment was made incorrect time was used, or the correction for the equation of time was forgotten, or applied with a wrong sign.

The adjustment for latitude is tested by seeing whether the image of the sun falls at the proper height on the card corresponding to the day of the year.

Once well adjusted, the instrument need not be disturbed, and it may be fixed in its place by cement or otherwise. It is possible, however, that at some stations, from the positions of buildings, &c., one place might be best for the instrument in summer, and another in winter. In such cases there is no objection to making the change. Of course the instrument will have to be re-adjusted after each change of position.

For the sake of those who wish to make use of the mathematical expressions for the errors of time and parallelism produced by given small errors of level and azimuth, the expressions are here subjoined.

Let l be the latitude, δ the sun's declination, both reckoned positive when north, α the small error of azimuth, λ that of level east and west, h the error of hour angle entailed, p the error of parallelism, α, λ, h, p being respectively reckoned positive when in the direction of the hands of a watch to an observer looking vertically downwards for the first, horizontally northwards for the second, downwards in the direction of the earth's axis for the third, downwards in the direction of the sun's rays at apparent noon for the fourth; then

$$h = \{\alpha \sin (l - \delta) - \lambda \cos (l - \delta)\} \sec \delta,$$
$$p = (\alpha \cos l + \lambda \sin l) \sec \delta,$$
and
$$\alpha = p \cos (l - \delta) + h \sin l,$$
$$\lambda = p \sin (l - \delta) - h \cos l.$$

7. *Choice and Insertion of the Cards.*—Cards are provided of three patterns, rectangular for the equinoxes, and curved for summer and winter. The summer and winter cards are alike except as to length (the summer cards being the longer), and as to having the hour figures printed so as to be seen erect when in the

one case the convex and in the other the concave edge of the card is held uppermost.

It will be noticed that the bowl is undercut inside so as to leave six grooves roofed in by flanges which are destined to confine the edges of the cards. The grooves or their flanges will here be numbered from the top downwards. The winter cards are inserted, concave upwards, with their edges under flanges Nos. 1, 3 and slid along till the noon hour line is against the line marked on the brass. Should the two marks on the brass not exactly agree, that nearest to the equator of the instrument had best be used. Nothing more is required in general till next day, when the card is pulled out and a fresh one put in. In time of snow, however, when there is any chance of sunshine, the snow should be removed from between the ball and the card.

If the sun should be shining when a fresh card is being put in, the observer should stand on the south side of the instrument, or otherwise shade the ball, lest a false score should be made on the card before it gets into the proper position.

The equinoctial cards are similarly inserted, with the hour figures erect, under flanges Nos. 2, 5, and the summer cards, convex uppermost, under flanges Nos. 4, 6.

The equinoctial cards are to be used during March and the first 12 days of April, and again during September and the first 12 days of October; the summer or winter cards, as the case may be, are to be used during the remainder of the year.

8. *Shortening of the Equinoctial Cards.*—If the ends of the equinoctial cards were left projecting above the brass frame, they would intercept the sun's rays near sunrise and sunset. The parts projecting above the horizontal top of the frame should therefore be cut off. If the observer chooses, he may cut off the ends before inserting the cards, by cutting one in the instrument, and using it as a pattern by which to cut the others. It would be unnecessary to remove the ends of the summer and winter cards, as they are not in the way.

A large number of experiments were made by means of two similar instruments placed side by side, as well as more roughly in other ways, on the effect of different modes of darkening the cards. It might, perhaps, have been expected beforehand that black cards would have been the most sensitive. Such, however, did not prove to be the case. With blackened cards the earliest indication of an effect of the sun's rays consisted in a slight alteration of the texture, visible only when the card was held so as to catch reflected light; whereas, with a moderately darkened card an alteration of colour produced by the heat could be seen before there was any visible alteration of texture; and this singe could be seen simultaneously with the determinate burns without the necessity for holding the card in any particular direction. And though the first change would most probably be produced on a black card, experience proved that the first *visible* change was produced on a suitably darkened card.

The difference between different kinds of cards was, however, far less than might perhaps have been anticipated. It was only in catching a few minutes more or less of a very feeble sunshine that, with the exception, perhaps, of a few pale and unsuitable kinds, one card differed from another. Cards darkened to a grey with carbon were amongst the best. It was decided, however, ultimately to employ prussian blue only moderately dark. Such cards when viewed through a red glass, which transmits all the visible rays which are strongest in heating effect, looks almost black, while the rays of high refrangibility which it freely reflects enable the record to be easily seen, while entailing but little loss of absorption of rays powerful in their heating effect. It is needless to remark that if a pigment were chosen *merely* from *à priori* considerations, its behaviour with respect to the invisible rays lying beyond the red would have to be taken into consideration. But to do this experimentally would involve an expenditure of time which the value of the result would hardly justify, since the suitability of a pigment may be ascertained by direct trial.

POSTSCRIPT BY THE AUTHOR, *March* 16*th*, 1880.

The instrument was designed for use in the United Kingdom, but there appears no reason why it should not be used even in

extreme latitudes. At the North Pole, for instance, the plane
of section of a complete ring would be the equatorial plane, and
a half ring would go completely round the polar axis. It would
merely be necessary to remove half-an-inch or an inch of the
flanges confining the summer cards, in order to permit of the
introduction of the end of a card. The card would then be slipped
along in its grooves. A summer card for the North Pole, as it lay
flat, would form an arc of an annulus with its ends in the direction
of radii, and as it lay in position would form a complete annulus
of a right circular cone, with a division down one generating line.
where the edges of the slip of card would meet, without either
overlapping or leaving a gap, if the card had been properly cut.
The equinoctial cards would form complete annuli of a cylinder
divided along one generating line. They would, it is true, have a
flange to hold them in along one edge only, but that would be
sufficient. What was said in the paper as to the liability to mis-
place a slip in a simple hemispherical bowl, such as that employed
at the Greenwich Observatory, was intended only to apply to the
case of observers of little or no experience. The cards actually
used fit very easily into their grooves, so that it would require
a good deal of dirt to make them jam. The grooves would tend
to be kept clean by the daily removal and insertion of a card;
and when the time came for shifting from one pair of grooves
to another, a change made only four times in the year, it would,
apparently, be no great trouble to clean sufficiently the grooves
coming into use should they be found to require it.

On a Simple Mode of Eliminating Errors of Adjustment in Delicate Observations of Compared Spectra.

[From the *Proceedings of the Royal Society*, Vol. XXXI, pp. 470—473. Received *February* 12, 1881.]

WHEN the identity or difference of position of two lines, bright or dark, in the spectra of two lights from different sources has to be compared with the utmost degree of accuracy, they are admitted simultaneously into different but adjacent parts of the slit of a spectroscope and viewed together. It was thus, for instance, that Dr Huggins proceeded in determining the radial component of the velocity of the heavenly bodies relatively to the earth. It is requisite that the two lights that are to be compared should fall in a perfectly similar manner on the slit: and it will be seen, from a perusal of his paper, how careful Dr Huggins was in this respect.

In a paper read before the Royal Society on the 3rd instant, Mr Stone has proposed to make the observation independent of a possible error in the exact coincidence of the lights compared, by constructing a reversible spectroscope, by which the light should be refracted alternately right and left, supposing for facility of explanation the slit to be vertical.

The idea is an elegant one, but I apprehend that there would be considerable difficulty in carrying it out. For a spectroscope giving large dispersion is of considerable weight, and the reversal of so heavy an apparatus would be liable to introduce possible errors arising from flexure*. It would be difficult to make sure that such did not exist, at any rate, unless the instrument were constructed with great nicety and firmness, which would add

---

* After the present paper was sent in to the Society, I was informed by Mr Stone that the spectroscope he had in his mind was a direct-vision one, which could be turned in its socket, the slit and cylindrical lens remaining fixed. To such an instrument the objection as to flexure would not apply.

considerably to the cost; and even then the care and time required for the reversal would help to obliterate the observer's memory of what he had seen in the first position of the instrument.

A method has occurred to me of effecting the reversal without reversing the spectroscope, but merely giving a lateral push to a little apparatus which need not weigh more than a few grains.

If the base of an isosceles prism be polished as well as the sides, and a ray of light parallel to the base and in a plane perpendicular to the edge fall on one of the equal sides of the prism so as to emerge from the other, after suffering an intermediate reflection (which will necessarily be total) at the base, its course after refraction will be parallel to its course before incidence; and there will, moreover, be no lateral displacement, provided the lateral distance of the base from the incident ray be such that the point of reflection is at the middle of the base.

If the slit of the spectroscope be covered by such a prism, placed close to the slit and facing the collimating lens, to the axis of which its base is parallel, it will not disturb the general course of the light incident on the spectroscope, nor even produce a lateral displacement provided the lateral position be that mentioned above; but in consequence of the reflection there will be a reversal as regards right and left, and any error in the placing of the lights to be compared will thus be detected and eliminated, by comparing the spectra seen with the light from the slit direct or reflected. If the prism be placed quite close to the slit it may be made very minute in section, though it should be long enough to cover the slit, and then the change of focus which it produces will be insignificant.

There will be no need, however, to make the prism so very minute, nor to place it so close to the slit, provided it be associated with a plate to take its place in the direct observation, and compensate for the change of focus which is produced by its introduction.

Let $ABCD$ be a section of the prism, let M be the middle point of the base AB, $KLMNO$ the course of a ray passing as above described, which is supposed to be the axis of the pencil coming through the middle of the slit. Let ϕ be the angle of incidence, which will be half the angle of the prism, and the complement of either angle A or B, ϕ' the angle of refraction, μ the index of refraction, b the base AB, l the length of path, $LM + MN$, of the ray within the glass, $p = LN$. In spectroscopic

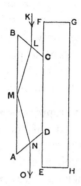

work it is the focus of rays in the primary plane that we have to deal with; and we get for the shortening (s) of the focus, or, in other words, the distance by which the slit is virtually brought nearer to the collimating lens,

$$s = p - l \, \frac{\cos^2 \phi}{\mu \cos^2 \phi'}.$$

But since $MBL = 90° - \phi$ and $MLC = 90° - \phi'$ we have

$$l = b \, \frac{\cos \phi}{\cos \phi'}; \quad \text{also } p = l \cos(\phi - \phi');$$

whence $s = b \left\{ \cos(\phi - \phi') \dfrac{\cos \phi}{\cos \phi'} - \dfrac{\cos^3 \phi}{\mu \cos^2 \phi'} \right\} = t \left(1 - \dfrac{1}{\mu} \right),$

where t is the thickness of a compensating plate which shall produce the same shortening of focus. In the figure, the part of the prism which is out of use is represented as cut away, to make the instrument more compact, and $EFGH$ represents the compensating plate. The faces CD of the truncated prism, and EF, HG, of the plate, of course need not be polished, and had better perhaps be blackened.

In the figure I have taken 80° for the angle of the prism, and supposed μ to be 1·52, which data give $t = 1·225\,b$, nearly. A blunter angle would have made the instrument a little more compact in the direction AB, but I wished to avoid needless loss of light by the two reflections that accompany the refractions. The size of the prism and compensating plate must depend upon its distance from the slit, and the angle subtended at the slit by the objective of the collimator. It should be a little larger than what is just sufficient to take in the largest pencil that is to be observed, but not beyond that. The object in keeping it as small as conveniently may be, is that only a trifling change of focus may be required when the instrument is pushed aside altogether, and the slit viewed directly through the spectroscope, without the slight loss of light due to the two reflections.

The compensating plate is represented as placed at the narrow end of the prism, which permits of the two being cemented together, thereby facilitating the support. I do not think that the minute quantity of light which is reflected at L, and scattered at the surface (even though blackened) FC in such a direction as to mingle with the direct light would be any inconvenience, being too faint to be visible at all. If it were wished to avoid this, or to get more easy access to the surfaces AD, BC, for cleaning if requisite, the plate might be placed at the other side; but in that case it must not be cemented to AB, as that surface is wanted for total reflection.

The little instrument I have suggested may conveniently be called a *slit-reverser*, to distinguish it from other arrangements which have been proposed, and in which the spectrum itself is reversed.

P.S. Feb. 21.—The method proposed above is more directly applicable to such an object as the comparison of really or apparently coincident lines in the spectra of two elements than to astronomical measurements, because in the latter case a great part of the difficulty arises from a want of perfect accuracy in the clockwork movement of the equatoreal. Yet I cannot help thinking that even for astronomical work the method will be found useful; for we can pass in a moment from the direct to the reflected image of the slit, and *vice versâ*, and by taking the measures alternately in the two modes, and combining them exactly as in weighing with a balance that is still swinging, any error progressive with the time would tend to be eliminated.

Discussion of the Results of some Experiments with Whirled Anemometers.

[From the *Proceedings of the Royal Society*, Vol. XXXII, pp. 170—188.
Received *April* 26, 1881.]

In the course of the year 1872, Mr R. H. Scott, F.R.S., suggested to the Meteorological Committee the desirability of carrying out a series of experiments on anemometers of different patterns. This suggestion was approved by the Committee, and in the course of the same year a grant was obtained by Mr Scott from the Government Grant administered by the Royal Society, for the purpose of defraying the expenses of the investigation. The experiments were not, however, carried out by Mr Scott himself, but were entrusted to Mr Samuel Jeffery, then Superintendent of the Kew Observatory, and Mr G. M. Whipple, then First Assistant, the present Superintendent.

The results have never hitherto been published, and I was not aware of their nature till on making a suggestion that an anemometer of the Kew standard pattern should be whirled in the open air, with a view of trying that mode of determining its proper factor, Mr Scott informed me of what had already been done, and wrote to Mr Whipple, requesting him to place in my hands the results of the most complete of the experiments, namely, those carried on at the Crystal Palace, which I accordingly obtained from him. The progress of the enquiry may be gathered from the following extract from Mr Scott's report in returning the unexpended balance of the grant.

" The comparisons of the instruments tested were first instituted in the garden of the Kew Observatory. This locality was found to afford an insufficient exposure.

" A piece of ground was then rented and enclosed within the Old Deer Park. The experiments here showed that there was a considerable difference in the indications of anemometers of different sizes, but it was not possible to obtain a sufficient range of velocities to furnish a satisfactory comparison of the instruments. Experiments were finally made with a rotating apparatus, a steam merry-go-round, at the Crystal Palace, which led to some results similar to those obtained by exposure in the Deer Park.

"The subject has, however, been taken up so much more thoroughly by Drs Dohrandt and Thiesen (*vide Repertorium für Meteorologie*, Vols. IV. and V.) and by Dr Robinson in Dublin, that it seems unlikely that the balance would ever be expended by me. I, therefore, return it with many thanks to the Government Grant Committee.

"The results obtained by me were hardly of sufficient value to be communicated to the Society."

On examining the records, it seemed to me that they were well deserving of publication, more especially as no other experiments of the same kind have, so far as I know, been executed on an anemometer of the Kew standard pattern. In 1860 Mr Glaisher made experiments with an anemometer whirled round in the open air at the end of a long horizontal pole*, but the anemometer was of the pattern employed at the Royal Observatory, with hemispheres of 3·75 inches diameter and arms of 6·725 inches, measured from the axis to the centre of a cup, and so was considerably smaller than the Kew pattern. The experiments of Dr Dohrandt and Dr Robinson were made in a building, which has the advantage of sheltering the anemometer from wind, which is always more or less fitful, but the disadvantage of creating an eddying vorticose movement in the whole mass of air operated on; whereas in the ordinary employment of the anemometer the eddies it forms are carried away by the wind, and the same is the case to a very great extent when an anemometer is whirled in the open air in a gentle breeze. Thus, though Dr Robinson employed among others an anemometer of the Kew pattern, his experiments and those of Mr Jeffery are not duplicates of each other, even independently of the fact that the axis of the anemometer was vertical in Mr Jeffery's and

---

* *Greenwich Magnetical and Meteorological Observations*, 1862, Introduction, p. li.

horizontal in Dr Robinson's experiments; so that the greater completeness of the latter does not cause them to supersede the former.

In Mr Jeffery's experiments the anemometers operated on were mounted a little beyond and above the outer edge of one of the steam merry-go-rounds in the grounds of the Crystal Palace, so as to be as far as practicable out of the way of any vortex which it might create. The distance of the axis of the anemometer from the axis of the " merry " being known, and the number of revolutions (n) of the latter during an experiment counted, the total space traversed by the anemometer was known.   The number (N) of *apparent* revolutions of the anemometer, that is, the number of revolutions *relatively to the merry*, was recorded on a dial attached to the anemometer, which was read at the beginning and end of each experiment. As the machine would only go round one way, the cups had to be taken off and replaced in a reverse position, in order to reverse the direction of revolution of the anemometer.   The *true* number of revolutions of the anemometer was, of course, $N + n$, or $N - n$, according as the rotations of the anemometer and the machine were in the same or opposite directions.

The horizontal motion of the air over the whirling machine during any experiment was determined from observations of a dial anemometer with 3-inch cups on 8-inch arms, which was fixed on a wooden stand in the same horizontal plane as that in which the cups of the experimental instrument revolved, at a distance estimated at about 30 feet from the outside of the whirling frame. The motion of the centres of the cups was deduced from the readings of the dial of the fixed anemometer at the beginning and end of each experiment, the motion of the air being assumed as usual to be three times that of the cups.

The experiments were naturally made on fairly calm days, still the effect of the wind, though small, is not insensible.   In default of further information, we must take its velocity as equal to the mean velocity during the experiment.

Let V be the velocity of the anemometer (*i.e.*, of its axis), W that of the wind, θ the angle between the direction of motion

of the anemometer and that of the wind. Then the velocity of
the anemometer relatively to the wind will be

$$\sqrt{V^2 - 2VW \cos\theta + W^2} \dots\dots\dots\dots\dots\dots(a).$$

The mean effect of the wind in a revolution of the merry will be
different according as we suppose the moment of inertia of the
anemometer very small or very great.

If we suppose it very small, the anemometer may be supposed
to be moving at any moment at the rate due to the relative
velocity at that moment, and therefore the mean velocity of
rotation of the cups in one revolution of the merry will be that
corresponding to the mean relative velocity of the anemometer
and the air. If, as is practically the case, W be small as compared
with V, we may expand (a) in a rapidly converging series according to ascending powers of W. All the odd powers will disappear
in taking the mean, and if we neglect the fourth and higher
powers we shall have for the mean

$$V + \frac{W^2}{4V},$$

so that $W^2 \div 4V$ is the small correction to be added to the
measured velocity of the anemometer in order to correct for the
wind.

On the other hand, if the moment of inertia of the anemometer be taken as very great, the rate of rotation of the cups
during a revolution of the merry will be sensibly constant. If
V' be the velocity of the anemometer relatively to the air, v the
velocity of the centre of one of the cups, and if we suppose the
rotation of the anemometer resisted by a force of which the
moment is F, then, according to Dr Robinson's researches, we
have approximately

$$F = AV'^2 - 2BvV' - Cv^2.$$

In the present case friction is not taken into account, and instead
of F we must take the moment of the effective moving force.
Furthermore, it appears from the experiments of Dr Robinson, in
Dublin, that the observations were almost as well satisfied by
taking the first two terms only of the above expression for F as
by taking all three, and this simplification may be employed with

abundantly sufficient accuracy in making the small correction for the wind.   We have, therefore

$$F = A V'^2 - 2Bv V' \quad\quad\quad\quad\quad\quad (b),$$

where V' is given by (a).   In order that the anemometer may be neither accelerated nor retarded from one revolution of the merry to another, the mean effective force must be *nil*; and taking the means of both sides of the above equation, observing that, in consequence of the supposed largeness of the moment of inertia, v is sensibly constant during one revolution of the merry, we have on employing the approximate value of the mean of V' or (a) already used

$$0 = A (V^2 + W^2) - 2Bv \left(V + \frac{W^2}{4 \bar{V}} \right).$$

But if U be the constant velocity of air relatively to the anemometer which would make the cups turn round at the same rate, we have similarly

$$0 = A U^2 - 2Bv U.$$

Eliminating Bv/A between these two equations we get

$$U \left(V + \frac{W^2}{4V} \right) = V^2 + W^2 \quad\quad\quad\quad (c),$$

and as the fourth and higher powers of W have been neglected all along, we get from the last

$$U = V + \frac{3W^2}{4V} \quad\quad\quad\quad\quad\quad\quad (d),$$

so that, on this supposition, the mean correction for the wind is $3W^2/4V$, or three times the correction of the former supposition.

The mean value of the radical (a) is given by an elliptic function; but even in an extreme case among the experiments, when the ratio of the velocity of the wind to that of the anemometer is as great as 3 to 5, the error of the approximate expression $V + W^2/4V$ amounts only to about 0·01 mile an hour, which may be quite disregarded.   The error in employing (d) for the determination of U instead of (c) is of about similar amount.

Three anemometers were tried, namely, one of the old Kew standard pattern, one by Adie, and Kraft's portable anemometer. Their dimensions will be found at the heads of the respective tables below.   With each anemometer the experiments were

made in three groups, with high, moderate, and low velocities
respectively, averaging about 28 miles an hour for the high, 14
for the moderate, and 7 for the low. Each group again was
divided into two subordinate groups, according as the cups were
direct, in which case the directions of rotation of the merry and
of the anemometer were opposite, or reversed, in which case the
directions of the two rotations were the same.

The data furnished by each experiment were: the time occupied by the experiment, the number of revolutions of the merry,
the number of *apparent* revolutions of the anemometer, given by
the difference of readings of the dial at the beginning and end
of the experiment, and the space S passed over by the wind,
deduced from the difference of readings of the fixed anemometer
at the beginning and end of the experiment.

The object of the experiment was, of course, to compare the
mean velocity of the centres of the cups with the mean velocity
of the air relatively to the anemometer. It would have saved
some numerical calculation to have compared merely the spaces
passed through during the experiment; but it seemed better to
exhibit the velocities in miles per hour, so as to make the experiments more readily comparable with one another, and with those
of other experimentalists. In the reductions I employed 4-figure
logarithms, so that the last decimal in V in the tables cannot
quite be trusted, but it is retained to match the correction for
W, which it seemed desirable to exhibit to 0·01 mile.

On reducing the experiments with the low velocities, I found
the results extremely irregular. I was subsequently informed by
Mr Whipple, that the machine could not be regulated at these
low velocities, for which it was never intended, and that it sometimes went round fast, sometimes very slowly. He considered
that the experiments in this group were of little, if any, value,
and that they ought to be rejected. They were besides barely
half as numerous as those of the moderate group. I have accordingly thought it best to omit them altogether.

In the following tables the first column gives the group, H
standing for high velocities, M for moderate; the subordinate
group, − standing for rotation of the anemometer opposite to that
of the machine, + for rotations in the same direction; and lastly

the reference number of the experiment in each subordinate group. T gives the duration of the experiment in minutes; n the number of revolutions of the machine; N the number of *apparent* revolutions of the anemometer; S the space passed over by the natural wind, in miles. These form the data. From them are calculated: V, the velocity of the anemometer, in miles per hour; W, the velocity of the wind; $W^2/2V$ the mean of the two corrections to be added to V on account of the wind, according as we adopt one or other of the extreme hypotheses as to the moment of inertia of the anemometer, namely, that it is very small or very large. The actual correction will be half the number in this column on the first supposition, and once and a half on the second. V_1, V_2 denote the velocity of the anemometer, or, in other words, of the artificial wind, corrected for the natural wind on these two suppositions respectively, so that the last two columns give 100 times the ratio of the registered velocity to the true velocity, or the registered as a percentage of the true, the registered velocity meaning that deduced from the velocity of the cups on employing the usual factor 3.

The dials of the first two anemometers read only to 10 revolutions, which is the reason why all the numbers N end with a 0.

The Old Kew Standard.

Diameter of Arms between Centres of Cups 48 inches; Diameter of Cups 9 inches. Fixed to Machine at 22·3 feet from the Axis of Revolution.

Group and No.	T	n	N	S	V	W	$\dfrac{W^2}{2V}$	$\dfrac{300v}{V_1}$	$\dfrac{300v}{V_2}$
$H-$									
1	15	303	1690	0·7	31·17	2·80	0·13	126·9	126·3
2	18	301	1690	0·5	26·63	1·67	0·05	124·1	123·8
3	16	301	1580	0·6	29·96	2·25	0·08	114·2	113·9
4	17	300	1710	1·1	28·11	3·88	0·27	125·9	124·7
5	17	300	1720	1·1	28·11	3·88	0·27	126·8	125·5
6	22	400	2210	1·3	28·96	3·27	0·18	121·4	120·6
7	23	400	2220	1·1	27·70	2·87	0·15	122·1	121·5
8	19	300	1670	0·7	25·14	2·21	0·10	122·6	122·2
9	19	300	1640	0·8	25·14	2·53	0·13	119·8	119·3
10	17	301	1670	0·8	28·20	2·71	0·13	122·1	121·5
11	19	300	1670	0·8	25·14	2·53	0·13	122·6	122·0
Mean...	...	...	...	...	26·75	2·78	0·15	122·6	121·9
$H+$									
1	17	302	980	0·0	28·29	0·00	0·00	114·2	114·2
2	17	300	960	1·0	28·11	3·53	0·22	112·6	111·7
3	15½	300	1000	1·4	30·82	5·42	0·48	115·7	113·9
4	22½	300	1080	1·8	21·23	4·80	0·54	122·3	119·2
5	19	300	1020	0·7	25·14	2·21	0·10	118·1	117·7
6	16	300	1030	0·9	29·86	3·37	0·19	119·0	118·2
7	18	300	1050	0·8	26·54	2·67	0·13	120·8	120·2
8	18	301	1060	0·6	26·63	2·20	0·09	121·4	121·0
9	18	300	1000	0·7	26·54	2·43	0·11	121·7	121·3
Mean...	...	...	...	...	27·02	2·96	0·21	118·4	117·5
$M-$									
1	30	300	1650	0·8	15·92	1·60	0·08	120·8	120·2
2	31	300	1670	1·6	15·41	3·10	0·31	121·7	119·3
3	34	300	1570	1·9	14·05	3·01	0·32	112·6	110·1
4	36	300	1540	1·7	13·26	2·83	0·30	110·0	107·6
5	36	300	1540	1·3	13·26	2·17	0·18	110·5	109·0
Mean...	...	...	...	...	14·38	2·54	0·24	115·1	113·2
$M+$									
1	28	301	880	0·0	17·63	0·00	0·00	102·6	102·5
2	38	300	940	2·0	12·57	3·15	0·38	109·5	106·4
3	38	300	890	1·6	12·57	2·52	0·25	105·7	103·7
4	36	300	990	0·8	13·26	1·33	0·07	115·4	114·9
5	35	300	990	1·0	13·66	1·71	0·07	115·3	114·8
Mean...	...	...	...	...	13·94	1·74	0·15	109·7	108·5

Adie's Anemometer.

Diameter of Arms between Centres of Cups 13·4 inches; Diameter of Cups 2·5 inches. Fixed to Machine at 20·7 feet from the Axis of Revolution.

Group and No.	T	n	N	S	V	W	$\frac{W^2}{2V}$	$\frac{300v}{V_1}$	$\frac{300v}{V_2}$
$H-$									
1	17	300	3860	1·0	26·16	3·53	0·24	95·2	94·4
2	15½	300	3650	1·4	28·61	5·45	0·52	89·5	87·9
3	22¼	300	3940	1·8	19·70	4·80	0·58	96·7	94·0
4	19	300	3760	0·7	23·33	2·21	0·10	93·1	92·7
5	16	300	3780	0·9	27·71	3·37	0·20	93·5	92·8
6	18	300	3890	0·9	24·63	2·67	0·14	96·4	96·0
7	18	301	3980	0·7	24·72	2·33	0·11	98·6	98·2
8	18	300	3940	0·8	24·63	2·67	0·14	97·8	97·3
Mean...	...	...	...	...	24·94	3·38	0·25	95·1	94·2
$H+$									
1	17	300	3240	1·1	26·08	3·89	0·29	94·9	93·9
2	17	300	3330	1·1	26·08	3·89	0·29	97·3	96·3
3	19	300	3760	0·7	23·33	2·21	0·10	109·2	108·7
4	16	300	3780	0·9	27·71	3·37	0·20	109·6	108·8
5	19	300	3060	0·8	23·33	2·53	0·14	90·3	89·8
6	17	301	3120	0·8	26·17	2·82	0·15	91·6	91·1
7	19	300	3160	0·8	23·33	2·53	0·14	93·0	92·5
Mean...	...	...	...	...	25·15	3·03	0·19	98·0	97·3
$M-$									
1	38	300	3620	2·0	11·67	3·16	0·43	87·7	84·9
2	38	300	3500	1·6	11·67	2·53	0·43	84·7	81·7
3	36	300	3910	0·8	12·26	1·33	0·07	97·5	96·9
4	35	300	3430	1·0	12·67	1·71	0·07	84·1	83·7
Mean...	...	...	...	...	12·07	2·18	0·25	88·5	86·8
$M+$									
1	31	300	3250	1·6	14·30	3·10	0·34	94·3	93·7
2	34	300	2920	1·7	13·04	3·00	0·35	85·7	85·5
3	34	300	2940	1·9	13·04	3·06	0·36	86·1	83·1
4	36	300	2760	1·7	12·31	2·83	0·33	81·4	79·2
5	36	300	2780	1·3	12·31	2·17	0·19	65·5	64·7
Mean...	...	...	...	...	13·00	2·83	0·31	82·6	81·0

Kraft's Portable Anemometer.

Diameter of Arms between Centres of Cups 8·3 inches; Diameter of Cups 3·3 inches. Fixed to Machine at 19·10 feet from the Axis of Revolution.

Group and No.	T	n	N	S	V	W'	$\dfrac{W^2}{2V}$	$\dfrac{300v}{V_1}$	$\dfrac{300v}{V_2}$
$H-$									
1	19	300	5594	0·6	21·53	1·89	0·08	95·6	95·4
2	17½	303	5681	0·7	23·60	2·69	0·15	96·2	95·4
3	15	303	5990	0·7	27·22	2·80	0·14	102·8	102·4
4	18	301	6086	0·5	22·79	1·67	0·06	104·2	104·0
5	16	301	6116	0·6	25·65	2·25	0·10	104·7	104·3
6	17	300	6143	1·1	24·06	3·88	0·31	105·2	103·6
7	17	300	6240	1·1	24·06	3·88	0·31	106·9	105·5
8	22	400	7896	1·3	24·79	3·27	0·21	101·4	100·4
9	23	400	7900	1·1	23·71	2·78	0·16	101·5	100·9
10	19	300	5966	0·7	21·53	2·21	0·11	102·3	101·7
11	19	300	5751	0·8	21·53	2·53	0·15	98·3	97·7
12	17	301	5842	0·8	24·14	2·82	0·16	99·6	99·0
13	19	300	5892	0·8	21·53	2·53	0·15	100·9	100·1
Mean...	...	...	...	...	23·55	2·71	0·16	101·5	100·8
$H+$									
1	17	300	5372	1·0	24·06	3·53	0·26	102·2	101·0
2	15½	300	5265	1·4	26·39	5·42	0·56	99·7	97·5
3	22⅔	300	5460	1·8	18·18	4·80	0·63	102·5	98·9
4	19	300	5093	0·7	21·53	2·21	0·11	97·4	96·8
5	16	300	5282	0·9	25·57	3·37	0·22	100·6	99·8
6	18	300	5274	0·8	22·72	2·67	0·16	100·6	99·8
7	18	301	5300	0·6	22·79	2·00	0·09	100·9	100·5
8	18	300	5363	0·8	22·72	2·67	0·16	102·2	101·4
Mean...	...	...	...	...	22·99	3·33	0·27	100·8	99·4
$M-$									
1	30	300	5488	0·8	13·63	1·60	0·09	93·3	92·5
2	31	300	5880	1·6	13·19	3·10	0·36	99·2	96·6
3	34	300	5168	1·6	12·03	2·53	0·27	86·8	84·8
4	34	300	5320	1·9	12·03	3·35	0·47	88·8	85·4
5	36	300	5030	1·7	11·36	2·83	0·35	84·0	81·4
6	36	300	4910	1·3	11·36	2·17	0·21	82·3	80·9
Mean...	...	...	...	...	12·27	2·60	0·29	89·1	86·9
$M+$									
1	38	300	4508	2·0	10·76	3·16	0·46	84·8	81·4
2	38	300	4250	1·6	10·76	2·53	0·30	80·9	78·7
3	36	300	5006	0·8	11·36	1·33	0·08	95·3	94·7
4	35	300	4743	1·0	11·69	1·71	0·12	90·3	89·3
Mean...	...	...	...	...	11·14	2·18	0·24	87·8	86·0

The mean results for the high and moderate velocities, contained in the preceding tables, are collected in the following table, in which are also inserted the mean errors.

Anemometer	Directions of Rotation	High Velocities				Moderate Velocities			
		Mom. inert. small		Mom. inert. large		Mom. inert. small		Mom. inert. large	
		p. c.	m. e.	p. c.	m. e.	p. c.	m. e.	p. c.	m. e.
Kew	Opposite...	122·6	2·4	121·9	2·3	115·1	4·9	113·2	5·2
	Alike	118·4	2·9	117·5	2·8	109·7	4·5	108·5	5·1
	Mean.......	120·5	...	119·7	...	112·4	...	110·8	
Adie	Opposite...	95·1	2·3	94·2	2·3	88·5	4·5	86·8	5·0
	Alike	98·0	6·5	97·3	6·5	82·6	7·3	81·0	7·3
	Mean.......	96·5	...	95·7	...	85·5	...	83·9	
Kraft	Opposite...	101·5	2·6	100·8	2·5	89·1	4·8	86·9	5·1
	Alike	100·8	1·2	99·4	1·3	87·8	5·0	86·0	6·0
	Mean.......	101·1	...	100·1	...	88·4	...	86·4	

The mean errors exhibited in the above table show no great difference according as we suppose the moment of inertia of the anemometer small or large in correcting for the wind. There appears to be a slight indication, beyond what may be merely casual, that the errors are a little greater on the latter supposition than on the former, which is what we should rather expect; for an anemometer would get pretty well under way in a fraction of a revolution of the whirling instrument. However, the difference is so small that it will suffice to take the mean of the two as the mean error belonging to the particular anemometer, class of velocity, and character of rotation under consideration. From the mean errors we may calculate nearly enough, by the usual formulæ, the probable errors of the various mean percentages for rotations opposite and alike. The probable errors of these mean percentages come out as follows:

Kew, 1·0 for high velocities; 2·7 for moderate velocities.

Adie, 1·5   „   „   2·0   „   „

Kraft, 0·9   „   „   1·8   „   „

These probable errors are so small that it appears that for the high and even for the moderate velocities the experiments are extremely trustworthy, except in so far as they may be affected by *systematic* sources of error.

If we compare the registered percentages of the true velocity of the air relatively to the anemometer according as the rotations are in opposite directions or in the same direction, we see that in five out of the six cases they are slightly greater when the rotations are opposite. The sole exception is in the group "Adie, high velocities," which is made up of the groups "Adie $H-$" and "Adie $H+$." On referring to the principal table for the Adie, we see that Experiments 3 and 4 in group $H+$ give percentages usually high, depending on the high values of N. These raise the mean for the group, and make the mean error far greater than those of the other five groups for high velocities. There appears little doubt, therefore, that the excess of percentages obtained for rotations opposite is real, and not merely casual. It is, however, so small as to give us much confidence in the correctness of the mean result, unless there were causes to vitiate it which apply to both directions of rotation alike.

It may be noticed that the difference is greatest for the Kew, in which the ratio of r to R is greatest, r denoting the radius of the arm of the anemometer, and R the distance of its axis from the axis of revolution of the machine, and appears to be least (when allowance is made for the two anomalous experiments in the group "Adie $H+$") for the Kraft, for which r/R is least. In the Kraft, indeed, the differences are roughly equal to the probable errors of the means. In these whirling experiments r/R is always taken small, and we might expect the correction to be made on account of the finiteness of R to be expressible in a rapidly converging series according to powers of r/R, say

$$A' \frac{r}{R} + B' \left(\frac{r}{R}\right)^2 + C' \left(\frac{r}{R}\right)^3 + \dots$$

We may, in imagination, pass from the case of rotations opposite that of rotations alike, by supposing R taken larger and larger in successive experiments, altering the angular velocity of revolution so as to preserve the same linear velocity for the anemometer, and supposing the increase continued until R changes sign in passing through infinity, and is ultimately reduced in magnitude to what it was at first. The ideal case of $R = \infty$ is what we aim at, in order to represent the motion of a fixed anemometer acted on by perfectly uniform wind by that of an anemometer uniformly impelled in a rectilinear direction in perfectly still air. We may judge of the magnitude of the leading term in the above correction, provided it be of an odd order, by that of the difference of the results for the two directions of rotation. Unless, therefore, we had reason to believe that A' were 0, or at least very small compared with B', we should infer that the whole correction for the finiteness of R is very small, and that it is practically eliminated by taking the mean of the results for rotations opposite and rotations alike.

We may accept, therefore, the mean results as not only pretty well freed from casual irregularities which would disappear in the mean of an infinite number of experiments, but also, most probably, from the imperfection of the representation of a rectilinear motion of the anemometer by motion in a circle of the magnitude actually employed in the experiments.

Before discussing further the conclusions to be drawn from the results obtained, it will be well to consider the possible influence of systematic sources of error.

1. *Friction.*—No measure was taken of the amount of friction, nor were any special appliances used to reduce it; the anemometers were mounted in the merry just as they are used in actual registration. Friction arising from the weight is guarded against as far as may be in the ordinary mounting, and what remains of it would act alike in the ordinary use of the instrument and in the experiments, and as far as this goes, therefore, the experiments would faithfully represent the instrument as it is in actual use. But the bearings of an anemometer have also to sustain the lateral pressure of the wind, which in a high wind is very considerable; and the construction of the bearing has to be attended to in order that this may not produce too much friction. So far the whirled instrument is in the same condition as the fixed. But besides the friction arising from the pressure of the artificial wind, a

pressure which acts in a direction tangential to the circular path of the whirled anemometer, there is the pressure arising from the centrifugal force. The highest velocity in the experiments was about 30 miles an hour, and at this rate the centrifugal force would be about three times the weight of the anemometer. This pressure would considerably exceed the former, at right angles to which it acts, and the two would compound into one equal to the square root of the sum of their squares. The resulting friction would exceed a good deal that arising from the pressure of the wind in a fixed anemometer with the same velocity of wind (natural or artificial), and would sensibly reduce the velocity registered, and accordingly raise the coefficient which Dr Robinson denotes by m, the ratio, namely, of the velocity of the wind to the velocity of the centres of the cups. It may be noticed that the percentages collected in the table on p. 83 are very distinctly lower for the moderate velocities than for the high velocities. Such an effect would be produced by friction; but how far the result would be modified if the extra friction due to the centrifugal force were got rid of, and the whirled anemometer thus assimilated to a fixed anemometer, I have not the means of judging, nor again how far the percentages would be still further raised if friction were got rid of altogether.

Perhaps the best way of diminishing friction in the support of an anemometer is that devised and employed by Dr Robinson, in which the anemometer is supported near the top on a set of spheres of gun-metal contained in a box with a horizontal bottom and vertical side which supports and confines them. For vertical support, this seems to leave nothing to be desired, but when a strong lateral pressure has to be supported as well as the weight of the instrument, it seems to me that a slight modification of the mode of support of the balls might be adopted with advantage. When a ball presses on the bottom and vertical side of its box, and is at the same time pressed down by the horizontal disk attached to the shaft of the anemometer which rests on the balls, it revolves so that the instantaneous axis is the line joining the points of contact with the fixed box. But if the lateral force of the wind presses the shaft against the ball, the ball cannot simply roll as the anemometer turns round, but there is a slight amount of rubbing.

This, however, may be obviated by giving the surfaces where the ball is in contact other than a vertical or horizontal direction.

Let *AB* be a portion of the cylindrical shaft of an anemometer; *CD*, the axis of the shaft; *EFGHI*, a section of the fixed box or cup containing the balls; *LMN*, a section of a conical surface fixed to the shaft, by which the anemometer rests on its balls; *FIKM*, a section of one of the balls; *F, I*, the points of contact of the ball with the box; *M*, the point of contact with the supporting cone; *K*, the point of contact or all but contact

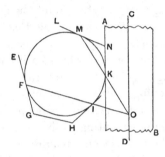

of the ball with the shaft. The ball is supposed to be of such size that when the anemometer simply rests on the balls by its own weight, being turned perhaps by a gentle wind, there are contacts at the points *M, F, I*, while at *K* the ball and shaft are separated by a space which may be deemed infinitesimal. Lateral pressure from a stronger wind will now bring the shaft into contact with the ball at the point *K* also, so that the box on the one hand and the shaft with its appendage on the other, will bear on the ball at four points. The surface of the box as well as that on the cone *LN* being supposed to be one of revolution round *CD*, those four points will be situated in a plane through *CD*, which will pass of course through the centre of the ball.

If the ball rolls without rubbing at any one of the four points *F, I, K, M* as the anemometer turns round, its instantaneous axis must be the line joining the points of contact, *F, I*, with the fixed box. But as at *M* and *K* likewise there is nothing but rolling, the instantaneous motion of the ball may be thought of as one in which it moves as if it were rigidly connected with the shaft and its appendage, combined with a rotation over *LNAB* supposed fixed. For the two latter motions the instantaneous axes are *CD*,

MK, respectively. Let *MK* produced cut *CD* in *O*. Then since the instantaneous motion is compounded of rotations round two axes passing through *O*, the instantaneous axis must pass through *O*. But this axis is *FI*. Therefore, *FI* must pass through *O*. Hence the two lines *FI*, *MK*, must intersect the axis of the shaft in the same point, which is the condition to be satisfied in order that the ball may roll without rubbing, even though impelled laterally by a force sufficient to cause the side of the shaft to bear on it. The size of the balls and the inclinations of the surfaces admit of considerable latitude subject to the above condition. The arrangement might suitably be chosen something like that in the figure. It seems to me that a ring of balls constructed on the above principle would form a very effective upper support for an anemometer whirled with its axis vertical. Possibly the balls might get crowded together on the outer side by the effect of centrifugal force. This objection, should it be practically found to be an objection, would not of course apply to the proposed system of mounting in the case of a fixed anemometer. Below, the shaft would only require to be protected from lateral motion, which could be done either by friction wheels or by a ring of balls constructed in the usual manner, as there would be only three points of contact.

2. *Influence on the Anemometer of its own Wake.*—By this I do not mean the influence which one cup experiences from the wake of its predecessor, for this occurs in the whirling in almost exactly the same way as in the normal use of the instrument, but the motion of the air which remains at any point of the course of the anemometer in consequence of the disturbance of the air by the anemometer when it was in that neighbourhood in the next preceding and the still earlier revolutions of the whirling instrument.

It seems to me that in the open air where the air impelled by the cups is free to move into the expanse of the atmosphere, instead of being confined by the walls of a building, this must be but small, more especially as the wake would tend to be carried away by what little wind there might be at the time. On making some enquiries from Mr Whipple as to a possible vorticose movement created in the air through which the anemometer passed, he wrote as follows:—" I feel confident that under the circumstances the tangential motion of the air at the level of the cups was so small as not to need consideration in the discussion of the results. As in one or two points of its revolution the anemometer passed close by some small trees in full leaf, we should have observed any eddies or artificial wind had it existed, but I am sure we did not."

3. *Influence of the Variation of the Wind; first, as regards Variations which are not Rapid.*—During the 20 or 30 minutes that an experiment lasted, there would of course be numerous fluctuations in the velocity of the wind, the mean result of which is alone recorded. The period of the changes (by which expression it is not intended to assert that they were in any sense regularly periodic), might be a good deal greater than that of the merry, or might be comparatively short. In the high velocities, at any rate, in which one revolution took only three or four seconds, the supposition that the period of the changes was large compared with one revolution is probably a good deal nearer the truth than the supposition that it is small.

On the former supposition, the correction for the wind during two or three revolutions of the merry would be given by the formulæ already employed, taking for W its value at the time. Consequently, the total correction will be given by the formulæ already used, if we substitute the mean of W^2 for the square of mean W. The former is necessarily greater than the latter; but how much, we cannot tell without knowing the actual variations. We should probably make an outside estimate of the effect of the variations, if we supposed the velocity of the wind twice the velocity during half the duration of the experiment, and nothing at all during the remainder. On this supposition, the mean of W^2 would be twice the square of mean W, and the correction for the wind would be doubled. At the high velocities of revolution, the whole correction for the wind is so very small, that the uncertainty arising from variation as above explained is of little importance, and even for the moderate velocities it is not serious.

4. *Influence of Rapid Variations of the Wind.*—Variations of which the period is a good deal less than that of the revolutions of the whirling instrument act in a very different manner. The smallness of the corrections for the wind hitherto employed depends on the circumstance that with uniform wind, or even

with variable wind, when the period of variation is a good deal greater than that of revolution of the merry, the terms depending on the first power of W, which letter is here used to denote the momentary velocity of the wind, disappear in the mean of a revolution. This is not the case when a particular velocity of wind belongs ·only to a particular part of the circle described by the anemometer in one revolution. In this case there will in general be an outstanding effect depending on the first power of W, which will be considerably larger than that depending on W^2. Thus suppose the velocity of whirling to be 30 miles an hour, and the average velocity of the wind 3 miles an hour; the correction for the wind supposed uniform, or if variable, then with not very rapid variations, will be comparable with 1 per cent. of the whole; whereas, with rapid variations, the effect in any one revolution may be comparable with 10 per cent. There is, how- ever, this important difference between the two: that whereas the correction depending on the square leaves a positive residue, however many experiments be made, the correction depending on the first power tends ultimately to disappear, unless there be some cause tending to make the average velocity of the wind different for one azimuth of the whirling instrument from what it is for another. This leads to the consideration of the following conceivable source of error.

5. *Influence of Partial Shelter of the Whirling Instrument.*— On visiting the merry-go-round at the Crystal Palace I found it mostly surrounded by trees coming pretty near it, but in one direction it was approached by a broad open walk. The con- sequence is, that the anemometer may have been unequally sheltered in different parts of its circular course, and the cir- cumstances of partial shelter may have varied according to the direction of the wind. This would be liable to leave an un- compensated effect depending on the first power of W. I do not think it probable that any large error was thus introduced, but it seemed necessary to point out that an error of the kind may have existed.

The effect in question would be eliminated in the long run if the whirling instrument were capable of reversion, and the experiments were made alternately with the revolution in one direction, and the reverse. For then, at any particular point of the course at which the anemometer was more exposed to wind than on the average, the wind would tend to increase the velocity of rotation of the anemometer for one direction of revolution of the whirling instrument just as much, ultimately, as to diminish it for the other.   Mere reversion of the cups has no tendency to eliminate the error arising from unequal exposure in different parts of the course.   And even when the whirling instrument is capable of reversion, it is only very slowly that the error arising from partial shelter is eliminated compared with that of irregularities in the wind; of those irregularities, that is to say, which depend on the first power of W.   For these irregularities go through their changes a very great number of times in the course of an experiment lasting perhaps half-an-hour; whereas, the effect of partial shelter acts the same way all through one experiment.   It is very desirable therefore, that in any whirling experiments carried on in the open air, the condition of the whirling instrument as to exposure or shelter should be the same all round.

The trees, though taller than the merry when I visited the place last year, were but young, and must have been a good deal lower at the time that the experiments were made.   Mr Whipple does not think that any serious error is to be apprehended from exposure of the anemometer during one part of its course and shelter during another.

From a discussion of the foregoing experiments, it seems to me that the following conclusions may be drawn:—

1.   That, at least for high winds, the method of obtaining the factor for an anemometer, which consists in whirling the instrument in the open air is capable, with proper precautions, of yielding very good results.

2.   That the factor varies materially with the pattern of the anemometer.   Among those tried, the anemometers with the larger cups registered the most wind, or in other words required the lowest factors to give a correct result.

3.   That with the large Kew pattern, which is the one adopted by the Meteorological Office, the register gives about 120 per cent. of the truth, requiring a factor of about .2·5, instead of 3. Even 2·5 is probably a little too high, as friction would be intro-

duced by the centrifugal force, beyond what occurs in the normal use of the instrument.

4. That the factor is probably higher for moderate than for high velocities; but whether this is solely due to friction, the experiments do not allow us to decide.

Qualitatively considered, these results agree well with those of other experimentalists. As the factor depends so much on the pattern of the anemometer, it is not easy to find other results with which to compare the actual numbers obtained, except in the case of the Kew standard. The results obtained by Dr Robinson, by rotating an anemometer of this pattern without friction purposely applied, are given at pp. 797 and 799 of the *Phil. Trans.* for 1878. The mean of a few taken with velocities of about 27 miles an hour in still air gave a factor 2·36, instead of 2·50, as deduced from Mr Jeffery's experiments. As special anti-friction appliances were used by Dr Robinson, the friction in Mr Jeffery's experiments was probably a little higher. If such were the case, the factor ought to come out a little higher than in Dr Robinson's experiments, which is just what it does. As the circumstances of the experiments were widely different with respect to the vorticose motion of the air produced by the action of the anemometer in it, we may I think conclude that no very serious error is to be apprehended on this account.

In a later paper (*Phil. Trans.* for 1880, p. 1055), Dr Robinson has determined the factor for an anemometer (among others) of the Kew pattern by a totally different method, and has obtained values considerably larger than those given by the former method. Thus the limiting value of the factor m, corresponding to very high velocities, is given at p. 1063 as 2·826, whereas the limiting value obtained by the former method was only 2·286. Dr Robinson has expressed a preference for the later results. I confess I have always been disposed to place greater reliance on the results of the Dublin experiments, which were carried out by a far more direct method, in which I cannot see any flaw likely to account for so great a difference. It would be interesting to try the second method in a more favourable locality.

I take this opportunity of putting out some considerations respecting the general formula of the anemometer, which may perhaps not be devoid of interest.

The problem of the anemometer may be stated to be as follows:—Let a uniform wind with velocity V act on a cup anemometer of given pattern, causing the cups to revolve with a velocity v, referred to the centre of the cups, the motion of the cups being retarded by a force of friction F; it is required to determine v as a function of V and F, F having any value from 0, corresponding to the ideal case of a frictionless anemometer, to some limit F_1, which is just sufficient to keep the cups from turning. I will refer to my appendix* to the former of Dr Robinson's papers (*Phil. Trans.* for 1878, p. 818), for the reasons for concluding that F is equal to V^2 multiplied by a function of V/v. Let

$$V/v = \xi, \qquad F/V^2 = \eta,$$

then if we regard ξ and η as rectangular co-ordinates, we have to determine the form of the curve, lying within the positive quadrant $\xi O \eta$, which is defined by these co-ordinates.

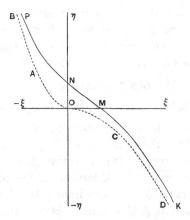

We may regard the problem as included in the more general problem of determining v as a function of V and F, where v is positive, but F may be of any magnitude and sign, and therefore, V also†. Negative values of F mean, of course, that the cups instead of being retarded by friction, are acted on by an impelling

[* Reprinted *infra*, p. 95.]

† Of course v must be supposed not to be so large as to be comparable with the velocity of sound, since then the resistance to a body impelled through air, or having air impinging on it, no longer varies as the square of the velocity.

force making them go faster than in a frictionless anemometer, and values greater than F_1 imply a force sufficient to send them round with the concave sides foremost.

Suppose now F to be so large, positive or negative, as to make v so great that V may be neglected in comparison with it, then we may think of the cups as whirled round in quiescent air in the positive or usual direction when F is negative, in the negative direction when F is greater than F_1. When F is sufficiently large the resistance may be taken to vary as v^2. For equal velocities v it is much greater when the concave side goes foremost, than when the rotation is the other way. For air impinging perpendicularly on a hemispherical cup, Dr Robinson found that the resistance was as nearly as possible four times as great when the concave side was directed to the wind as when the convex side was turned in that direction*. When the air is at rest and the cups are whirled round, some little difference may be made by the wake of each cup affecting the one that follows. Still we cannot be very far wrong by supposing the same proportion, 4 to 1, to hold good in this case. When F is large enough and negative, F may be taken to vary as v^2, say to be equal to $- Lv^2$. Similarly, when F is large enough and positive, F may be taken equal to $L'v^2$, where in accordance with the experiment referred to, L' must be about equal to $4L$. Hence we must have nearly

$$\eta = - L\xi^2, \text{ when } \xi \text{ is positive and very large;}$$
$$\eta = 4L\xi^2 \quad \text{,,} \quad \text{negative} \quad \text{,,} \quad \text{,,}$$

Hence if we draw the semi-parabola OAB corresponding to the equation $\eta = 4L\xi^2$ in the quadrant $\eta, O, - \xi$, and the semi-parabola OCD with a latus rectum four times as great in the quadrant $\xi, O, - \eta$, our curve at a great distance from the origin must nearly follow the parabola OAB in the quadrant $\eta, O, - \xi$, and the parabola OCD in the quadrant $\xi, O, - \eta$, and between the two it will have some flowing form such as $PNMK$. There must be a point of inflexion somewhere between P and K, not improbably within the positive quadrant ξ, O, η. In the neighbourhood of this point the curve NM would hardly differ from a straight line. Perhaps this may be the reason why Dr Robinson's experiments in the paper published in the *Phil. Trans.* for 1878 were so nearly represented by a straight line.

---

* *Transactions of the Royal Irish Academy*, Vol. xxii, p. 163.

ON THE DETERMINATION OF THE CONSTANTS OF THE CUP ANEMOMETER BY EXPERIMENTS WITH A WHIRLING MACHINE. BY T. R. ROBINSON, D.D., F.R.S.

Appendix*.

[From the *Philosophical Transactions*, 1878, pp. 818—821.]

THE object of the experiments being to determine the relation between the velocity of actual wind supposed uniform (the air also being at, or reduced to, a normal density), the velocity of the cups, and the friction, I assume in the first instance as correct, the values of those two quantities given by the experiments with the whirling machine, and proceed to consider the relation.

Let V' be the velocity with which the air passes the anemometer, that is, in the case of the actual experiments, the velocity of the centre of the anemometer itself corrected for the velocity of the wind produced by it; let v be the velocity of the centre of the cups, F the moment of the total friction. Then supposing the density of the air normal for a given anemometer, v will depend only on V' and F, that is, there will be a functional relation between the three variables V', v, F, leaving two of them independent.

In investigating experimentally the relation between two variables, it is often very useful to plot the results of experiment, as the general character of the relation sought, and the allowance to be made for errors of observation can thus be estimated. The relation between three variables would be expressed graphically by a surface instead of a curve, and it is troublesome to model a surface. If, however, we can find a relation between the variables which is satisfied, *provided* some other relation is satisfied, we can thereby reduce the number of independent variables from two to one, and employ ordinary plotting in investigating the relation

[* Cf. footnote, *supra*, p. 93.]

between the variables. In fact, the relation sought is reduced from one of the form $V' = \phi(v, F)$ to one of the form

$$f(V', v, F) = \psi\{f_1(V', v, F)\},$$

where ϕ, ψ denote unknown, and f, f_1 known functions.

In the present case, since by hypothesis the anemometer is in a permanent state, the moment of the friction is equal to the *total* impelling force of the air, *i.e.*, the total pressure arising from the motion of the air, without distinction of impelling or retarding, but reckoning the latter as a negative impelling force. Now, in cases of rapid motion like that of the air passing the cups of the anemometer, it is well ascertained that the resistance varies as the square of the velocity, all other circumstances being the same. Hence, with a given anemometer, when only the scale of the velocities changes, *i.e.*, when V varies as v, the moment of the total impelling force may be expected to vary as the square of the velocity. When the density changes it may be expected also to vary as the density. Hence we may expect that when v varies as V', then F varies as $\rho V'^2$, or in other words that

$$F = \rho V'^2 \phi(v/V') \dots\dots\dots\dots\dots\dots(1),$$

where ϕ denotes some function the form of which is not at present under consideration.

Let $v/V' = \xi$, $F/\rho V'^2 = \eta$, and for each observation let the point whose coordinates are ξ, η be laid down on paper. If $F\rho^{-1}$ were merely some arbitrary function of V' and v, the points so laid down would be spread out over the paper, but if equation (1) be true they will lie in a definite curve.

The actual experiments were executed in series, in each of which only one independent variable was changed, so that if the experiments were infinitely numerous and infinitely exact the locus of the point whose coordinates are ξ, η would be a definite continuous curve. And the test of the truth of (1) is that the curves belonging to the different series shall coincide, instead of being arranged in some order of sequence.

Plate 68* shows the result of plotting the observations taken with anemometer No. III. On inspecting the figure it will be seen that the different series fit very well into one another.

[* Not here reproduced.]

Departures there are no doubt in the individual observations from a mean curve, but these appear to be casual, not methodical and depending upon the order of the series.

The result of the observations then is confirmatory of the fundamental supposition made hitherto, that when the friction is so arranged that the velocity of the air passing the anemometer bears a given ratio to the velocity of the cups, the moment of the total impelling force varies as the square of either velocity.

Assuming then the truth of equation (1)*, we have next to inquire what is the form of the function ϕ?

A complete hydrodynamical solution of the problem is altogether beyond our power. On the other hand, the irregularities of the observations prevent us from going, by observation alone, more than a certain way towards the determination. We must, therefore, endeavour to combine as best may be the indications of mechanical theory with the results of experiments.

In his paper "On the Cup Anemometer," in the *Transactions of the Royal Irish Academy*, Dr Robinson has shown that (supposing the density constant, say = 1) the relation between the moment of the impelling force and the moment of the friction is either accurately or approximately of the form

$$F = \alpha V'^2 - 2\beta V'v - \gamma v^2 \dots\dots\dots\dots (2),$$

which would give for the locus of the point whose coordinates are ξ, η the parabola

$$\eta = \alpha - 2\beta\xi - \gamma\xi^2 \dots\dots\dots\dots\dots(3).$$

If now we turn to the plotting of the observations, we see that the best smooth curve to represent the observations, free from sinuosities which the observations do not warrant us in supposing real, is either accurately or approximately a straight line,

$$\eta = \alpha' - 2\beta'\xi \dots\dots\dots\dots\dots\dots(4);$$

in fact, so nearly does a straight line represent the observations that it is not easy to say to which side the concavity of the line, if curved it be, should lie. On the whole there appears to be a slight indication of a gentle concavity *towards* the origin.

---

* It formed no part of the object of the experiments to determine the relation of the impelling force to ρ, which merely comes in as a small correction for reducing observations made on different days to a common standard. It is the dependence of F on V' and v that is contemplated in the text.

It may be remarked in passing that the formula (4) which the experiments show to be at least very approximately true, leads to a very simple expression for v in terms of V', namely—

$$v = aV' - b/V',$$

where a and b are constants.

The figure shows that there cannot be much doubt as to the distance from the origin at which the curve intersects the axis of ξ, nor as to the direction of the curve at that point; and generally that the curve is well determined in its right-hand half, though it becomes more uncertain towards the left. If λ be the value of ξ at the point of intersection, and $-t$ the tangent of the inclination at that point, the equation of the curve, assumed to be a parabola, will be

$$\eta = t(\lambda - \xi) - C(\lambda - \xi)^2 \dots\dots\dots\dots (5);$$

or again, if we suppose known two points (p, h) and (q, k) lying in the well-determined part of the curve, its equation will be

$$\eta = (q - p)^{-1} \{ h(q - \xi) + k(\xi - p) - C'(\xi - p)(\xi - q) \} \dots (6);$$

and as $(\lambda - \xi)^2$ or $(\xi - p)(\xi - q)$ will be small throughout the well-determined part of the curve, the constant C or C' will admit of considerable latitude of variation without much affecting the satisfaction of the observations. Conversely, if we attempt to determine the constant C or C' from the observations, in addition to the two elements λ, t, or h, k, the determination will be extremely precarious. And if we arrange the formula (5) or (6) according to powers of ξ, so as to throw it into the form (3), the precariousness of the determination of C or C' will more or less affect all the three constants α, β, γ.

Accordingly, if we take this formula, and attempt to determine the three constants α, β, γ, from the observations, it may be that by different processes we shall arrive at results differing considerably, not only as regards γ, but even, though to a less degree, as regards α and β. It is not until we use the values of α, β, γ, so obtained for the determination of two out of the three elements of the parabola which are well or fairly determined by the observations, that we perceive the accordance which underlies the apparent discrepancy.

If the simple formula (4) so nearly fits the observations, it is by no means *merely* as an empirical formula of interpolation presenting two arbitrary constants whereby an approximate accordance may be brought about, or in the way that a small arc of an arbitrary curve may be approximately represented by a straight line. The observations were also plotted by taking for coordinates $V' : v$ and $F : v^2$ instead of $v : V'$ and $F : V'^2$, and in this case the curvature of the curve was very decided. Accordingly, though the observations may be satisfied by the first two terms of the formula (2) almost as well as by the three, that is by no means true of the last two, though in both cases alike we have two arbitrary constants at our disposal.

Note on the Reduction of Mr Crookes's Experiments* on the Decrement of the Arc of Vibration of a Mica Plate oscillating within a Bulb containing more or less Rarefied Gas.

[From the *Philosophical Transactions*, Vol. CLXXII, 1881, pp. 435—446. Received and read *February* 17, 1881.]

(Abstract from *Proceedings of the Royal Society*, Vol. XXXI, pp. 458—460.)

THE determination of the motion of the gas within the bulb, which would theoretically lead to a determination of the coefficient of viscosity of the gas, forms a mathematical problem of hopeless difficulty. Nevertheless we are able, by attending to the condition of similarity of the motion in different cases, to compare the viscosities of the different gases for as many groups of corresponding pressures as we please. Setting aside certain minute corrections, which would have vanished altogether had the moment of inertia of the vibrating body been sufficient to make the time of vibration sensibly independent of the gas, as was approximately the case, the condition of similarity is that the densities shall be as the log. decrements of the arc of vibration, and the conclusion from theory is that when that condition is satisfied, then the viscosities are in the same ratio. Pressures which satisfy the condition of similarity are said to "correspond."

It was found that on omitting the high exhaustions, the experiments led to the following law:—

The ratios of the viscosities of the different gases are the same for any two groups of corresponding pressures. In other words, if the ratios of the viscosities of a set of gases are found (they are given by the ratios of the log. decrements) for one set of corresponding pressures, these pressures may be changed in any given ratio without disturbing the ratios of the viscosities.

This law follows of course at once from Maxwell's law, according to which the viscosity of a gas is independent of the pressure. It does not, however, by itself alone prove Maxwell's law, and might be satisfied even were Maxwell's law not true. The constancy, however, of the log. decrement, when the circumstances are such that the molar inertia of the gas may presumably

[* Sir W. Crookes's experiments are described in the memoir preceding, "On the viscosity of gases at high exhaustions," *Phil. Trans.* Vol. CLXXII, 1881, pp. 387—434.]

be neglected, proves that at any rate when the density is not too great that law is true ; and the variability of the log. decrement at the higher pressures in all but the very light gas hydrogen is in no way opposed to it, though Mr Crookes's experiments do not enable us to test it directly, but merely establish a more general law, which embraces Maxwell's as a particular case.

The viscosities referred to air as unity which came out from Mr Crookes's experiments were as follows :—

Oxygen	1·117
Nitrogen and carbonic oxide	0·970
Carbonic anhydride	0·823
Hydrogen	0·500

The viscosity of kerosoline vapour could not be accurately deduced from the experiments, as the substance is a mixture, and the vapour density therefore unknown. Assuming the relative viscosity to be 0·0380, the vapour density required to make the experiments fit came out 3·408 referred to air, or 49·16 referred to hydrogen.

When once the density is sufficiently small, the log. decrement may be taken as a measure of the viscosity. Mr Crookes's tables show how completely Maxwell's law breaks down at the high exhaustions, as Maxwell himself foresaw must be the case. Not only so, but if we take pressures at those high exhaustions which are in the same ratios as "corresponding" pressures, the log. decrements in the different gases are by no means in the ratios of the densities.

It would appear as if the mechanical properties of a gas at ordinary pressures and up to extreme exhaustions (setting aside the minute deviations from Boyle's law, &c.), were completely defined by two constants, suppose the density at a given pressure and the coefficient of viscosity, but at the high exhaustions at which phenomena of "ultra-gas" begin to appear, specific differences came in, to include which an additional constant, or perhaps more than one, requires to be known.

———————

In the course of his long series of researches " On Repulsion resulting from Radiation," Mr Crookes had frequently occasion to observe the deflections of a light bar or lamina of some substance delicately suspended and oscillating by torsion. When such a bar was set in vibration, the vibrations tended more or less rapidly to subside, in consequence, no doubt, of the viscosity of the gas enclosed in the apparatus. At first it seemed as if the rate of subsidence tended to reach a constant value which remained the same at all higher exhaustions. But as methods of exhaustion were improved, and the gases were so rarefied that the effect of a candle in causing repulsion distinctly fell off, the rate of subsidence of the oscillations was found greatly to fall off too. This falling off at extreme exhaustions seemed to present a very interesting field of study in connexion with the molecular condition of gases *. The inquiry would naturally involve the observation of the nearly constant rate obtained at somewhat lower exhaustions; and the same apparatus would serve for experiments on the rate of subsidence at higher densities, up to that corresponding to atmospheric pressure.

A comparison of the rates of subsidence in different gases at great but not extreme exhaustions was further interesting as a new means of determining the ratios of the viscosities of different gases. In fact, at high exhaustions, the motion of the gas tends to a condition of ideal simplicity from which a comparison of the viscosities of different gases would immediately result. The effect of the viscosity of a gas on its own motion is regulated by the value of a constant which I have elsewhere† called the index of friction of the gas, namely, the coefficient of viscosity divided by the density. According to Maxwell's law, the coefficient of viscosity is independent of the density, and therefore the index of friction varies inversely as the density. Hence as the exhaustion proceeds the motion of the gas tends to become what it would be if the viscosity were infinite, and the bounding surfaces had their actual motion. In the limit, the instantaneous motion of the gas depends only on that of the vibrating plate, to which it is proportional, except in so far as the finiteness of the angle of oscillation

[* In the experiments of Kundt and Warburg, *Ann. der Phys.*1875–6, referred to in Sir W. Crookes's memoir, this stage of exhaustion was attained. These investigations were conducted according to Maxwell's plan of a disc vibrating torsionally between fixed discs, but without definite measurement of the vacua employed; they revealed a relative reduction of the amount of damping when the discs were brought close together, which was ascribed by the authors to a slight frictional slip, theoretically indicated, of the gas over the surfaces of the discs, to an extent varying inversely as the pressure and thus sensible only at high exhaustions. This slip, as extended by O. Reynolds to include gradient of temperature as well as velocity, is, according to Maxwell's later theory, *Phil. Trans.* 1879, *Collected Papers*, II. p. 701, essential to the action of Crookes's radiometer. On the other hand, Sir W. Crookes (*loc. cit.* p. 425) on the basis of his experience of high vacua, suggests a connexion of this result, and also of some anomalies encountered in the case of hydrogen, with the presence of aqueous vapour condensed on the plates; of this an effect had been traced by the authors in a gradual change of viscosity after the exhaustion had been stopped, but according to them the present effect appears at an earlier stage in the exhaustion.]

† "On the Effect of the Internal Friction of Fluids on the Motion of Pendulums," *Camb. Phil. Trans.* Vol. IX, p. [8]. [*Ante*, Vol. III, p. 1.]

entails a difference of position of the plate relatively to the bounding wall of the bulb, a difference however which is trifling on account of the smallness of the angle through which the plate oscillates. The forces therefore arising from the viscosity tend in the limit to vary entirely as the angular velocity, and not, as is the case when the index of friction is comparatively small, partly also as the angular acceleration. The result therefore will be that the oscillations are retarded by a force varying as the angular velocity and producing therefore a subsidence of the motion such that the proportionate rate of change of the arc is proportional to the coefficient of viscosity.

Mr Crookes's experiments were carefully made from pressures as high as the atmospheric pressure downwards. At first there is a very evident decrease of subsidence as the pressure decreases, except in the case of hydrogen, in which it is very small, we may say insensible. Then it remains very nearly constant for a considerable range of exhaustion, and at last, for extreme exhaustions, it rapidly fades away.

In the second of these three stages the condition of ideal simplicity above mentioned is doubtless approximately attained. If however we confined our attention to this part only of the series, the lower part, although so carefully made, would remain unutilised; and further, we should remain uncertain whether in taking the logarithmic decrement as proportional to the viscosity our approximation was not too rough.

The determination of the motion of the gas corresponding to a given motion of the vibrating solid, and thereby the determination of the forces which the gas exerts on the solid, forms a perfectly definite problem, the solution of which, if it could be effected, would lead to a determination of the coefficient of viscosity from the observed influence of the gas on the motion of the plate. But although in these slow motions the terms in the hydrodynamical equations which involve the squares of the velocities are insensible, so that the equations may be taken as linear, the problem is one of hopeless difficulty except in a few simple cases. In the paper referred to, I have given the solution in the case of a sphere vibrating in a mass of fluid either unlimited or confined by a concentric spherical envelope, and in that of a long cylindrical rod vibrating in an unconfined mass of fluid. In the latter especially

of these cases the solution involves functions of a highly complicated form.   For a lamina, such as that employed by Mr Crookes,
the problem could not even be solved if the fluid were regarded as
perfect, much less when the viscosity is taken into account.

But though we are baffled in the attempt to give an absolute
solution of the problem, theory indicates the conditions of similarity
of the motion of the gas in two different cases, and enables us
thereby to compare the viscosities when those conditions are
satisfied.

The bulb, vibrating plate, and torsion thread being always the
same, the two things that varied from one experiment to another
were the nature and the pressure of the gas, not the temperature,
which for the present was kept constant.   The moment of inertia
of the lamina was sufficiently large to allow the time of vibration
to be nearly the same in the different experiments.   For the
present I will suppose it constant, reserving to a later stage the
consideration of the correction to be made for its variation.

Let ρ be the density of the gas, p the observed pressure, D the
density under a standard pressure, μ the coefficient of viscosity,
and let accented letters refer to another gas.   The dimensions of
the terms in the equations of motion show that in comparing two
cases in which the nature and pressure of the gases alone differ,
the motions will be similar provided

$$\frac{\mu}{\rho} = \frac{\mu'}{\rho'}, \text{ or } \frac{\mu}{pD} = \frac{\mu'}{p'D'} \dots\dots\dots\dots\dots(1).$$

This condition being satisfied, the resultant pressures of the gas on
the solid will vary as μ or as ρ; and as the logarithmic decrements
(l) will vary in the same proportion, we shall have

$$\frac{l}{\rho} = \frac{l'}{\rho'}, \text{ or } \frac{l}{pD} = \frac{l'}{p'D'} \dots\dots\dots\dots\dots(2).$$

The equations (1) and (2) are such that when one is satisfied
so is the other.   It will be convenient to regard (2) as giving the
condition of similarity, and then (1) or

$$\frac{\mu}{l} = \frac{\mu'}{l'} \dots\dots\dots\dots\dots\dots(3)$$

gives the ratio of the viscosities at the two corresponding pressures
in the two gases.

The times of vibration were practically constant when once the exhaustion was pretty high, at least until the very highest exhaustions were reached, when it fell off a very little; but at atmospheric pressure and at low exhaustions it was somewhat greater, though not much. Its variability will not affect the results obtained by the above method, provided only the times are the same in the two experiments of each pair, which was very approximately the case. Nevertheless it may be well to consider the correction to be made in consequence of the inequality of the times.

Let τ be the time of vibration from rest to rest, then in comparing two similar systems the time-scale must be varied, so as always to be proportional to τ, and the hydrodynamical equations show that for the condition of similarity we have, in place of (1), the equation

$$\frac{\mu\tau}{\rho} = \frac{\mu'\tau'}{\rho'} \quad\ldots\ldots\ldots\ldots\ldots\ldots\ldots\ldots(4).$$

As the two dynamical systems are not similar as a whole, but only the gaseous parts of them, we must have recourse to the equation of motion of the vibrating lamina. Let θ be the angle of torsion, I the moment of inertia, $n^2 I \theta$ the force of restitution, which will be proportional to the angle of torsion, provided at least the glass fibre be treated as perfectly elastic, as it doubtless may be in considering the correction to be made for the inequality of the times of vibration, even though its defect of elasticity might not, possibly, be absolutely insensible in its influence on the main motion. Then if there were no fluid the equation of motion of the lamina would be

$$I \frac{d^2\theta}{dt^2} + n^2 I \theta = 0.$$

As in the cases treated of in the paper of mine already referred to, the resultant pressure of the fluid on the lamina (the term "pressure" here including the tangential action), will partly agree in phase with the displacement or force of restitution, partly with the velocity of the lamina. The first part will have the effect of adding to the mass of the lamina an ideal mass depending on the density, viscosity, and time of vibration. From the dimensions of the quantities involved with respect to time and density, this ideal mass must be of the form $\rho f\left(\dfrac{\mu\tau}{\rho}\right)$. There is no need to express the dependence of the function f on the scale of lengths. In a similar manner the part of the resultant pressure which is multiplied by $d\theta/dt$ must be expressed by ρ multiplied by some other function of $\mu\tau/\rho$ and divided by a time. We may express it therefore by $2\rho n F\left(\dfrac{\mu\tau}{\rho}\right)$, where n is $\dfrac{\tau}{\pi}$. Denoting the two functions of $\mu\tau/\rho$ by A and B respectively, we have accordingly for the equation of motion

$$(I + A\rho)\frac{d^2\theta}{dt^2} + 2B\rho n\frac{d\theta}{dt} + a^2 I\theta = 0 \quad\ldots\ldots\ldots\ldots(5).$$

The integral of this equation is

$$\theta = e^{-qt}(c\cos mt + c'\sin mt),$$

where

$$q = \frac{B\rho n}{I + A\rho}, \quad m^2 = \frac{a^2 I}{I + A\rho} - \frac{B^2\rho^2 n^2}{(I + A\rho)^2};$$

and since by definition of n, $m = n$, we have

$$n^2 + q^2 = \frac{a^2 I}{I + A\rho},$$

and then by eliminating A between the last equation and the last but two we have

$$B = \frac{a^2 I q}{\rho n (n^2 + q^2)} \quad\ldots\ldots\ldots\ldots\ldots\ldots(6).$$

Now n is π/τ, and $q\tau$ is the Napierian logarithmic decrement, or l/M, M being the modulus of the common system. Hence (6) becomes

$$B = \frac{a^2 M I l \tau^2}{\pi\rho (M^2\pi^2 + l^2)},$$

and as B is the same in the two systems compared we have

$$\frac{l}{\rho} \cdot \frac{\tau^2}{M^2\pi^2 + l^2} = \frac{l'}{\rho'} \cdot \frac{\tau'^2}{M^2\pi^2 + l'^2} \quad\ldots\ldots\ldots\ldots(7),$$

an equation which takes the place of (2), and serves to define corresponding densities, and then (4) gives the ratio of the viscosities at those densities, or say, at the corresponding pressures. If we eliminate the ratio of ρ to ρ' between (4) and (7) we get

$$\frac{\mu}{l\tau}(\pi^2M^2 + l^2) = \frac{\mu'}{l'\tau'}(\pi^2M^2 + l'^2) \quad \ldots\ldots\ldots\ldots\ldots(8),$$

which takes the place of (3).

The ratio of the factor $\pi^2M^2 + l^2$ to π^2M^2 alone but little exceeds unity; thus even for oxygen at 760 millims. pressure it is barely 1·0085, and of course the ratio of that factor for one gas to the factor for another gas at the corresponding pressure will differ from unity still less. Hence it is almost a needless refinement to keep in this factor at all. However, even if we retain it in (8) it is quite superfluous in (7), which merely determines what densities are to be deemed to correspond in seeking the logarithmic decrements. For until extreme rarefactions are reached, to which the above investigation no longer applies, the logarithmic decrement changes so slowly that a small error in the density of one gas which is deemed to correspond to a given density in another will make no sensible error in the logarithmic decrement. And not only may the factors above mentioned be omitted, but as the ratio of τ to τ' will differ but little from a ratio of equality, the formula (7) may be dispensed with altogether, and the simpler formula (2) employed. But when the logarithmic decrements have been found, in determining the ratio of the viscosities from (8) it is better not to disregard the quantities by which the ratios of τ to τ' and of $\pi^2M^2 + l^2$ to $\pi^2M^2 + l'^2$ differ from ratios of equality. And if we now wish to know more precisely what densities or pressures do correspond, we may obtain them from (4).

In the numerical calculations which follow, the difference in the times of vibration (τ) at *corresponding* pressures in the different gases is neglected, and likewise the difference between the ratios of the factors $M^2\pi^2 + l^2$ and a ratio of equality. The general effect of this omission, which is very minute, will be considered in the end.

The values of D adopted were, hydrogen, 1; air, 14·42; oxygen, 16; nitrogen, 14; carbonic anhydride, 22; carbonic oxide, 14.

Tables were first formed of the logarithms of l/pD for the different gases for the various pressures given by Mr Crookes down to 0·76 millim. The pressures standing against equal numbers in these tables for the different gases would be " corresponding" pressures. The pressures corresponding to a given number may be obtained from the tables by interpolation; and as the experiments were made at close intervals it seemed sufficient to regard only the two adjacent numbers and use proportional parts.

Corresponding pressures					Logs. of ratios to corresponding pressures in air			
Air	O	N	CO_2	CO	O	N	CO_2	CO
760	767·3	760·8	413·3	760·8	+ ·0042	+ ·0005	− ·2645	+ ·0005
660	666·6	660·9	359·2	663·0	+ ·0044	+ ·0006	− ·2642	+ ·0020
560	565·5	559·6	306·2	563·8	+ ·0042	− ·0003	− ·2622	+ ·0029
460	463·7	459·0	250·0	459·8	+ ·0034	− ·0010	− ·2649	− ·0002
360	365·9	361·6	194·7	359·5	+ ·0071	+ ·0019	− ·2669	− ·0006
260	263·2	261·8	141·9	258·3	+ ·0053	+ ·0030	− ·2630	− ·0029
160	161·2	159·5	86·4	166·7	+ ·0033	− ·0013	− ·2676	− ·0022
Mean of log p'/p					+ ·0046	+ ·0005	− ·2648	− ·0005
Number, or p'/p					1·010	1·001	0·540	0·999
Log $(p'D' \div pD)$, from mean					·0497	$\bar{1}$·9876	$\bar{1}$·9186	$\bar{1}$·9870
$p'D' \div pD$					1·121	0·972	0·829	0·971

I do not think it worth while to give these tables at length, but I subjoin a small table calculated from them, giving for oxygen, nitrogen, carbonic anhydride, and carbonic oxide the pressures corresponding to air pressures decreasing by 100 millims. from 760 to 160. It will be seen from Mr Crookes's tables that below these pressures there is little variation in l until very high exhaustions are reached. Hydrogen does not enter into the table, as the highest pressure (760 millims.) in the experiments corresponds to a pressure of only about 106 millims. in air. The table contains also the logarithms of the ratios of the pressures in the different gases to the corresponding pressures in air.

An inspection of the numbers in the same vertical column in the right-hand portion of the above table shows that the logarithm in question is constant as nearly as the observations can show. This leads to the following law.

If any pressure be taken in one gas and the pressures found in other gases for which the coefficients of viscosity are as the densities (pressures which have been defined as " corresponding "), then if another system of pressures be taken proportional to the former the pressures in the new system will also correspond; and consequently the ratios of the coefficients of viscosity of the different gases will be the same for the pressures in one such system as in another.

This law is in accordance with Maxwell's law, but does not by itself alone prove Maxwell's law. It leaves the functional relation between the coefficient of viscosity and the density for any one gas arbitrary, and deduces from thence the relation for all other gases, this relation introducing merely one unknown constant for each. What the law gives may be put in a clear form by a geometrical illustration. I assume Boyle's law, so that for any one gas the ratios of the densities in different cases or the ratios of the pressures may be used indifferently.

Let, then, the relation between the viscosity and density be represented graphically by taking abscissæ to represent densities and ordinates to represent coefficients of viscosity. Then the law found above may be enunciated by saying that the curves for all gases are geometrically similar, the origin being the centre of similitude. Maxwell's law would give a particular case of such similar curves, namely, a system of straight lines parallel to the axis of abscissæ.

The deviation from uniformity of the logarithmic decrements for any one of these gases at these comparatively speaking high pressures is not therefore in any way inconsistent with Maxwell's law, but is fully accounted for by the very natural supposition that the rarefaction is not yet sufficient to render the molar inertia of the gas insensible as regards its influence on the gas's own motion, a supposition which can be shown to be true when we employ the approximately known absolute value of the coefficient of viscosity. The same consideration shows, moreover, that we have only to carry the exhaustion further in order to render the effect of that inertia insensible, and accordingly, *if* Maxwell's *law be true*, to make the logarithmic decrement sensibly independent of the pressure.

That such is actually the case is shown by Mr Crookes's tables or the diagram A, in which they are graphically represented. We observe a manifest tendency for the logarithmic decrement to become constant till the law is interrupted by the breaking down of viscosity attending extreme exhaustions, or by certain deviations which in some cases (as in those of oxygen and kerosoline vapour) show themselves a little earlier: these deviations will be referred to further on; for the present I merely avoid exhaustions high enough to introduce them. This approximate constancy of logarithmic decrement is observed in hydrogen from the first, which is accounted for by the high index of friction of that gas as compared with the others at equal pressure.

This evident constancy or tendency towards constancy in the viscosity as the rarefaction goes on supplies the missing link, and establishes Maxwell's law on the basis of Mr Crookes's experiments even taken by themselves. It is not, of course, *directly proved* for the higher pressures in the gases other than hydrogen; its extension to such pressures is a matter of inference, derived from observing, first, that it is found to be true within such limits of density that the condition of ideal simplicity supposed at the commencement of this note is presumably sensibly attained; and, secondly, that above those limits, though we are unable from mathematical difficulties to examine its truth directly, yet we are able to deduce from theory one inference on the supposition of its truth, which is found to be in accordance with the results of experiment.

Hitherto the ratios of the coefficients of viscosity have been deduced from a part of Mr Crookes's tables, in which the logarithmic decrements changed very evidently with the pressure. We may now deduce those ratios by what may almost be deemed an independent method, namely, by attending only to the part of the tables at which the logarithmic decrement is all but constant. If the condition of ideal simplicity supposed at the outset were quite attained, we might disregard the pressures in the comparison, which would entitle that method to be considered quite distinct from the former; but as that condition is not absolutely reached, it will be proper not wholly to neglect the condition of correspondence of pressure, though a rough determination of correspondence will suffice.

Suppose, then, we take the air pressures from 120 to 26 millims. The ratio of the corresponding pressures in oxygen, &c., is given in the last line but two of the preceding table. The corresponding limits are for oxygen 132 to 29; for nitrogen and carbonic oxide the same (sensibly) as for air; for carbonic anhydride 65 to 14; for hydrogen the limits are not given in the table, but they are 864 and 187 nearly. In strictness each pressure should be considered separately; but as the intervals were not intentionally divided in a different manner for the different gases, and as the logarithmic decrements are very nearly constant between the specified limits, it seems sufficient to take the mean for each gas of those corresponding to pressures that lie between the assigned limits. We thus get for air, ·1002; oxygen, ·1120; nitrogen, ·971; carbonic anhydride, ·822; carbonic oxide, ·971; hydrogen, ·500. Reducing to air = 1, and writing down for comparison the numbers expressing the ratios of the coefficients of viscosity to that of air given in the previous table, we have for the ratios in question—

	O	N	CO_2	CO	H
From air pressures, 760 to 160...	1·121	0·972	0·829	0·971	...
„ „ 120 to 26...	1·118	0·969	0·820˙	0·969	0·499
Values adopted	1·120	0·970	0·822	0·970	0·499

We see that almost identically the same numbers are obtained whether they are deduced from the higher pressures, for which the logarithmic decrements notably diminish with the pressure, or from the part of the tables in which they are nearly independent of the pressure. The greatest difference is in the case of carbonic anhydride, where it is rather more than one per cent. This difference is in part accounted for by the omission of the correction for the time of vibration. If the times of vibration at *corresponding* pressures as determined by Mr Crookes be taken, they will be found to be very nearly the same; indeed, the differences are quite comparable with the errors of observation of those times. Perhaps the differences could be got with greater certainty from theory than from observation. According to theory the effect of the gas on the time of vibration turns mainly on the term $A\rho$. Now, though A is a function which we cannot calculate, yet we know that it is the same at corresponding pressures in two gases. The effect at such pressures varies therefore from one gas to another as the density, and therefore as the coefficient of viscosity.

It is here supposed (as is practically true) that the term is so small compared with I, with which it is associated, that its square, &c., may be neglected.

Taking the time of two oscillations (or of one complete oscillation) for air at an exhaustion at which the effect of the molar inertia has ceased to be sensible, but the slight decrease due to the removal of the viscosity has not yet come in, at $10^{s}\cdot76$, we get for the mean effect at pressures 760, 660, ... 160 about $0^{s}\cdot20$. The coefficient of viscosity for carbonic anhydride being 18 per cent. less than for air, we get $0^{s}\cdot036$ for the average difference of times in air and that gas, which is $\frac{1}{300}$th of the average time; and since according to (8) $\mu \propto \tau$ we must deduct $\cdot003$ from the $\cdot829$ given above, leaving $\cdot826$, the mean of which and $\cdot820$ gives $\cdot823$, nearly the number adopted. Similarly the number $1\cdot121$ for O in the first line should be raised about $\cdot002$.

There is still a small correction to make depending on the factor $\pi^2 M^2 + l^2$. Since by (8) μ varies as this factor, and l^2 is very small compared with $\pi^2 M^2$, and moreover $\mu \propto l$ nearly, it will suffice to deduct $l^3/\pi^2 M^2$ from l, and use the l's so corrected. The correction being, however, very small, it will suffice to take an average l and make the deduction for it. The deductions for the six gases came to about $\cdot005$, $\cdot009$, $\cdot005$, $\cdot003$, $\cdot005$, $\cdot001$. Deducting these numbers from the relative viscosities given above, and reducing afresh to the scale air $= 1$, we get the following final numbers:

Air	O	N	CO$_2$	CO	H
1·000	1·117	0·970	0·823	0·970	0·500

I have left kerosoline vapour to the last on account of the uncertainty as to its vapour density. It is a mixture of different substances, being the more volatile part of petroleum. I am informed by Mr Greville Williams that it contains much pentane, the theoretical vapour density of which on the hydrogen scale would be 36. Taking at a venture $D = 36$, and choosing suppose the pressure 54 millims., for which $l = \cdot0404$, and further assuming the limiting logarithmic decrement for air before the breakdown to be $0\cdot1000$, as it seems to be from Mr Crookes's table, we find $0\cdot392$ for the relative viscosity of kerosoline vapour. This is pretty certainly too high. If we suppose the true number to be

0·380, we get for the air number corresponding to $l' = ·0425$, $l = ·1129$, which from the table of results for air belongs to $p = 740$. This would give for the vapour density of kerosoline

$$D' = \frac{·0380 \times 740}{·1000 \times 82·5} \; D = 3·408D = 49·16 \text{ if } D = 14·42.$$

For the air pressure corresponding to 54 millims. in kerosoline we should have $740 \times 54 \div 82·5 = 484·3$. For the corresponding logarithmic decrement we get from the table of experiments for air ·1059. The logarithmic decrement for kerosoline at 54 millims. calculated from that for air at 484·3 millims. is ·1059 × ·0390, or ·04024, which comes very near the observed number ·0404. It is curious that we should apparently be able to calculate very approximately an unknown vapour density from observations on the decrement of the arc of vibration of a vibrating lamina.

Before considering the falling off of viscosity at high exhaustions, I would point out a result of theory which is of some interest in connexion with the forms of Mr Crookes's curves.

At p. [34]* in the paper of mine already referred to I have given in equation (61) an expression (which has to be subsequently multiplied by ρ) for the resistance to a sphere vibrating in a viscous fluid within a concentric spherical envelope. When the index of friction is sufficiently great, as will be the case when the exhaustion is high enough, this expression may be developed according to ascending powers of m, which will be convergent even from the first. It will be found that the successive terms will be multiplied by

$$m^{-2}, \; m^0, \; m^2, \; m^4 \ldots,$$

where m^2 is a pure imaginary multiplied by ρ/μ; and from the mode of treatment there adopted it will readily be seen that the terms fall alternately on the arc and on the time.

The same thing may, however, be shown to be true generally, independently of the form of the vibrating body. It is here supposed, as has been all along, that the motion is sufficiently slow to allow us to neglect squares and products of the components of the velocity, or of their differential coefficients. For if μ/ρ be very great, we may imagine the hydrodynamical equations solved

[* *Ante*, Vol. III, p. 36.]

by successive substitution. First we should neglect the terms in ρ, and solve the equations; then substitute in the terms multiplied by ρ the result of the first approximation and solve again, and so on. And though we cannot actually solve the equations, still this consideration shows that the solution must be of the form

$$a\mu + b\rho + c\frac{\rho^2}{\mu} + d\frac{\rho^3}{\mu^2} + \dots,$$

where $a, b, c, d \dots$ involve neither μ nor ρ. And by adopting the artifice for the introduction of the time employed in the paper it readily appears that the terms fall alternately on the arc and on the time. Hence taking the two most important terms only in the expression for the effect on the arc we shall have for l an expression of the form

$$a'\mu + c'\frac{\rho^2}{\mu}, \quad \text{or} \quad a'\mu + c'\frac{D^2}{\mu}p^2.$$

Hence in a curve plotted with l and p for coordinates, the tangent as p diminishes would tend to become parallel to the axis of p, and the curvature where the curve cuts the axis of l would vary from one gas to another directly as the square of the density at a standard pressure and inversely as the coefficient of viscosity. For the gases examined the curvatures near the axis of l would therefore vary

for	Air	O	N and CO	CO_2	H
as	$14\cdot42^2 \div 1\cdot000$,	$16^2 \div 1\cdot117$,	$14^2 \div 0\cdot970$,	$22^2 \div 0\cdot823$,	$1^2 \div 0\cdot500$
or as	$1\cdot000$,	$1\cdot102$,	$0\cdot971$,	$2\cdot828$,	$0\cdot010$

It will be seen in Mr Crookes's diagram A, that if we imagine the curves continued upwards on their old lines, cutting off the changes which take place at very high exhaustions, the tangent tends to become vertical, and, moreover, the rate at which the direction of the tangent changes as we go down agrees well, as far as the eye can judge, with the above figures. We may notice in particular the extreme flatness of the hydrogen curve.

As we proceed upwards to the higher exhaustions, the first thing that strikes us (first in order of occurrence, very far from first in order of magnitude) is the curious *increase* which is observed in the logarithmic decrement in the case of oxygen and of kerosoline vapour. Small as this is, Mr Crookes considers it real. It puzzled me at first, since it occurs while the pressure is still comparatively high, such as 15 millims. or 20 millims., so that the mean free path must still be extremely small, and might, one would naturally suppose, be treated as infinitely small considering the dimensions of the apparatus. It occurred to me afterwards that it is probably referable to the thinness of the vibrating body. As the lamina moves there must in the immediate neighbourhood of the edge be a thin stratum or cushion of gas in which there is a very intense shearing motion. The intensity of the shearing makes up in good measure for the narrowness of the cushion, and renders the effect of the cushion a not insignificant fraction of the whole. The narrowness of the stratum may well be such as to forbid us to treat the mean free path as infinitesimal long before we are prohibited from so regarding it in comparison with the dimensions of the vessel, or the lateral dimensions of the lamina. That among the four unmixed gases examined oxygen should be the one to show this effect, seems to be connected with the fact that at comparatively low exhaustions (such as 0·76 millim.) it shows repulsion effects much the most strongly*; and both phenomena seem to indicate that for oxygen the length of path (I do not say *free* path) is comparatively large, "path" here meaning the space throughout which a molecule preserves approximately its original direction of motion.

When we come to those high exhaustions at which the decrement of arc gives way, the law of proportional logarithmic decrements at corresponding (and those proportional) pressures, which we hitherto found to be so accurately obeyed, breaks down altogether. A single example will suffice to show this. Take hydrogen at 330 **M**, for which $l = 0·0495$. The ratios of corresponding pressures resulting from the numbers for the relative viscosities given on p. 107, and the numbers to which they lead, are as follows:—

Gas	Air	O	N	CO_2	CO	H
Ratios of corresponding pressures	0·1387	0·1396	0·1386	0·0748	0·1386	1
Pressures corresponding to 330 M in H	45·8	46·1	45·7	24·7	45·7	330·0
l calculated from l in H	·0990	·1106	·0961	·0815	·0961	
l observed	·0758	·0829	·0722	·0525	·0734	0·0495

[* In this connexion the footnote, p. 102, may be referred to.]

8—2

It would seem as if when a gas may be treated as a continuous mass with continuously varying conditions of pressure, velocity, &c., as is done in the application of the hydrodynamical equations, a gas is completely defined* as to its mechanical action by two constants, suppose the density at a standard pressure and the coefficient of viscosity; but when the conditions are such as oblige us to take account of the finiteness of path of the molecules, specific differences are manifested which oblige us to introduce at least one constant more in order that the gas may be even mechanically defined; for of course I am not contemplating the chemical properties. It is worthy of note in this connexion that the two gases, oxygen and kerosoline vapour, which showed the phenomenon of a rate of decrease of arc of vibration *increasing* with a decreasing density, are just those which lie at the two extremes as regards viscosity, while as regards density at a given pressure they are separated by carbonic anhydride, which nevertheless does not show the phenomenon in question. It may well be that the mode of encounter of such complex structures as the molecules of a gas varies from one gas to another; and that while some of the laws of gases admit of explanation when the molecules are regarded as elastic spheres, or as particles repelling one another according to some definite law of force, other properties fail to receive an explanation when such a simplification of conception is adopted.

* I here leave out of account such differences as the small deviations from Boyle's law which have been observed with different gases.

On the Cause of the Light Border frequently noticed
in Photographs just outside the Outline of a Dark
Body seen against the Sky: with some Introductory
Remarks on Phosphorescence.

[From the *Proceedings of the Royal Society*, Vol. XXXIV, pp. 63—68.
Received *May* 20, 1882.]

An observation I made the other day with solar phosphori,
though not involving anything new in principle, suggested to me
an explanation of the above phenomenon which seems to me very
likely to be the true one. On inquiring from Captain Abney
whether it had already been explained, he wrote: "The usual
explanation of the phenomenon you describe is that the silver
solution on the part of the plate on which the dark objects fall has
nowhere to deposit, and hence the metallic silver is deposited to
the nearest part strongly acted upon by light." As this explanation seems to me to involve some difficulties, I venture to offer
another.

1. I will first mention the suggestive experiment, which is
not wholly uninteresting on its own account, as affording a pretty
illustration of what is already known, and furnishing an easy and
rapid mode of determining in a rough way the character of the
absorption of media for rays of low refrangibility.

The sun's light is reflected horizontally into a darkened room,
and passed through a lens*, of considerable aperture for its focal
length. A phosphorus is taken, suppose sulphide of calcium giving
out a deep blue light†, and a position chosen for it which may be

---

* The lens actually used was one of crown glass which I happened to have;
a lens of flint glass would have been better, as giving more separation of the
caustic surfaces for the different colours.

† The experiments were actually made, partly with a tablet painted with
Balmain's luminous paint, partly with sulphide of calcium and other phosphori in
powder.

varied at pleasure, but which I will suppose to be nearer to the lens than its principal focus, at a place where a section of the pencil passing through the lens by a plane perpendicular to its axis shows the caustic surface well developed. A screen is then placed to intercept the pencil passing through the lens, and the phosphorus is exposed to sunlight or diffuse daylight, so as to be uniformly luminous, and is then placed in position; the screen is then removed for a very short time and then replaced, and the effect on the phosphorus is observed.

Under the circumstances described there is seen a circular disk of blue light, much brighter than the general ground, where the excitement of the phosphorus has been refreshed. This is separated by a dark halo from the general ground, which shines by virtue of the original excitement, not having been touched by the rays which came through the lens.

2. The halo is due to the action of the less refrangible rays, which, as is well known, discharge the phosphorescence. Their first effect, as is also known, is however to cause the phosphorus to give out light; and if the exposure were very brief, or else the intensity of the discharging rays were sufficiently reduced, the part where they acted was seen to glow with a greenish light, which faded much more rapidly than the deep blue, so that after a short time it became relatively dark.

3. This change of colour of the phosphorescent light can hardly fail to have been noticed, but I have not seen mention of it. In this respect the effect of the admission of the discharging rays is quite different from that of warming the phosphorus, which as is known causes the phosphorus to be brighter for a time, and then to cease phosphorescing till it is excited afresh. The difference is one which it seems important to bear in mind in relation to theory. Warming the phosphorus seems to set the molecules more free to execute vibrations of the same character as those produced by the action of the rays of high refrangibility. But the action of the discharging rays changes the character of the molecular vibrations, converting them into others having on the whole a lower refrangibility, and being much less lasting.

4. Accordingly when the phosphorus is acted on simultaneously by light containing rays of various refrangibilities,

the tint of the resulting phosphorescence, and its more or less lasting character, depend materially on the proportion between the exciting and discharging rays emanating from the source of light. Thus daylight gives a bluer and more lasting phosphorescence than gaslight or lamplight. I took a tablet which had been exposed to the evening light, and had got rather faint, and, covering half of it with a book, I exposed the other half to gaslight. On carrying it into the dark, the freshly exposed half was seen to be much the brighter, the light being, however, whitish, but after some considerable time the unexposed half was the brighter of the two.

Again, on exposing a tablet, in one part covered with a glass vessel containing a solution of ammonio-sulphate of copper, to the radiation from a gas flame, the covered part was seen to be decidedly bluer than the rest, the phosphorescence of which was whitish. The former part, usually brighter at first than the rest, was sure to be so after a very little time. The reason of this is plain after what precedes.

A solution of chromate of potash is particularly well suited for a ray filter when the object is to discharge the phosphorescence of sulphide of calcium. While it stops the exciting rays it is transparent for nearly the whole of the discharging rays. The phosphorescence is accordingly a good deal more quickly discharged under such a solution than under red glass, which along with the exciting rays absorbs also a much larger proportion than the chromate of the discharging rays.

5. I will mention only one instance of the application of this arrangement to the study of absorption. On placing before excited sulphide of calcium a plate of ebonite given me by Mr Preece as a specimen of the transparent kind for certain rays of low refrangibility, and then removing the intercepting screen from the lens, the transmission of a radiation through the ebonite was immediately shown by the production of the greenish light above mentioned. Of course, after a sufficient time, the part acted on became dark.

6. I will mention two more observations as leading on to the explanation of the photographic phenomenon which I have to suggest.

In a dark room, an image of the flame of a paraffin lamp was thrown by a lens on to a phosphorescent tablet. On intercepting the incident rays after no great exposure of the tablet, the place of the image was naturally seen to be luminous, with a bluish light. On forming in a similar manner an image of an aperture in the window-shutter, illuminated by the light of an overcast sky reflected horizontally by a looking-glass outside, this image of course was luminous; it was brighter than the other. On now allowing both lights to act simultaneously on the tablet, the image of the flame being arranged to fall in the middle of the larger image of the aperture, and after a suitable exposure cutting off both lights simultaneously, the place of the image of the aperture on which the image of the lamp had fallen was seen to be *less* luminous than the remainder, which had been excited by daylight alone. The reason is plain. The proportion of rays of lower to rays of higher refrangibility is much greater in lamp-light than in the light of the sky; so that the addition of the lamplight did more harm by the action of the discharging rays which it contained on the phosphorescence produced by the daylight, than it could do good by its own contribution to the phosphorescence.

7. The other observation was as follows:—The same tablet was laid horizontally on a lawn on a bright day towards evening, when the sun was moderately low, and a pole was stuck in the grass in front of it, so as to cast a shadow on the tablet. After a brief exposure the tablet was covered with a dark cloth, and carried into a dark room for examination.

It was found that the place of the *shadow* was *brighter* than the general ground, and also a deeper blue. For a short distance on both sides of the shadow the phosphorescence was a little feebler than at a greater distance.

This shows that, though the direct rays of the sun by them-selves alone would have strongly excited the phosphorus, yet acting along with the diffused light from all parts of the sky, they did more harm than good. They behaved, in fact, like the rays from the lamp in the experiment of § 6. The slightly inferior luminosity of the parts to some little distance on both sides of that on which the shadow fell, shows that the loss of the diffuse light corresponding to the portion of the sky cut off by the pole was quite sensible when that portion lay very near the sun.

All this falls in very well with what we know of the nature of the direct sunlight and the light from the sky. In passing through the atmosphere, the direct rays of the sun get obstructed by very minute particles of dust, globules of water forming a haze too tenuous to be noticed, &c. The veil is virtually coarser for blue than for red light, so that in the unimpeded light the proportion of the rays of low to those of high refrangibility goes on continually increasing, the effect by the time the rays reach the north increasing as the sun gets lower, and has accordingly a greater stretch of air to get through. Of the light falling upon the obstructing particles, a portion might be absorbed in the case of particles of very opaque substances, but usually there would be little loss this way, and the greater part would be diffused by reflection and diffraction. This diffused light, in which there is a predominance of the rays of higher refrangibility, would naturally be strongest in directions not very far from that of the direct light; and the loss accordingly of a portion of it where it is strongest, in consequence of interception by the pole in front of the tablet, accounts for the fact that the borders of the place of the shadow were seen to be a little less luminous than the parts at a distance.

8. The observations on phosphorescence just described have now prepared the way for the explanation I have to suggest of the photographic phenomenon.

It is known that, with certain preparations, if a plate be exposed for a very short time to diffuse daylight, and be then exposed to a pure spectrum in a dark room, on subsequently developing the image it is found that while the more refrangible rays have acted positively, that is, in the manner of light in general, a certain portion of the less refrangible have acted in an opposite way, having undone the action of the diffuse daylight to which the plate was exposed in the first instance.

It appears then that in photography, as in phosphorescence, there may in certain cases be an antagonistic action between the more and less refrangible rays, so that it stands to reason that the withdrawal of the latter might promote the effect of the former.

Now the objective of a photographic camera is ordinarily chemically corrected; that is to say, the minimum focal length is made to lie, not in the brightest part of the spectrum, as in a telescope, but in the part which has strongest chemical action. What this is, depends more or less on the particular substance acted on; but taking the preparations most usually employed, it may be said to lie about the indigo or violet. Such an objective would be much under-corrected for the red, which accordingly would be much out of focus, and the ultra-red still more so.

When such a camera is directed to a uniform bright object, such as a portion of overcast sky, the proportion of the rays of different refrangibilities to one another is just the same as if all the colours were in focus together; but it is otherwise near the edge of a dark object on a light ground. As regards the rays in focus, there is a sharp transition from light to dark; but as regards rays out of focus, the transition from light to dark though rapid is continuous. It is, of course, more nearly abrupt the more nearly the rays are in focus. Just at the outline of the object there would be half illumination as regards the rays out of focus. On receding from the outline on the bright side, the illumination would go on increasing, until on getting to a distance equal to the radius of the circle of diffusion (from being out of focus) of the particular colour under consideration the full intensity would be reached. Suppose, now, that on the sensitive plate the rays of low refrangibility tend to oppose the action of those of high refrangibility, or, say, act negatively, then just outside the outline the active rays, being sharply in focus, are in full force, but the negative rays have not yet acquired their full intensity. At an equal distance from the outline on the dark side the positive rays are absent, and the negative rays have nothing to oppose, and therefore simply do nothing.

9. I am well aware that this explanation has need of being confronted with experiment. But not being myself used to photographic manipulation, I was unwilling to spend time in attempting to do what could so much better be done by others. I will, therefore, merely indicate briefly what the theory would lead us to expect.

We might expect, therefore, that the formation of the fringe of extra brightness would depend:—

(1)   Very materially upon the chemical preparation employed. Those which most strongly exhibit the negative effect on exposure to a spectrum after a brief exposure to diffuse light might be expected to show it the most strongly.

(2)   Upon the character of the light. If the light of the bright ground be somewhat yellowish, indicating a deficiency in the more refrangible rays, the antagonistic effect would seem likely to be more strongly developed, and, therefore, the phenomenon might be expected to be more pronounced.

(3)   To a certain extent on the correction of the objective of the camera. An objective which was strictly chemically corrected might be expected to show the effect better than one in which the chemical and optical foci were made to coincide, and much better than one which was corrected for the visual rays.

It is needless to say that on any theory the light must not be too bright or the exposure too long; for we cannot have the exhibition (in the positive) of a brighter border to a ground which is white already.

P.S.—Before presenting the above paper to the Royal Society I submitted it to Captain Abney, as one of the highest authorities in scientific photography, asking whether he knew of anything to disprove the suggested explanation. He replied that he thought the explanation a possible one, encouraged me to present the paper, and kindly expressed the intention of submitting the question to the test of experiment.

I have referred to the photographic action of the more and less refrangible rays as antagonistic. This is practically true so far as the explanation I have ventured to offer is concerned, inasmuch as the more refrangible rays convert a salt of silver which is not developed into one which is developable, while the less refrangible convert the latter into one which is not developable. But Captain Abney has pointed out to me that though the first and third salts cannot be distinguished by appearance, nor by the action of the developing solution, they are nevertheless not the same, so that the two actions of the rays are not, rigorously speaking, antagonistic, inasmuch as the one is not strictly the reverse of the other. Thus with bromide of silver the explanation of the observed phenomena, according to Captain Abney, is that the undevelopable bromide is converted, chiefly by the action of the more refrangible rays, into a sub-bromide, which is developable; and this again is converted, chiefly by the action of the less refrangible rays, into an oxy-bromide, which is undevelopable. As however under the ordinary circumstances for obtaining a good picture the action of the light is chiefly of the first kind, and a much longer exposure would be required to bring out prominently the second kind of action, the explanation I have suggested is not virtually affected, though the two actions could not be prolonged indefinitely, as in the illustrative experiment in phosphorescence described in § 6.

EXTRACTS FROM REPORTS OF THE COMMITTEE ON
SOLAR PHYSICS.

[From First Report, *Nov.* 22, 1882, issued as Parliamentary Paper C, 3411,
signed by G. G. Stokes (Chairman), Balfour Stewart, J. Norman Lockyer,
W. de W. Abney, R. Strachey, W. H. M. Christie, J. F. D. Donnelly,
with F. R. Fowke, Secretary.]

SECTION IV; CONCLUSIONS.

(1) *Solar Phenomena.*

1. *Sun pictures.*—It appears to us probable that the varying
phenomena of the sun's disc represent the play of a huge system
of convection currents, the down rushes of which are indicated
by the darker patches and up rushes by the brighter patches.
These currents appear to be always present on the sun's surface,
and to give rise to the mottled appearance which it presents
under high magnifying powers. It would seem, however, that
the scale of these phenomena, and the rapidity of the indicated
motions, are on certain occasions greatly increased, presenting to
the observer that complex appearance which is associated with
the outbreak of spots. Thus sun spots may be supposed to denote
gigantic down currents of comparatively cold matter from above,
while the faculæ and red flames may denote the corresponding
up rush of hot matter from beneath.

We are also of opinion that it is in virtue of this convection
system that our luminary is able to bring to the surface the
intensely heated matter requisite to supply the enormous quantity
of radiant energy which it is known continuously to give out.
It would seem to follow that when this convection system is
peculiarly vigorous the radiation from the sun's disc must be
peculiarly vigorous also. Now all observations tend to convince
us that the sun's atmosphere is most agitated at epochs of
maximum sun spot frequency, and perhaps we may likewise
conclude from Mr Lockyer's investigations that certain definite
levels present the spectral characteristics of a peculiarly high
temperature at such times. We must look to future spectroscopic observations to settle these points by giving us evidence
from year to year of the velocity of motion in the up rushes

and down rushes of the sun's atmosphere, as well as of the heights to which the red flames are carried; and also by presenting us with a continuous record of the spectral lines exhibited by certain selected portions of the solar disc.

On the whole the evidence, judging solely from the sun itself, seems to us to be in favour of the view that our luminary is most energetic in its radiation at times of maximum sun spot frequency.

Should this inequality in the sun's power prove to be of practical influence on the meteorology of the earth, it would become of great importance to be able to analyse sun spot records in such a way as approximately to predict the state of the sun for any future year.

The ability to do this will depend on the possession of a series of accurate sun spot records sufficiently extensive to enable us to arrive at a true knowledge of the law of the sun's variability.

At present, therefore, all attempts in the direction of analysis must be regarded as merely preparatory, and our great aim must be to bring together and collate the various scattered observations of sun spots, faculæ, and prominences, so as to produce a trustworthy and sufficiently complete record to which some method of analysis should then be applied. This, therefore, forms one chief branch of the work we hope to see carried out, and we are glad to think that all solar observers are anxious to do what they can to hasten forward its completion.

Another object, of equal importance, is the establishment of solar photographic observations for the future on such a basis as may reasonably be expected to give a picture for every day, by utilising the results of observations in different parts of the earth, and avoiding thereby those interruptions of the record that must occur in any single locality.

The back work in sun pictures consists of two parts. First, the pictures mentioned in the catalogue appended to this report, which ends with the year 1877, and, secondly, the Indian and other sun pictures from the beginning of 1878 to the end of 1881. It is probable that the total cost of reducing and collating this back work will not exceed £1,000, which might be spread over five years; and we recommend that this work be undertaken.

With respect to the future, we recommend that steps be taken to secure a reasonable prospect of daily pictures, the Astronomer Royal having undertaken to reduce such pictures as may be necessary to form a complete series.

2. *Spectroscopic Work.*—When we consider that the spectroscope has been in our possession since 1869 as an instrument for

continuously observing various localised phenomena of the solar atmosphere, we cannot but express our regret that more has not been done to employ this new instrument of research; for, even yet, daily and systematized observations are almost unknown.

As there can be no question of the importance of a daily record of spectroscopic solar phenomena, we proceed next to discuss those special points to which, according to our present knowledge (and in a new subject it is more than ever important to make this qualification), our studies should be directed.

Foremost among these inquiries we would insist upon the importance of obtaining records of all phenomena visible during total eclipses of the sun. It must not be forgotten, that with regard to the sun's atmosphere, more information can be obtained during an eclipse than from a year's work on the uneclipsed sun; and with regard to the form and extent of the atmosphere generally, we already know that the changes are so great from eclipse to eclipse that inquiries may be greatly hampered in the future if we fail to obtain these records whenever opportunity occurs.

We believe that these opinions are shared by physical astronomers in other countries, and this being so, concerted action by civilized Governments may render the part to be taken by each comparatively inexpensive.

Coming to the uneclipsed sun, we find that our great needs at present are more observations of the chromospheric and prominence lines in climates where these can be easily and continuously made.

Observations of the lines widened in solar spots rank, perhaps, next in order of importance, as the facts show that the vapours which produce them are not those which give rise to the appearance of prominences. At present this work is limited to Greenwich and South Kensington, and at the latter place at all events the observations are confined to a restricted part of the spectrum.

We recommend that communications should be opened with observatories where such work would be likely to be prosecuted, suggesting simultaneous observations on a definite plan for a limited period, say for the next three years; and that the reduction of these observations should be undertaken at South Kensington.

In this way we believe results of the highest importance in their bearing upon solar theory would be obtained in the shortest possible time, and with a minimum of expense.

We now come to another part of the inquiry—that which refers to combined laboratory and observatory work; to the comparison of lines seen in the spectrum of the sun with those seen in the spectra of terrestrial bodies.

Here, unfortunately, the work is almost entirely confined to South Kensington, where its progress is very slow, partly in consequence of the limited facilities, partly in consequence of the bad climate.

As such work is necessarily the keystone of the arch which may unite celestial and terrestrial chemistry, we recommend that certain steps should be taken to accelerate the rate of progress, and we have indicated these to the Science and Art Department.

(2) *Solar Radiation.*

The best proof of solar variability would be the direct one given by an actinometer or instrument so constructed as to measure with accuracy the amount of radiant energy given out by the sun, but as yet hardly any such observations have been made.

We attach great importance to the use of such instruments in the future, and we are in hopes that in India and elsewhere much information may soon be obtained by their means.

But even assuming the possession of a perfect actinometer there are considerable difficulties in the way of obtaining by its means the true solar radiation.

Allowing, for the sake of argument, a variability in the sun's power, it seems probable that during those years when the sun is increasing in power the quantity of aqueous vapour suspended in the air should also continue to increase. But it is known from the researches of Prof. Tyndall and others that this would imply a continually increasing atmospheric absorption, which would stop each year an increasing proportion of the rays of the sun, and prevent these from reaching the earth. It might thus happen that observations of the solar radiation made near the earth's surface might give something much less than the true increase of solar power, inasmuch as a continually increasing proportion of the solar rays would have been year after year intercepted by the atmosphere. The most obvious way of getting out of this difficulty would be to make observations at high altitudes, and we hope in addition to the records of the heat-actinometer now in operation at the Alipore Observatory, near Calcutta, to obtain records from one at Lé (at a considerable elevation, about 11,000 feet, above the level of the sea). We trust also that Dr Roscoe's chemical actinometer will be established at Lé. Furthermore, with the view of throwing light on the condition of fluctuations in the received solar radiation caused by the atmosphere, we would suggest that these actinometers should be used in conjunction with some qualitative instrument which gives an immediate graphical and visible indication of the power of the sun. A modification of an instrument devised by Mr Winstanley would appear to be very suitable for this purpose.

Presuming that we are thus able to obtain unexceptional observations at Lé, still it is certain that there must remain an appreciable quantity of aqueous vapour in the air above.

It has been suggested by General Strachey that a travelling observer carrying with him an actinometer might ascend to a considerable height by a series of stages, making observations at each. We might thus be able to obtain a more exact estimate of the absorption of the air and moisture respectively, and thence to deduce what the true solar radiation would be if we could altogether escape the atmosphere.

We ought to mention that Professor Langley of the Allegheny Observatory is at present devoting a good deal of attention to this problem, and we are induced to hope that by the united efforts of observers in the elevated portions of both hemispheres a great deal of light may be thrown upon this important subject.

This is, perhaps, the place in which to notice another species of observation which it may soon be possible to make.

Captain Abney has discovered a method of photographing the infra-red spectrum, and it is hoped that light may be thrown by this method on solar radiation. Indeed, recent observations which he undertook at an altitude of 9,000 feet tend to show that for a qualitative and partially quantitative estimate of atmospheric absorption the method promises results of high value.

It should here be remarked that, by comparing with a standard certain definite regions of the spectrum unabsorbed by any of the constituents of the earth's atmosphere, we might be able to ascertain any variation in the quantity or in the quality of the true solar radiation.

We recommend that the heat actinometer by Prof. Stewart, the chemical actinometer by Prof. Roscoe, and the supplementary instrument by Mr Winstanley be established at Lé, it being understood that the Indian Government is ready to bear the expense of the instruments and of the observations with them.

We recommend also that the suggestions of General Strachey already mentioned should, if possible, be carried into effect.

(3) *Influence of the State of Sun on the Meteorology of the Earth.*

While observations of atmospheric and of underground temperature appear to give evidence of a fluctuation in the temperature of the air having the same period as that of sun spots, yet on the whole they appear to show that a maximum of sun spots corresponds to a low and not to a high temperature.

But if we bear in mind that temperature is an exceedingly complex phenomenon, and that an excessive rainfall generally produces a low temperature, we cannot receive this as evidence tending to show that the sun is least powerful at epochs of maximum sun spots. It seems possible, however, that something might be done by confining ourselves to short-period inequalities, if the existence of such inequalities should be made out with tolerable certainty.

Professor Stewart has pointed out the apparent existence at Toronto of fluctuations of diurnal temperature range, having periods very nearly corresponding to those of sun spots, and of a nature which leads him to infer that a maximum of sun spots is associated with a maximum of solar power. We think that this investigation might be extended so far as to take three or four prominent solar inequalities, and see whether they correspond to similar inequalities of temperature range. This would appear to be a method likely to show if there be any real connexion between the sun spots and terrestrial meteorology, waiving for the present the question of true periodicity in such inequalities.

With respect to rainfall, while the observations appear to indicate that at certain localities we have a maximum amount of rain about the time of maximum sun spot frequency, yet we cannot say that taking the rainfall stations as a whole there is incontestable evidence of a single rainfall period corresponding to that of sun spots. This subject is one that requires further investigation; and observation of the heights of rivers appears to be a hopeful direction in which to look for evidence bearing on this question.

In the next place, with respect to storms, the evidence appears to be favourable to an increase in the number of great atmospheric disturbances, corresponding to times of maximum sun spots, but the systematic discussion of anemometric observations has hitherto received far too little attention.

It seems to us that a study of isobaric lines may throw light upon the problems now before us. The relatively low summer barometer in the middle of continents, and high barometer at sea, and the opposite disposition of pressure which holds for the winter months, are problems which deserve further investigation.

We hope that this subject may be advanced by the discussion of a lengthened series of those excellent meteorological charts which the American Government are now publishing.

We recommend that communications should be entered into with the Meteorological Council, with the view of concerting a plan for investigating with sufficient thoroughness the nature and extent of the supposed relation between solar variability and the meteorology of the Earth.

We have no hesitation in expressing our belief that the continued careful study of solar phenomena will prove to be of the greatest scientific value, and that there is no reason for doubting that the advance of true knowledge in this direction will, in some form or other, and sooner or later, prove to be of real practical value also, as all experience has shown that it has been in other branches of human knowledge.

Whether or not we shall ever possess the power of foreseeing the character of the seasons in this country, or to what extent they may in truth be related to those changes in the condition of the sun to which our attention is specially directed, it is of course impossible for us to say. But of the extreme importance of doing all that lies in our power to advance a sound knowledge of the laws of climate which so directly affect the well-being of the whole human race there can be no question.

(4) *Influence of the State of Sun on the Magnetism of the Earth.*

There can be no doubt that the diurnal range of the earth's magnetism is greatest when there are most sun spots, and that on such occasions there is likewise an unusually large number of magnetic storms with their accompaniments in the form of earth currents and auroral displays.

There are also strong indications that the fluctuations in the diurnal ranges lag, as a rule, in point of time behind the solar influences which produce them.

This would most naturally lead us to suppose that these diurnal magnetic effects are not directly caused by solar magnetic influences, but indirectly by solar radiation *.

Again, the large amount of variation in the declination range, which increases nearly in the ratio of *two* to *three* between times

[* The insufficiency of any possible amount of direct magnetic change in the Sun is now clearly recognised: cf. Larmor, *Phil. Mag.*, Jan. 1884, p. 23; Lord Kelvin, Address to Royal Society, Nov. 30, 1892, in *Popular Addresses*, III, p. 509.]

of minimum and times of maximum sun spots, might perhaps induce us to locate the solar influence which brings about this result in the upper regions of the earth's atmosphere where there is reason to think inequalities in solar radiation would be particularly felt.

In order to investigate this subject we recommend: First, a more extended comparison of the declination curves at Kew and Stonyhurst, after the manner of the preliminary comparison already referred to in this report. Secondly, a more extended comparison between the meteorological and the magnetical weather of the British Isles, after the manner of the preliminary comparison also referred to in this report. Thirdly, a more extended investigation with the object of deciding whether declination range inequalities do really appear to travel from west to east as far at least as phase is concerned.

Finally, with the view of carrying out the valuable suggestions of Senhor Capello, contained in a letter published in an appendix to this report, page 239, namely, that the directors of observatories possessing self-recording magnetographs should arrange together some uniform plan of utilising the curves produced by such instruments, we recommend that communications should be entered into with the Kew Committee of the Royal Society in order to concert some method of action in this respect.

[From same Report, *Nov.* 22, 1882, p. 57.]

CONNEXION BETWEEN SOLAR AND TERRESTRIAL PHENOMENA.

Sun and Magnetism: New Theory.

With regard to the connexion between solar activity and auroræ, magnetic disturbances, and earth currents, Professor Stokes has proposed a new theory[*]. That auroræ consist in electric discharges taking place, usually at any rate, in the higher regions of the atmosphere, is allowed on all hands; the only question is, how are these discharges occasioned, and what is the nature of the connexion between phenomena apparently so remote? Professor Stokes contends that the source of the discharge is to be sought in atmospheric electricity, which not being relieved by the thunderstorms which take place in low and moderately low latitudes accumulates in the higher latitudes till it has sufficient power to occasion a discharge. He regards the earth as analogous to a Leyden jar, the lower portion of the

---

[*] *Nature*, Vol. xxiv, 1881, pp. 593—598 and 613—618. [Lecture on Solar Physics.]

atmosphere forming the dielectric, while the higher portion bears some analogy to the exterior metallic coating of a jar; only, the air in the higher regions is far from being a good conductor, like the tinfoil of the jar, and merely opposes a very much smaller resistance to a disruptive discharge than does the denser portion below. Thus when the difference of tension between the upper atmosphere over one region and that over some other more or less distant region becomes sufficient, a discharge takes place. The opposite electricities, previously bound by induction at the surface of the earth, being thus set free, a redistribution takes place, giving rise to earth currents; and the assemblage of currents partly terrestrial, partly atmospheric, form very nearly . closed circuits, and exercise magnetic influence at a distance, giving rise to magnetic disturbances. When the sun is unusually disturbed, the intensely heated portions of matter which come up from below to the sun's surface cause increased radiation, especially as regards rays of high refrangibility, and this, being in part absorbed in the upper regions of the earth's atmosphere, is supposed to render them better conductors, or rather to cause them to oppose less resistance than before to disruptive discharges, facilitating thereby displays of auroræ, and occasioning the earth currents and magnetic disturbances, which on this theory have their origin in auroral discharges.

[From same Report, *Nov.* 22, 1882, Appendix A, by Prof. Stokes, pp. 75—77.]

MEMORANDUM AS TO ONE MODE OF DEALING WITH THE
INDIAN SOLAR PHOTOGRAPHS.

I assume the available data to be a set of photographs, each taken at a known time, and presenting one fiducial line, whether it be the shadow of an equatorially mounted wire, or the common chord of two solar images taken on the same plate after a suitable interval without disturbing the telescope in the interim.

Taking the photographs in chronological order it will be convenient in the first instance to attach reference numbers to the spots, which might be written on a waste positive, or on a hand sketch. In successive photographs of what is *unquestionably* the same spot, it would be well to use the same reference number followed by a distinctive letter *A, B, C.*

For anything beyond a statistical enumeration of numbers and areas the first step is to measure the position of a spot relatively to the sun's disc and the fiducial line.   For this I have thought of two plans, the first the simpler and cheaper, the second the more accurate and complete.

The first is to rule with diamond on glass, or better perhaps etch on glass, a network of cross-lines, the interval being, say, the 100th part of the average diameter of the photographed image. The photograph being placed with the collodion uppermost, the scale is laid on it, with the ruled lines against the collodion, and one set parallel to the fiducial line. The scale might be centred on the photograph, but this would involve some trouble unless the scale were provided with slow motions in two directions, and then the photograph might be injured unless a little space were left between, in which case errors of parallax might come in. Instead of attempting more than a rough centering, it would seem best to read both limbs and the spots just as they lie, when by a simple calculation we get the abscissæ of the spots referred to the centre of the disc, and the diameter of the disc. I will suppose that the latter plan is adopted. In that case the figuring of the graduation had best go from zero on the left, instead of having the zero in the middle. Supposing the photographs about 4 inches in diameter each interval would be about the 25th of an inch, and might be divided by estimation to 10ths.

We should commence with the readings taken in a direction perpendicular to the fiducial line. Using a lens we should read the first limb, the spots in succession, the second limb. A thread moving parallel to the fiducial line would be required for guiding the eye down a ruled line to the lower edge where the graduation is figured. Precisely the same process would have to be gone through for the readings in the perpendicular direction.

An exceedingly simple calculation would then give the two coordinates of a spot measured from the centre of the image, and referred to radius 100.

It may be noticed in passing that the effect of terrestrial refraction is eliminated by referring each set of coordinates to the measured diameter in its own direction. It is only, however, in case the photograph were taken when the sun was pretty low that the effect of refraction would be sensible.

The most convenient process is next to calculate the polar coordinates (or in other words the distance from the centre and angle of position) from the rectangular coordinates. Let S, E, P be the centre of the sun, the earth regarded as a point, the spot. The radius vector inferred from the rectangular coordinates is very nearly that belonging to the orthogonal projection of SP on a plane through S perpendicular to SE. A very minute correction, which depends on the non-parallelism of SE, PE, and which can be made at once by inspection, suffices to reduce it to the orthogonal projection. The distance from the centre referred to the scale radius = 1 is the sine of the inclination of SP to SE.

The time of taking the photograph being known, we get the sun's longitude L from the *Nautical Almanac,* and thence the inclination, I, of circles of latitude and declination passing through the sun from the formula

$$\tan I = \tan O \cos L,$$

where O is the obliquity of the ecliptic. The fiducial line, which is applied primarily to the equator, is hereby referred to the ecliptic; and by applying to this angle the angle of position, which forms one of the polar coordinates, we get the inclination of the plane ESP to a plane through ES perpendicular to the ecliptic. Two formulæ equally simple with the above then give the heliocentric latitude and longitude of the spot.

The other method, which I should prefer, is to measure at once the polar coordinates, using a microscope of very low power with cross-lines. The microscope slides along a horizontal rest, graduated so as to read with a vernier, say to 0·001 inch, or else 0·001 of the average radius. The rest and microscope are supported on a horizontal circle graduated to degrees, with a vernier reading to 5′ and by estimation to 1′. The plate is laid on a support underneath this, which is provided with slow motions in two rectangular directions for centering.

If readings taken by applying a scale to the photograph and estimating the fractions of intervals be deemed sufficiently accurate, a glass scale graduated by concentric circles and radial lines might be used instead of the rectangular network. The estimation would be a little more troublesome and less accurate, but the trouble of subsequent reduction to polar coordinates would be saved.

We should thus get with very little trouble the heliocentric places of the various spots in each photograph, referred to the celestial sphere. But we should also wish to know the places referred to the sun itself, treated as if it all revolved together like a solid globe, which Carrington has shown to be only approximately true. This might be done by calculation, using Carrington's elements. But the calculation is considerably longer than those hitherto mentioned, and it would be desirable to avoid it by a graphical process.

The simplest process of this kind that I have thought of, and one which would, I believe, give the reduced places even more accurately than is required considering errors of observation, is to use a large globe. This would have the further advantage of presenting to the eye the combined results in a digested form. The globe should be provided with two axes at an inclination of 7° 15'. The meridian circle should be graduated to degrees, better perhaps subdivided to 20'. The globe is a blank, save as to two circles, the equators of the two axes respectively. These circles are divided into degrees, or better to 20'. One pole represents the pole of the ecliptic, the other that of the sun's equator.

The globe being mounted so as to turn for the ecliptic, the places of the spots referred to the celestial sphere are entered on it by means of the latitudes and longitudes found as above. When a sufficient number are entered to make it worth while, the globe is mounted so as to turn about the other axis. An arbitrary origin of time is taken, say 1879, January 1, Greenwich mean noon. The dates of the photographs, reckoned from the beginning of the year in days and fractions of a day, are divided by the assumed time of the sun's rotation, and the integral rotations being omitted, the fraction of a rotation is converted into degrees. Call this angle A. The spots marked on the globe for a given photograph are then all set back by the value of the angle A for that photograph, and marked on the globe in ink of a different colour.

We shall thus have depicted, not only the places of the spots on the celestial sphere, but their places on a sphere turning with the sun.

It would be desirable to lay down in the ecliptic position or rather mounting of the sphere, not only the heliocentric places of the spots, but also those of the earth in the ecliptic, and to have provided a brass limb, a little more than a quadrant, which is loose and intended merely for a ruler. Perhaps a mere flexible strip of metal would be better. When a spot is observed a good way from the centre of the disc, the quadrant rule should be laid down between the places of the earth and the spot, and a line, longer or shorter according as the spot was observed more or less near the edge, drawn through the place of the spot. This may be afterwards useful in comparing results, since the position of the spot along the drawn line is more or less uncertain.

When the globe gets too full, the entries on it can be copied out on two plane maps, one for celestial and the other for solar places, the old entries on the globe painted out, and the globe used afresh.

[From same Report, *Nov.* 22, 1882, Appendix H, pp. 212—214.]

MEMORANDUM FOR THE USE OF OBSERVERS WITH PROFESSOR
BALFOUR STEWART'S ACTINOMETER, PREPARED BY PROFESSOR
G. G. STOKES, SEC. R.S., AND ADOPTED BY THE COMMITTEE
ON SOLAR PHYSICS.

The Committee regard the instrument herein described as an
apparently good instrument, though it has not yet been tried
under conditions more favourable to actinometric observation than
can be obtained in the United Kingdom.   Partly on this account,
but even more on account of the still immature condition of the
whole subject of actinometry, they deem it premature to attempt
to draw out anything like a code of instructions for regular
observation with it, and think it best to place the instrument in
the hands of one or more intelligent observers interested in the
subject, and residing in suitable localities at very considerable
elevations above the level of the sea, indicating to them the
objects which it is sought to attain, but leaving it in great
measure to their own judgment, and to the experience they will
gain in using the instrument, how best to carry out the observations in detail.

Let it be understood then that the chief object which it
is sought *ultimately* to attain by the use of an actinometer is a
knowledge of the variations (if any) in the heat radiation from the
sun itself.

The first great obstacle to the attainment of this object is
that arising from variations in the heat-intercepting power of the
earth's atmosphere.   To reduce this to a minimum a station is in
the first instance chosen which, while favourably conditioned in
other respects as regards climate, is well elevated above the level
of the sea, so as to get rid of the denser and more dusty and hazy
portions of the atmosphere.   A suitable station having been
chosen, the next point is to select proper occasions for observation.
A cloud covering the sun would of course make the observation
impossible; but even a slight veil of cirrus is found to interfere
materially with the amount of heat coming directly (that is,
without any deflection) from the sun.   Hence the observer should
choose times when the blue sky is to all appearance free from
haze.   Detached clouds need not prevent an observation, unless
during some part of the time of exposure they come so near the
sun that the rays they reflect are liable to pass through the lens
in such a direction as to fall on the bulb of the thermometer.

Vitiation of the observation by *visible* causes of interception of the heat rays having been thus guarded against, there still remains the possibility of casual fluctuations being produced by the invisible constituents of the atmosphere. For the detection of these, and for learning the conditions of their absence, we can only have recourse to a comparison of the results of observations made on different occasions. To render such a comparison effective, memoranda should be made at the time of the observations of the condition of the atmosphere, so far as can be judged by the eye, and by readings of the ordinary meteorological instruments, and the altitude of the sun should be measured (no great accuracy being required in the measures) and recorded, or else subsequently calculated from the known time of day and year.

With a view to throwing light on the conditions of atmospheric fluctuations in the radiation received at the surface of the earth, the Committee would suggest that the actinometer above described should be observed in conjunction with some qualitative instrument which gives an immediate graphical and visible indication of the power of the sun. An instrument devised by Mr Winstanley, and modified by Captain Abney, would appear to be very suitable for this purpose. A complete accordance between the two instruments is not to be expected, because the thermometer in Mr Winstanley's instrument is exposed to radiation from various directions. The difference between the two instruments in this respect may be useful in throwing light on the causes of atmospheric fluctuation.

When the observer has learned how to avoid at least the grosser forms of atmospheric fluctuation, he may attack the problem of the effect of the sun's altitude on the amount of heat radiation intercepted. For this purpose specially favourable days should be chosen, and observations made at frequent intervals, from shortly after sunrise to near sunset. The condition of the atmosphere on these days should be carefully recorded. The days used for the purpose should not be confined to one season, as it is possible that the normal condition of the atmosphere at a given place, and with it the amount of absorption for a given altitude of the sun, may vary with the season. All through the observations above referred to, or at least after he has learned to recognise and avoid the more serious atmospheric fluctuations, the observer must bear in mind that the instrument itself (the actinometer) is on its trial, and he must be alive to the possibility of variations in the readings, due merely to different conditions of exposure, or to other purely instrumental sources. For testing the instrument itself, times should be chosen when, as far as the observer can judge, there is a freedom from casual atmospheric fluctuations, and it would be well to take a good number of consecutive observations with the screen alternately on and off.

When the observer has learned how to avoid, as far as may be, merely casual atmospheric fluctuations, and considers that the instrument has been sufficiently tested, he may commence observations taken with a view to their possibly forming part of a permanent record. For this it would be proper to get a result for each day, so far as atmospheric conditions permit; but how many observations it would be desirable to take, whether they should be taken at stated hours, or with stated altitudes of the sun, or whether the most favourable opportunities as to atmospheric conditions should be seized which present themselves, not too far from noon, so that the sun has a high altitude, are questions which cannot well be answered till the preliminary experiments above mentioned have been made. The observations for permanent record can hardly be *reduced* until the effect of altitude has been determined, but they may be *made* as soon as the observer has made sufficient progress in learning to avoid casual sources of error.

When all has been done that can be done at one station, the discussion of the records obtained may lead to presumptive evidence in favour of a variation in the actual radiation emitted by the sun itself; but so important a conclusion could not be considered as established without corroborative evidence, arising from the comparison of simultaneous observations at at least one other favourably situated station widely separated from the former. As soon, therefore, as the method shall have been brought into thorough working order, especially as regards the rules to be followed for the avoidance of casual atmospheric fluctuations, it is desirable that the stations of observation should be multiplied.

On the Highest Wave of Uniform Propagation.
(Preliminary notice.)

[From the *Proceedings of the Cambridge Philosophical Society*, Vol. IV, Pt. 6, pp. 361—365. *May* 14, 1883.]

THERE is one particular case of possible wave motion, applicable to a fluid of practically infinite depth, in which all the circumstances of the motion admit of being expressed mathematically in finite terms, the necessary equations being satisfied exactly, and not approximately only; while the general expressions contain an arbitrary constant permitting of making the amplitude any whatsoever up to the extreme limit of cycloidal waves, coming to cusps at the crests. This possible solution of the equations was given first by Gerstner, near the beginning of the present century. The motion however to which it relates is not of the irrotational class, and could not therefore be excited in a fluid previously at rest by forces applied to the surface; nor could it be propagated into still water from a disturbance at first at a distance. In fact, the conditions requisite for its existence are of a highly artificial character; so that the chief interest of the solution is one arising from the imperfection of our mathematics, which makes it desirable to discuss a case of possible motion, however artificial the conditions may be, in which everything relating to the motion can be pretty simply expressed in finite terms.

There can be no question however that it is the irrotational class of possible wave motions which possesses the greatest, almost the only, intrinsic interest; since it is this kind alone which can be excited in a fluid previously at rest by means of forces applied to the surface, such for example as the unequal pressure of the wind on the surface, or propagated into previously still water from a distance.

In a paper read before the Society in 1847, and published in the *Transactions**, I have investigated the motion of oscillatory waves in which the motion is not very small, by the method of successive substitutions, proceeding to the second order in the case of an arbitrary depth, and to the third order in the simpler case in which the depth is infinite. In the latter case the terms of the third order were found to be very small even in the case of waves of very considerable magnitude. The series converge less rapidly when the depth is finite; and when the length is very great compared with the depth of the fluid the convergence becomes so slow that the method practically fails, and is not therefore applicable to solitary waves.

The circumstances of the motion of solitary waves of considerable height have been investigated by M. Boussinesq† and Lord Rayleigh‡.

The evidence of the existence of a type of oscillatory irrotational waves which are uniformly propagated is derived from the nature of the process of approximation, which is one of a systematic character that can certainly be carried on to as many orders as we please, all the conditions of the problem being satisfied to the order to which we step by step advance. If therefore the series that we are working with be convergent, there can be no question of the possible existence of uniformly propagated waves. But for a given depth and given wave-length there remains in our series a disposable constant on which the height depends, and on the value of which the degree of convergency depends. By taking this constant small *enough* the series will be convergent; though *what* the limit may be that separates convergency from divergency, the process of expansion does not show.

It seems to me pretty certain that the series will remain convergent until a singular point appears at the boundary of the fluid. Some years ago I was led by simple considerations to the conclusion that the occurrence of a singular point in the profile at which two branches meet at a finite angle (or as it might conceivably have been, touch, forming a cusp) entailed as a consequence the existence of two tangents inclined at angles of

[* *Ante*, Vol. I, pp. 197—229.]

† *Comptes Rendus*, Tom. LXXII (1871), p. 755.

‡ *Philosophical Magazine*, Vol. I (1876), p. 257.

± 30° to the horizon; so that the ridges of the waves came to
wedges of 120°*. In a supplement to my former paper† lately
published I have conducted the approximation in a different
manner, which is more convenient for proceeding to a high order.
In this latter method the coordinates x, y are expressed in terms
of the velocity-potential ϕ and stream-line function ψ, instead of ϕ
and ψ in terms of x and y. The approximation is carried to the
fifth order for deep water, and to the third when the depth is
finite. Still even in this method the labour of the approximation
rapidly increases with the order, so that the result of working out
a great number of terms would not repay the labour; and expansion by series is hardly applicable to the determination of the
circumstances of the highest possible wave. When a series whose
general term contains a power or other function of some parameter
a is convergent when a lies below a certain critical value, and
divergent when it lies above, it may be convergent when a has the
critical value, but if so its convergence is very slow. If we allow
that the highest possible wave comes to a ridge of 120°, that,
combined with our knowledge of the form of waves of very considerable height, would enable us to draw very approximately the
theoretical outline of the highest possible wave. But it is
tantalizing to get thus near it and not to be able to complete
the solution.

The expansion in series would be of little avail for the reason
I have mentioned; but it occurred to me that some method of
trial and error might succeed. I have devised one which promises
so well that a notice of the *method* may not be without interest to
the Society, though I am not at present in a condition to present
the *result*, except it were in a rough way, not having completed
the numerical calculations required, nor even begun them in the
second, and that the more interesting, of the two cases to which
the method applies. I have employed the series contained in the
supplement before referred to, not however directly, for the
purpose of numerical calculation, but merely as stepping-stones
enabling me to effect a certain analytical transformation in which
the use of series is got rid of.

The method is not confined to the case of the highest possible
wave, but may also be used for lower waves, though unless the

[* *Ante*, Vol. I (1880), p. 226.]

† *Mathematical and Physical Papers*, Vol. I, [1880, pp. 314—326].

wave is near the maximum it is better to have recourse to the series. In any case of uniform propagation, we may readily reduce the motion to a case of steady motion, and when that is done the velocity of a particle at the surface will be the same as that of a particle sliding along a smooth curve corresponding to the outline of the wave, and will accordingly be that due to the depth below a fixed straight line, which for the sake of a name may be called the *datum line*. In the case of the highest wave, since a particle at the vertex of a wedge must be momentarily at rest, the datum line will pass through the crest; in other cases its height above the wave must be assumed for trial as well as the outline of the wave.

The trial outline (and the trial datum line in the case of a wave short of the highest) being known, the velocity at any point of the surface is known, and therefore by an ordinary integration or by a quadrature the velocity-potential at the surface is known. Hence ϕ being known in terms of x, x is known in terms of ϕ. But the coordinates of points in the surface are given in terms of ϕ by equations (23), (24) of the supplement referred to on putting ψ, the parameter of a stream line, $= 0$. These equations have been simplified by choosing the units of space and time such as to make a wave-length, and also the change of ϕ in passing from one wave to the next, each equal to 2π, and k is the value of $-\psi$ at the bottom. The equations then become

$$x = -\phi + \Sigma A_n (e^{nk} + e^{-nk}) \sin n\phi \dots\dots\dots\dots(1),$$

$$y = \qquad \Sigma A_n (e^{nk} - e^{-nk}) \cos n\phi \dots\dots\dots\dots(2).$$

The negative sign of ϕ in the first of these equations arises from choosing the direction of x positive as that of the propagation of the waves in the first instance, and therefore the direction of x negative as that of the superposed velocity by which the motion is converted into steady motion.

x having been determined in terms of ϕ, $x + \phi$ is to be deemed a known function of ϕ, $f(\phi)$ suppose, which will have 2π for its period. The coefficients A_n may then be deemed known, and on substituting in (2) we shall have y expressed in terms of $f(\phi)$. Denoting this expression for y by $F(\phi)$ we shall have

$$F(\phi) = \frac{1}{\pi} \Sigma \int_0^{2\pi} \frac{e^{nk} - e^{-nk}}{e^{nk} + e^{-nk}} \cos n\phi \sin n\phi' f(\phi') \, d\phi' \dots(3),$$

where of course the integration with respect to ϕ' may stop at π if we double the coefficient, since $f(2\pi - \phi) = -f(\phi)$.

If the trial curve had been the true outline, the curve of which the ordinate is determined by (3) would have come out identical with the original, which would have proved the original to have been correct: otherwise the new curve will be a much closer approximation to the true form than the trial curve, and may be used as a fresh trial curve, and so on.

In (3) the integration is supposed to be performed first and then the summation. If we attempted to perform the summation first, we should encounter a series which is neither convergent nor divergent, but fluctuating. Such a series may however be summed by regarding it as the limit of the convergent series formed from it by multiplying the nth term by the nth power of a quantity less than 1, and which is supposed to become equal to 1 in the limit. The summation cannot however, so far as I know, be actually effected except in two cases.

The first case is that of a fluid of infinite depth, for which the fraction involving the exponentials becomes equal to 1, and the series divides into a pair of series of sines of arcs in arithmetical progression, which may be summed by regarding it as the limit of another series; a view to which we are naturally conducted by regarding the stream-line of the surface as the limit of a stream-line taken first a little below the surface. The other case is that in which the wave-length is regarded as infinitely great instead of infinitely small compared with the depth of the fluid. In this case we first take a crest for the origin of x and then make λ infinite, when the sum takes the form of a definite integral which may be evaluated according to the known formula

$$\int_0^\infty \frac{e^{ax} - e^{-ax}}{e^{ax} + e^{-ax}} \sin bx\, dx = \frac{\pi/a}{e^{\pi b/2a} - e^{-\pi b/2a}}.$$

As the other crests have moved off to infinity, we are in this case left with a solitary wave.

The results in the two cases are as follows :—

1.　Case of oscillatory waves in a fluid of infinite depth

$$F(\phi) = \frac{1}{2\pi} \int_0^\pi \{f(\phi + \chi) - f(\phi - \chi)\} \cot \tfrac{1}{2}\chi\, d\chi.$$

2. Case of a solitary wave

$$F(\phi) = \frac{1}{k} \int_0^\infty \{f(\phi + \omega) - f(\phi - \omega)\} \frac{d\omega}{e^{\pi\omega/2k} - e^{-\pi\omega/2k}}.$$

In the former case, partly as a severe test of the method, and partly for other reasons, I took as the trial outline a serrated line composed of straight lines inclined at angles* of $\pm 30°$ to the horizon, giving ridges formed of wedges of 120°, and troughs formed of similar wedges inverted. Even in this case, where the assumed form is so egregiously wrong as regards the troughs, the first approximation led to a form presenting ridges of 120°, which for a considerable way down hardly differed from the original, while in lieu of the angular troughs I got a curved outline, with a depth from crest to trough of about 0·22 of the wave length, instead of $1/2\sqrt{3}$ or 0·309† as in the serrated outline assumed for trial. The results of some other trials are encouraging as to the success of the method, but as I mentioned already the numerical calculations are not finished ‡.

---

* In the case of a simple serrated outline, $f(\phi)$, and therefore $F(\phi)$, is independent of the assumed inclination.

[† This should be 0·288.]

[‡ See the paper next following.]

[Not before published.]

On the Maximum Wave of Uniform Propagation. Being a Second Supplement to a Paper on the Theory of Oscillatory Waves.

Proposed method* of drawing the wave profile, and determining the velocity of propagation, for oscillatory waves on deep water, when the disturbance is too great to be mastered by the approximations, proceeding according to powers of b, given in last paper of Vol. I of my book.

From the approximations as far as they go, from our general knowledge of the forms of waves, and in the case of the maximum wave from the knowledge that it comes to a ridge of 120°, we can draw a flowing curve representing approximately the wave outline. Draw also a horizontal line above, at an estimated height so that the ordinate downwards from it to any point at the surface may be that to which the velocity is due when the wave motion is converted into steady motion. In the case of the maximum wave, the horizontal line passes through the crests. Let the unit of length be the one-2πth part of the length of a wave. Find graphically or numerically $\int \sqrt{y}\, ds$ as a function of x in the first instance. This will be ϕ, save as to a constant multiplier. Alter the scale so as to make the integral equal to π for a half-wave. Then we get $x - \phi$ (or $x + \phi$ if we take ϕ negatively) as a function, first of x, and then by re-plotting of ϕ. Let it be $f(\phi)$, then the relations

$$f(\phi) = A_1 \sin \phi + A_2 \sin 2\phi + A_3 \sin 3\phi + \ldots,$$
$$y = A_1 \cos \phi + A_2 \cos 2\phi + A_3 \cos 3\phi + \ldots$$

give

$$y = \frac{1}{2\pi} \int_0^\pi \{f(\phi + x) - f(\phi - x)\} \cot \frac{x}{2}\, dx,$$

which is to be found graphically or by tabulation. Plot the curve. This will coincide with the trial curve if the trial curve and trial line were right; otherwise the curve will probably indicate the character of the correction. We must then repeat the process on the better trial curve, and so on till we get sufficiently near the truth.

22 *September*, 1880.

[* This abstract was found type-written, uniform with the paper which follows.]

In a supplement to the above paper (*ante*, Vol. I, p. 314) I have given a method which is much more convenient than the original one for conducting the approximation when carried to a somewhat high order. Yet even by this method the analytical expressions become very laborious to work out when the order to which we proceed is high. And even in the most favourable case, namely, that in which the depth may be taken as infinite, the convergency of the series we have to deal with becomes, as will shortly be shown, very slow when the wave of greatest possible height is reached; while when the depth is small compared with the length of a wave, so that the waves approach in character to a succession of solitary waves, the convergency though ultimately assured is slow for a considerable way, and may even be exchanged for divergency at first, long before the highest possible wave is reached. It is therefore practically impossible to determine in this way the form of the highest possible wave, or even of a wave somewhat short of that.

Let us examine in the first instance the ultimate convergency in the case of the highest possible wave.

It has been shown at p. 226 of the first volume (not indeed by a method absolutely rigorous, since it involves the assumption that the increase of height may be pushed to the limit of giving singular points in the outline) that the greatest wave comes to a ridge of 120°, and that near the ridge the difference between the velocity-potential and that at the ridge varies as the $\frac{3}{2}$ power of the distance, and consequently the distance as the $\frac{2}{3}$ power of the increment of velocity-potential. If therefore y be expressed as a function, not of x but of ϕ, we shall obtain a wave-like curve presenting a series of cusps.

The values of the coordinates expressed in terms of the functions ϕ, ψ, and simplified by choosing the units of length and time such as to make $m = 1$ and $c = 1$, are given in equations (23), (24), on p. 320 of Vol. I. In these equations the origins of x and ϕ are taken at a trough and the condition to be satisfied at the free surface is not yet introduced. They are

$$x = -\phi + \Sigma A_i \{e^{i(\psi+k)} + e^{-i(\psi+k)}\} \sin i\phi \quad\ldots\ldots(1),$$
$$y = -\psi + \Sigma A_i \{e^{i(\psi+k)} - e^{-i(\psi+k)}\} \cos i\phi \quad\ldots\ldots(2),$$

where all integral values of i from 1 to ∞ are to be taken. The parameter ψ of the stream-line of the free surface being 0, we have for points in the outline of the wave

$$x = -\phi + \Sigma A_i (e^{ik} + e^{-ik}) \sin i\phi \ \dots\dots\dots\dots(3),$$

$$y = \qquad \Sigma A_i (e^{ik} - e^{-ik}) \cos i\phi \ \dots\dots\dots\dots(4).$$

Let y be denoted by $f(\phi)$. In terms of $f(\phi)$ supposed known we can determine A_i. We have

$$\pi (e^{ik} - e^{-ik}) A_i = \int_0^{2\pi} f(\phi') \cos i\phi' d\phi' \ \dots\dots\dots(5).$$

Suppose that at the point for which $\phi = \pi$ there is a cusp, and that elsewhere within the limits of integration $f(\phi)$ and its derivatives have values that vary continuously without becoming infinite. Divide the total interval 0 to 2π in (5) into a number of partial intervals, and treat each by integration by parts. The result for any interval will in general be

$$\frac{1}{i} f(\phi) \sin i\phi + \frac{1}{i^2} f'(\phi) \cos i\phi - \frac{1}{i^3} f''(\phi) \sin i\phi - \dots,$$

which has to be taken between limits corresponding to the beginning and end of the interval. For the particular interval however comprising the cusp (which it will be convenient to suppose situated within an interval and not at the edge) the above transformation cannot be made, as $f'(\phi)$ becomes infinite within the limits of the interval. Except as regards this interval, the terms of the above development at the inferior limit will be destroyed by those at the superior limit in the interval preceding, and the terms at the superior limit by those at the inferior limit of the interval following, which applies even to the first and last intervals if we regard, as we may, the function $f(\phi)$ as returning into itself. Hence we see that except as regards what is contributed by the immediate neighbourhood of the cusp, the function $A_i (e^{ik} + e^{-ik})$ of i decreases faster as i increases than any inverse power of i, it decreases in fact, in general, in geometric progression.

If $$f(\phi) = A - B\phi_1^{\frac{3}{2}}$$

in the immediate neighbourhood of the cusp, where $\phi_1 = \phi - \pi$, we have for the part of A_i due to that neighbourhood, omitting the constant term, for which the above transformation does apply,

$$\pi \left(e^{ik} - e^{-ik} \right) A_i = - B \int \phi_1{}^{\frac{3}{2}} \cos i\phi_1 d\phi_1,$$

and as we may contemplate this integral apart from the restriction that i is to be an integer we may put it

$$- B \frac{d}{di} \int \phi_1{}^{-\frac{1}{2}} \sin i\phi_1 d\phi_1.$$

Now when i is very large,

$$\int \phi_1{}^{-\frac{1}{2}} \sin i\phi_1 d\phi_1$$

taken from a given small negative to a given small positive quantity will not sensibly differ from the integral taken from $-\infty$ to ∞, which we know is $\Gamma\left(\frac{2}{3}\right) i^{-\frac{2}{3}}$. Hence for very large values of i we have for the part of A_i due to the neighbourhood of the cusp

$$- B\pi^{-1} e^{-ik} \frac{d}{di} \left\{ \Gamma\left(\tfrac{2}{3}\right) i^{-\frac{2}{3}} \right\}, \quad \text{or} \quad \tfrac{2}{3} \pi^{-1} \Gamma\left(\tfrac{2}{3}\right) B e^{-ik} i^{-\frac{5}{3}}.$$

It appears therefore from (4) that the coefficient of $\cos i\phi$ in a term of very high order will vary ultimately as $i^{-\frac{5}{3}}$. The series will therefore converge only very slowly.

It is however a point of no little interest to determine from theory, at least approximately, the form of the highest possible wave.

We have seen that analytical development of the coefficients of multiples of ϕ is of little avail, because on the one hand a large number of terms of the series would have to be taken, and on the other the labour of the approximation, even according to the method explained in the first supplement (Vol. I, p. 314) increases rapidly with the order of the term. We must therefore have recourse to some independent method.

I have devised a method, depending upon a sort of trial and error, by which the true form can rapidly be approximated to, though each approximation involves quadratures which would be laborious if great accuracy were demanded; if however only sufficient accuracy to arrive at a good figure is required, the labour is not very great.

Suppose an outline of the wave assumed for trial, the assumed curve consisting of a series of undulations all alike, and symmetrical with respect to both the maxima and minima ordinates, and everywhere of a flowing form, except (in the particular case of the maximum wave) at the crests, where it comes to wedges of 120°. I say "in the particular case of the maximum wave" because the method, though specially designed for the investigation of the maximum wave, is not confined to it, but is of general application, and might be usefully employed were it wished to investigate a wave short of the maximum, but still so high as to make the series used in Vol. I too slowly convergent to be practically useful. A tolerably near approximation to the true form may be guessed from our knowledge of the approximate forms of high though not excessively high waves. The assumed function may be either expressed by an empirical formula, or by a numerical table, or by a flowing curve which is drawn by eye with the aid of a flexible ruler.

Let the wave motion be reduced to steady motion by superposing the requisite horizontal velocity. Then if the assumed curve be true, the velocity along the surface will vary in the same manner as that of a particle sliding along a smooth curve. This follows from the equation of condition that has to be satisfied at the free surface. The velocity at any point will in fact be that due to the depth below a certain horizontal line, which for the convenience of a name I will call the *datum line*. If we are dealing with the maximum wave, a particle at a crest will be momentarily at rest, and therefore the datum line will pass through the crests. If we are dealing with a wave short of the maximum, the datum line will lie at a finite distance above the crests. The height will be approximately known from our knowledge of the approximate motion of high waves. This height must be assumed for trial as well as the form of the curve.

Let V be the whole velocity; s the length of the curve measured from the origin, which is placed at the lowest point; y_1 the depth of any point in the surface below the datum line. Then without specifying the units of length and time we have

$$V = \sqrt{2gy_1}, \quad \phi = -\int V ds = -\int \sqrt{2gy_1} \frac{ds}{dx} dx.$$

The sign is here −, because the waves are supposed propagated in the positive direction, so that in the steady motion the flow is backwards. Since the units have been so chosen as to make the length of a half-wave equal to π, and the increment of ϕ in passing in the positive direction from a trough to a crest equal to $-\pi$, we have accordingly

$$\phi = -\pi \int_0^x y_1^{\frac{1}{2}} \frac{ds}{dx}\, dx \Big/ \int_0^\pi y_1^{\frac{1}{2}} \frac{ds}{dx}\, dx \quad \dots\dots\dots(6).$$

From this equation we can by integration or quadratures determine ϕ as a function of x for any value of x from 0 to π, which determines ϕ generally on account of the symmetry and periodicity of our functions. We may observe that x and $-\phi$ are co-periodic, and have the same mean value. As ϕ is known in terms of x, we may regard x as known in terms of ϕ. Let $x + \phi = F(\phi)$, then $F(\phi)$ being deemed known, we can determine A_i from (3) by the usual formula, and then by substituting in (4) we shall get y or $f(\phi)$. If the ordinates thus obtained agree with those of the trial curve, it proves that the curve was right; otherwise we shall obtain a new curve, which will be found to be a much closer approximation to the true form. This may then be used as a new trial curve, and so on until the approximation is sufficient.

The process would however be futile from the slowness of convergence of the series if we had first to determine the A_i's and then sum the series. In the two extreme cases, however, of a very great or very small depth the summation may be effected in the first instance and the integration left to the end, and then the integral may be obtained by quadratures. In the general case we encounter, in the attempt to carry out this process, a series which so far as I know cannot be summed. At this point therefore the investigation takes one or other of two different directions which require to be followed separately.

Case of Oscillatory Waves in Deep Water.

Writing $e^{-ik}A_i$ for A_i, and then making k infinite, we have from (1) and (2), for the expressions of the coordinates in terms of the functions ϕ, ψ,

$$x = -\phi + \Sigma A_i e^{i\psi} \sin i\phi \quad \dots\dots\dots(7),$$
$$y = -\psi + \Sigma A_i e^{i\psi} \cos i\phi \quad \dots\dots\dots(8),$$

and from (8) and (4) we have, remembering the definitions of the functions $F(\phi), f(\phi)$,

$$F(\phi) = \Sigma A_i \sin i\phi \quad\text{......................(9)},$$

$$f(\phi) = \Sigma A_i \cos i\phi \quad\text{......................(10)}.$$

From these equations we can determine either of the functions $F(\phi)$, $f(\phi)$ supposing the other known. Supposing $F(\phi)$ known, we have

$$f(\phi) = \frac{1}{\pi} \Sigma \cos i\phi \int_0^{2\pi} F(\phi') \sin i\phi' d\phi' \quad\text{.........(11)}.$$

We cannot immediately transpose the order of the two processes summation and integration, as the series would not be convergent. But we may regard the expression (11) as the limit of the expression thence obtained by introducing r^i into the general term, where r is a quantity less than 1, and which is supposed to become 1 in the limit. We thus get by summing the series $\Sigma r^i \cos i\phi \sin i\phi'$

$$f(\phi) = \frac{1}{2\pi} \int_0^{2\pi} \left\{ \frac{r \sin(\phi' + \phi)}{1 - 2r \cos(\phi' + \phi) + r^2} + \frac{r \sin(\phi' - \phi)}{1 - 2r \sin(\phi' - \phi) + r^2} \right\} F(\phi') d\phi' \quad\text{.........(12)}.$$

When r becomes 1, the two terms within brackets become respectively $\frac{1}{2} \cot \frac{1}{2}(\phi' + \phi)$ and $\frac{1}{2} \cot \frac{1}{2}(\phi' - \phi)$, and may be replaced by them accordingly, unless possibly in the immediate neighbourhood of $\phi' \pm \phi = 0$ or π, when the cotangent becomes infinite. Reserving this point for consideration a little further on, and making the change, we have

$$f(\phi) = \frac{1}{4\pi} \int_0^{2\pi} \left\{ \cot \tfrac{1}{2}(\phi' + \phi) + \cot \tfrac{1}{2}(\phi' - \phi) \right\} F(\phi') d\phi'.$$

On account of the periodicity of the function $F(\phi)$ it does not signify at what point the integration begins provided the range be 2π. Putting then $\phi' + \phi$ or $\phi' - \phi$, as the case may be, $= \chi$, and integrating from $\chi = 0$ to $\chi = 2\pi$, and observing that $F(\phi')$ changes sign with ϕ', we have

$$f(\phi) = \frac{1}{4\pi} \int_0^{2\pi} \left\{ F(\phi + \chi) - F(\phi - \chi) \right\} \cot \tfrac{1}{2}\chi \, d\chi$$

$$= \frac{1}{2\pi} \int_0^{\pi} \left\{ F(\phi + \chi) - F(\phi - \chi) \right\} \cot \tfrac{1}{2}\chi \, d\chi \quad\text{.........(13)},$$

since the quantity under the integral sign in the first line remains unchanged when $2\pi - \chi$ is written for χ.

As the substitution of 1 for r might have been reserved to the end, and as the quantity under the integral sign in (13) does not become infinite, but merely takes the form $0 \times \infty$, remaining finite, we see that the validity of (13) is not affected by the circumstance that one of the terms within brackets in (12) is made infinite within the limits of integration by putting $r = 1$.

In just the same way $F(\phi)$ might have been deduced from $f(\phi)$ supposing the latter known. The result is

$$F(\phi) = -\frac{1}{2\pi} \int_0^\pi \{f(\phi + \chi) - f(\phi - \chi)\} \cot \tfrac{1}{2}\chi \, d\chi \ ...(14).$$

By way of trying the method in such a way as to put it to a pretty severe test, I took for the trial outline a serrated profile, composed of equal straight lines inclined at angles of $\pm 30°$ to the horizon. The quadratures were worked roughly, as it was intended merely to test the method, not to obtain a final result. In the following figure it will be observed that the assumed form satisfies the conditions of wave motion about the crests, while they are egregiously violated in the neighbourhood of the troughs. The figure* gives the outline obtained by the first approximation. The dotted lines represent the trial outline where it does not sensibly confound itself with the curve obtained.

The unit of length employed in the above investigation is the one-2πth part of λ, and to complete the solution we must determine the unit of time, or what comes to the same, the velocity of propagation, which hitherto has been taken as unity by properly choosing the unit of time. For this we must revert to the equation of condition at the surface, or, which is the same thing otherwise expressed, the condition that, in the steady motion, the motion of a particle at the free surface is the same as that of a particle gliding along a smooth curve. The method followed in the present and former supplements leads us naturally to define the velocity of propagation by the constant c, though in the general case of an arbitrary depth the relation of this constant to the physical circumstances of the motion is not immediately apparent, except to a first approximation, when the ambiguity

[* The rough figure has not been reproduced : see *infra*, p. 158.]

in the meaning of the term "velocity of propagation" noticed in
Art. 3 of the original paper (Vol. I, p. 202) does not occur. In
neither of the extreme cases however in which the ratio of the
length of a wave to the depth of the fluid is regarded as infinitely
small or infinitely great is there any ambiguity, whatever be the
magnitude of the motion.

We see from (7) that the increment of ϕ in passing from
a crest to a trough is $-\pi$, or expressed in general units $-\frac{1}{2}c\lambda$.
But by the condition at the surface it is also $-\int\sqrt{(2gy_1)}\,ds$ taken
from a crest to the next trough, y_1 being the depth below the
datum line.   Hence

$$c = \frac{2}{\lambda}\sqrt{2gy_1}\,ds = \gamma\sqrt{\frac{g\lambda}{2\pi}} \dots\dots\dots\dots\dots (15),$$

where γ is a numerical factor which must be found by quadratures.
This particular form of expression is of course chosen to exhibit
the ratio of the velocity of propagation to what it would be in
infinitely low waves of the same height.

The components of the velocity at any point are given by the
differential coefficients of x or y with respect to ϕ or ψ, by the
formulae on p. 315 of Vol. I, namely,

$$\left.
\begin{aligned}
u &= \frac{d\phi}{dx} = S^{-1}\frac{dx}{d\phi} = S^{-1}\frac{dy}{d\psi} \\
v &= \frac{d\phi}{dy} = -S^{-1}\frac{dx}{d\psi} = S^{-1}\frac{dy}{d\phi} \\
S &= \left(\frac{dx}{d\phi}\right)^2 + \left(\frac{dx}{d\psi}\right)^2
\end{aligned}
\right\} \dots\dots\dots\dots(16).$$

At a great or even moderate depth, the series (7) and (8)
converge rapidly, on account of the factor $e^{i\psi}$ in which ψ is
negative and not small, and $F(\phi)$ being supposed known the
easiest way would doubtless be to develop $F(\phi)$ in a series of sines
to a few terms, substitute the resulting values of A_1, A_2, A_3... in
(7) or (8), and differentiate. But in the case of the maximum
wave the series (7) and (8) would converge only slowly in the
neighbourhood of the surface; and in accordance with the spirit
of the method here employed we ought to dispense with the use
of the series.

This may be done just as before. Either coordinate will serve for the differentiation, and if we choose y we may employ the equation (12) already obtained, merely writing $y + \psi$ for $f(\phi)$ and e^ψ for r. Making the same further transformation as in that case we get

$$y + \psi = \frac{1}{\pi} \int_0^\pi \{F(\phi + \chi) - F(\phi - \chi)\} \frac{\sin \chi \, d\chi}{e^\psi + e^{-\psi} - 2 \cos \chi} \dots (17).$$

To sum up, the method here suggested is as follows:—

From our knowledge of the theoretical forms of high waves we must first assume a form as near the truth as we can guess. The choice is assisted by our knowledge that if we are dealing with the maximum wave the crests must be wedges of 120°, and if the wave be a little below the maximum the crests must be rounded in such a manner as to approach towards wedges of 120°. We must further assume for trial a datum line. If the waves be the highest possible, we know that the datum line passes through the wedge-formed crests. If though high they are short of the maximum, the datum line will lie above the crests. Its height above the mean elevation will not greatly differ from what it is for the maximum wave. We must now determine ϕ by quadratures from (6) for a sufficient number of points. The functional relation between ϕ and x being thus ascertained, to the degree of accuracy that we think expedient, $x + \phi$ is to be regarded as a known function, $F(\phi)$, of ϕ. The ordinate y or $f(\phi)$ is then to be calculated by quadratures for a sufficient number of points from (13). The difference between the corrected outline and that with which we started for trial will enable us to judge whether the former may be deemed sufficiently exact; if not, the process is to be repeated, using the corrected outline for a new trial outline. When we are satisfied with the outline, the equation (15) will give the velocity of propagation, and thereby make known the hitherto unknown unit of time to which the expressions are referred.

The evidently near approach of the first approximation to the true form even when we start with such an outlandish figure as a serrated line formed of a succession of equal straight lines inclined at angles of $\pm$ 30° to the horizon, indicates that we need not fear to use for the trial form a polygon of a moderate number of sides, so chosen as to make a tolerable approach to the true form. In the form got for a first approximation from the serrated outline, the troughs are in all probability still a little too low. Guided by these considerations I take for the trial form a polygon made up of lines inclined to the horizon at angles of 30° from 0 to 0·5 of the half wave's length, 20° from 0·5 to 0·8, 10° from 0·8 to 0·9, and 0° from 0·9 to 1. From the figure on p. 212 of Vol. I, [not] reproduced a little further on, it seemed probable that the hollows would be flatter towards their middles than towards their sides. A polygon for trial form presents great facilities for the first step in the calculation, namely determining ϕ in terms of x, as the relation for each side in succession is given by a simple formula.

In approximating to the maximum wave, it will be convenient if we can find an empirical formula of a tolerably simple character that shall give for a single wave a near approximation to the true form*. The result of the trial with the serrated outline, combined with what we know of the character of the wave, will enable us to come pretty close to the true form.

The empirical formula, giving y_1 as a function of x, is not meant to extend beyond the limits $\pm \pi$ or x. The symmetry requires that it be an even function of x. It must vanish at the two limits. We must also have at the limits $dy_1/dx = \pm \tan 30°$. Again the result of the trial with the serrated outline, as well as what was shown at p. 226 of Vol. I regarding the form of ϕ or ψ in terms of the coordinates in the immediate neighbourhood of a crest, leads to the conclusion that at the crest $d^2y_1/dx^2 = 0$. Assuming then the empirical formula

$$y_1 = A + Bx^2 + Cx^4 + Dx^6,$$

and introducing the three conditions above mentioned, we get

$$y_1 = A \left(1 - 3 \frac{x^2}{\pi^2} + 3 \frac{x^4}{\pi^4} - \frac{x^6}{\pi^6}\right) + \frac{\sqrt{3}\pi}{24} \left(9 \frac{x^2}{\pi^2} - 14 \frac{x^4}{\pi^4} + 5 \frac{x^6}{\pi^6}\right).$$

The remaining arbitrary constant A enables us to assign to y_1 at the centre what may appear a probable value. The trough would probably lie a little above that determined for a first

----

[* This passage was marked by the author to be omitted.]

approximation in starting with the serrated outline. We may accordingly assume as a near approximation $y_1 = 1\cdot33$. This gives

$$y_1 = 1\cdot330 - 1\cdot950\,\frac{x^2}{\pi^2} + 0\cdot816\,\frac{x^4}{\pi^4} - 0\cdot196\,\frac{x^6}{\pi^6}.$$

Case of a Solitary Wave.

When the length of a wave is very great compared with the depth of the water, it has been shown (Vol. I, pp. 209, 325) that unless the height be extremely small the series begin by diverging. Accordingly as the length tends to infinity, the initial terms must be very small, or else the sum would be very large, which is contrary to hypothesis. The terms will therefore get small and numerous, and the proper form of the expression when the limit of an infinite wave length is attained will be an integral. In the expressions (1) and (2) the origin of ϕ has been supposed to be a point where the wave is symmetrical right and left, and may be therefore either a crest or a trough. In the investigation for the waves in deep water it has been supposed to be a trough, but now it will be more convenient to suppose it a crest.

Assuming then the origin to be at a crest, and supposing the wave length to become infinite, the other elevations will move off to infinity, and we shall pass to a solitary wave. Replacing the sums in (1) and (2) by integrals, we shall have for the expressions for the coordinates

$$x = -\phi + \int_0^\infty \varpi\,(n)\,(e^{n\,(\psi+k)} + e^{-n\,(\psi+k)})\sin n\phi\,dx \;\;....(a),$$

$$y = -\psi + \int_0^\infty \varpi\,(n)\,(e^{n\,(\psi+k)} - e^{-n\,(\psi+k)})\cos n\phi\,dx \;\;....(b),$$

where ϖ denotes an arbitrary function, and for the coordinates of the surface

$$x + \phi = f(\phi) = \int_0^\infty \varpi\,(n)\,(e^{nk} + e^{-nk})\sin n\phi\,dx \;\;......(c),$$

$$y = F(\phi) = \int_0^\infty \varpi\,(n)\,(e^{nk} - e^{-nk})\cos n\phi\,dx......(d).$$

Comparing (c) with the known development of an arbitrary function in a series of that form we have

$$\varpi\,(n)\,(e^{nk} + e^{-nk}) = \frac{2}{\pi}\int_0^\infty f(\phi')\sin n\phi'\,d\phi',$$

and now substituting in (d) we get

$$F(\phi) = \frac{2}{\pi} \int_0^\infty \int_0^\infty \frac{e^{nk} - e^{-nk}}{e^{nk} + e^{-nk}} \cos n\phi \sin n\phi' dx\, d\phi' \quad \ldots\ldots(e).$$

But it is known that

$$\int_0^\infty \frac{e^{nk} - e^{-nk}}{e^{nk} + e^{-nk}} \sin n\alpha\, dx = \frac{\pi}{k}\, \frac{1}{e^{\pi a/2k} - e^{-\pi a/2k}},$$

and applying this formula to (e) we get

$$F(\phi) = \frac{1}{k} \int_0^\infty \left\{ \frac{1}{e^{\pi (\phi'+\phi)/2k} - e^{-\pi (\phi'+\phi)/2k}} \right. $$
$$\left. + \frac{1}{e^{\pi (\phi'-\phi)/2k} - e^{-\pi (\phi'-\phi)/2k}} \right\} f(\phi')\, d\phi'.$$

Since $f(\phi')$ changes sign with ϕ', retaining the same magnitude, the quantity under the integral sign remains unchanged when ϕ' changes sign; therefore we may take the integral from $-\infty$ to ∞ and divide by 2, and then by an easy transformation we get

$$F(\phi) = \frac{1}{k} \int_0^\infty \{f(\phi + \omega) - f(\phi - \omega)\} \frac{d\omega}{e^{\pi\omega/2k} - e^{-\pi\omega/2k}} \quad \ldots(f).$$

In a similar manner $f(\phi)$ might be expressed in terms of $F(\phi)$ supposed known, but this expression will not be required.......

[This investigation, of which the preceding paper was an abstract, has been found among Sir George Stokes' manuscripts in a form ready for press, along with a fair copy of the laborious calculations required for the three applications of the method to be mentioned presently. The following measurements of the ratio d of height of wave to length, taken from these three plotted approximations, just as they have been found, are supplied by Mr T. H. Havelock, Fellow of St John's College, who has kindly examined the manuscript calculations. They appear to show that the method is not so successful as Sir George Stokes anticipated.

i. The trial outline was a serrated profile composed of two equal straight lines inclined at angles of $\pm 30°$ to the horizon. For it $d = 1/2\sqrt{3} = \cdot288$; and for the next approximation deduced therefrom, $d = \cdot22$, as stated *supra*, p. 145.

ii. The trial figure was a polygon made up of lines inclined to the horizon, at an angle of $30°$ from 0 to 0·5 of the half wave length, $20°$ from 0·5 to 0·8, $10°$ from 0·8 to 0·9, and $0°$ from 0·9 to 1. In this case $d = \cdot21$; and from the next approximation deduced therefrom $d = \cdot165$.

iii. The trial curve was a circular arc of $60°$. In this case $d = \cdot134$; the next approximation gave $d = \cdot195$.

These calculations may be compared with the result given by Michell (*loc. cit. infra*), which makes d for the highest wave equal to ·142. It will be observed that the change in d is in each case in the right direction, but there is little sign of approximation to the true result.

The form of the highest possible waves, those with sharp summits of 120°, has been completely determined by J. H. Michell (*Phil. Mag.* (5) XXXVI, Dec. 1893, pp. 430—437) for deep water, by a very refined analysis ; while, as above mentioned, waves not near this limiting form are adequately treated by Sir George Stokes' previous method of approximation.

For this purpose Michell utilizes the function log dz/dw, where $w = \phi + \iota\psi$ and z is the complex variable $x + \iota y$: the consideration of this function had already led him (Michell, *Phil. Trans.* Jan. 1890, pp. 389—441 ; Love, *Proc. Camb. Phil. Soc.* VII, May 1891, pp. 175—201) to a general method of solution of problems of steady jets or free streams in two dimensions, issuing from reservoirs with straight boundaries. In the present problem, when reduced as above to a steady motion, the same function can be extended over the whole plane of z, and has no singularities except those corresponding to the sharp crests of the waves. If the function $\dfrac{d}{dz} \log \dfrac{dz}{dw}$ is used, these points are simple poles, as Stokes had virtually shown. The poles being thus known, the analytical type of this function is determined, and, introducing the surface condition, the form of the profile of the waves is deduced by a rapid approximation in a Fourier series. Cf. *supra*, p. 149.

The corresponding solution, by means of trains of image-poles, for waves in water of finite depth, is only briefly sketched at the end of Michell's paper. It would include as a limiting case that of the solitary wave.

This latter problem has also been treated by J. McCowan (*Phil. Mag.* (5) XXXI, 1891, pp. 45—58 and pp. 553—555, XXXVIII, 1894, pp. 351—365) by assuming, arbitrarily, suitable simple types of expression for w, namely

$$U\iota z + a \tan \tfrac{1}{2} \iota m z$$

for the general case, and

$$\iota U \left(1 - f k^2 \sec^2 \tfrac{1}{2} \iota m z\right) \left(1 - k^2 \sec^2 \tfrac{1}{2} \iota m z\right)^{\frac{1}{2}}$$

for the highest or sharp-crested wave, in which the constants can be adapted so as to secure approximate satisfaction of the surface conditions; the results are compared with those of Boussinesq and of Rayleigh. A further note by Sir George Stokes, written in this connexion, is reprinted *infra*, p. 160.

Although as regards its special problem the fragment here published has been superseded by Michell's investigation, it is still of interest, on account of the illustrations it affords, incidentally, of some refined points in analysis.]

NOTE ON THE THEORY OF THE SOLITARY WAVE.

[From the *Philosophical Magazine*, Vol. XXXI, Sept. 1891, pp. 314—316.]

IN a paper on the Solitary Wave by Mr J. McCowan, printed in the July number of the *Philosophical Magazine*, for a copy of which I am indebted to the kindness of the author, he refers to a conclusion which I advanced in a paper written long since, and reprinted full ten years ago, according to which a solitary wave could not be propagated without change of form. As I have known for the last ten years that this conclusion was erroneous, and have published a paper in which the motion of a uniformly propagated solitary wave was considered, I am not concerned to defend it; but it may be well to point out the true source of the error, respecting which I cannot agree with Mr McCowan.

While the first volume of my 'Collected Papers' was going through the press, I was led to the conclusion (see p. 227) that the highest possible waves of the oscillatory kind (the motion being irrotational) presented a form in which the crests came to wedges of 120°. On reflecting on the application of this to very long waves propagated in water of which the depth is small compared with the length of wave, I was led to perceive that the conclusion above mentioned was erroneous, and also that the source of the error was that it was not sufficient, even though a solitary wave were very long, to treat it as indefinitely long, and consequently to take the horizontal velocity as the same from the surface to the bottom. On speaking on the subject to Lord Rayleigh, he referred me to the previous papers on the solitary wave by M. Boussinesq and himself, with which I was not at the time acquainted. The conclusion of a supplement to my paper on oscillatory waves, which forms the last article in Vol. I, shows that I was then fully alive to the possibility of the propagation of a solitary wave without change; and in a short paper entitled

"On the highest wave of uniform propagation (preliminary notice),"
read before the Cambridge Philosophical Society in 1883, and
printed in the Proceedings (Vol. IV, p. 361), I have indicated
a new method, depending on a process of trial and error, for
determining numerically the circumstances of uniform propagation
of waves, whether of the oscillatory or solitary class, more especially
in the extreme case in which the crest comes to a wedge of 120°,
so that the wave is on the point of beginning to break*.

I cannot agree with Mr McCowan either that the form of
expansion which I used is inadmissible, or that the form which
he proposes at p. 58 to substitute, that of a series involving
exponentials in which the coefficient of x in the index is real, is
(at least for my purpose) admissible. It is not true that a non-
periodic function of x cannot be expressed by means of periodic
functions; for example, the non-periodic function e^{-x^2} may be
expanded in the definite integral

$$e^{-x^2} = \frac{2}{\sqrt{\pi}} \int_0^\infty e^{-a^2} \cos 2ax \,.\, da,$$

each element of which is periodic. On the other hand, the
form of expansion proposed by Mr McCowan is (at least for my
purpose) inadmissible, on account of the discontinuity of the
expression†.

I will now mention more particularly the step which led
me to a wrong conclusion. It is easily shown that in a very
long wave propagated in water the depth of which is small
compared with the length of the wave, the horizontal velocity
is nearly the same from the bottom to the surface for any vertical
section of the wave made by a plane perpendicular to the direction
of propagation. For a given depth of water and maximum height
of wave, this is so much the more nearly true as the length of the
wave is greater. The horizontal velocity tends indefinitely towards
constancy from top to bottom as the length of the wave increases
indefinitely. Now Sir George Airy has shown that for a wave
in which we may suppose the particles in a vertical plane to
remain always in a vertical plane, as they must do if the horizontal

[* See note appended *supra*, p. 158.]
[† Mr McCowan replied that the criticism of this paragraph arose from
a misapprehension of his meaning.]

162   NOTE ON THE THEORY OF THE SOLITARY WAVE.

velocity is the same from top to bottom, the form of the wave must gradually change as it progresses. It might seem therefore that, however small we take the height of a wave at the highest point, we have only to make the wave long enough and Airy's investigation will apply, and the wave will change its form in time as it travels along. Now in the solitary wave of Russell, the lower the wave, the longer it is; and therefore it might seem as if we had only to make the wave low enough and long enough and the length would be so great that Airy's investigation would apply, and the form would change, though slowly.

The answer to this is that, however small we take the height, we are not at liberty to increase the length indefinitely. There is in fact a relation between the height and the length in a solitary wave which can be propagated uniformly, which, though it is of such a nature that the length becomes infinite when the height becomes infinitely small, nevertheless forbids us for a given height, however small, to increase the length indefinitely.

The possibility of the existence of a solitary wave of uniform propagation is so far established by my investigation relating to oscillatory waves, as that it is made to depend on the principle that the infinite series by which the circumstances of the motion of oscillatory waves is expressed must remain convergent until the height is so far increased that the outline presents a singular point, namely the wedge of 120°. If this be conceded, we have only to make the wave-length (in the sense of the distance from crest to crest, not in that of the distance from the point where one of the swells begins to where it ceases to be sensible) infinite, in order to pass to the case of a solitary wave uniformly propagated.

THE OUTSKIRTS OF THE SOLITARY WAVE.

[From Prof. H. LAMB's *Hydrodynamics*, 1895, p. 421.
Communicated by Sir GEORGE STOKES*.]

THE motion at the outskirts of the solitary wave can be represented by a very simple formula. Considering a progressive wave travelling in the direction of x-positive, and taking the origin in the bottom of the canal, at a point in the front part of the wave, we assume

$$\phi = A e^{-m(x-ct)} \cos my \quad \dots\dots\dots\dots(\text{xiii}).$$

This satisfies $\nabla^2 \phi = 0$, and the surface-condition

$$\frac{d^2\phi}{dt^2} + g \frac{d\phi}{dy} = 0 \quad \dots\dots\dots\dots\dots (\text{xiv})$$

will also be satisfied for $y = h$, provided

$$c^2 = gh \frac{\tan mh}{mh} \quad \dots\dots\dots\dots\dots\dots(\text{xv}).$$

This will be found to agree approximately with Lord Rayleigh's investigation if we put $m = b^{-1}$.

[* A letter to Prof. Lamb of date July 25, 1900, contains the following passage : "I do not know whether you may have noticed that the equation I gave you (*Hydrodynamics* [1895], p. 421, eq. xv)

$$c^2 = gh \frac{\tan mh}{mh}$$

is an *exact* equation, that is to say it gives an exact relation between two constants relating to a solitary wave, namely, (1) the velocity of propagation, and (2) the logarithmic decrement (or increment) of the velocity [of flow at the surface, which is $A m e^{-m(h-ct)}$] at a great distance from the middle of the wave. It therefore serves as a useful test of the degree of accuracy of an approximate solution of the problem. Its application to a solitary wave only postulates two things, (1) the possibility of the existence of a solitary wave, (2) the fact that in such a wave the velocity [of flow] at a great distance from the middle of the wave ultimately changes in geometric progression. Of neither of these postulates is there, I think, any doubt."]

ON A REMARKABLE PHENOMENON OF CRYSTALLINE REFLECTION.

[From the *Proceedings of the Royal Society;*
received *Feb.* 25, 1885.]

Introduction.

IN a letter to me, dated March 29, 1854, the late Dr W. Bird
Herapath enclosed for me some iridescent crystals of chlorate of
potash, which he thought were worth my examination. He noticed
the intense brilliancy of the colour of the reflected light, the
change of tint with the angle of incidence, and the apparent
absence of polarisation in the colour seen by reflection.

The crystals were thin and fragile, and rather small. I did not
see how the colour was produced, but I took for granted that it
must be by some internal reflection, or possibly oblique refraction,
at the surfaces of the crystalline plates that the light was polarised
and analysed, being modified between polarisation and analysation
by passage across the crystalline plate, the normal to which I supposed must be sufficiently near to one of the optic axes to allow
colours to be shown, which would require no great proximity, as
the plates were very thin. To make out precisely how the colours
were produced seemed to promise a very troublesome investigation
on account of the thinness and smallness of the crystals: and
supposing that the issue of the investigation would be merely to
show in what precise way the phenomenon was brought about by
the operation of well-known causes, I did not feel disposed to
engage in it, and so the matter dropped.

But more than a year ago Professor E. J. Mills, F.R.S., was so
good as to send me a fine collection of splendidly coloured crystals
of the salt of considerable size, several of the plates having an area
of a square inch or more, and all of them being thick enough to

handle without difficulty. In the course of his letter mentioning the despatch of the crystals, Professor Mills writes: "They (the coloured crystals) are, I am told, very pure chemically, containing at most 0·1 per cent. foreign matter. They are rarely observed— one or two perhaps now and then, in a large crystallisation....I have several times noticed that small potassic chlorate crystals, when rapidly forming from a strong solution, show what I suppose to be interference colours; but the fully formed crystals do not show them."

Some time later I was put into communication with Mr Stanford, of the North British Chemical Works, Glasgow, from which establishment the crystals sent me by Professor Mills had come. Mr Stanford obligingly sent me a further supply of these interesting crystals, and was so kind as to offer to try any experiment that I might suggest as to their formation.

I am informed that at the recent Health Exhibition a stand was exhibited from the chemical works at Widnes, showing a fine collection of brilliantly coloured crystals of chlorate of potash. I did not see it. It would seem that the existence of these coloured crystals is pretty generally known, but I have not seen mention of them in any scientific journal, nor, so far as I know, has the subject been investigated.

On viewing through a direct-vision spectroscope the colours of the crystals which I had just received from Professor Mills, the first glance at the spectrum showed me that there must be something very strange and unusual about the phenomenon, and determined me to endeavour to make out the cause of the production of these colours. The result of my examination is described in the present paper.

SECTION I.—*Preliminary Physical Examination.*

1. It will be necessary to premise that chlorate of potash belongs to the oblique system of crystallisation. The fundamental form may be taken as an oblique prism on a rhombic base, the plane bisecting the obtuse dihedral angle of the prism being the plane of symmetry. Rammelsberg denotes the sides of the prism by P, and the base by C, and gives for the inclinations of the

faces $PP = 104° 22'$ and $CP = 105° 35'$. The face C, which is perpendicular to the plane of symmetry, is so placed as to bring three obtuse plane angles together at two opposite corners of the parallelopiped. The salt usually forms flat rhombic or hexagonal plates parallel to the C plane, the edges of the rhombus being parallel to the intersections of the P faces by the C plane, and the hexagons being formed from the rhombic plates by truncating the acute angles by faces parallel to the intersection of the C plane by the plane of symmetry.

The plane angles of the rhombic plates, calculated from the numbers given by Rammelsberg, are $100° 56'$ and $79° 4'$, while the hexagonal plates present end-angles of $100° 56'$ and four side-angles of $129° 32'$. These angles are sufficiently different to allow in most cases the principal plane of a plate, or even of a fragment of a plate, to be determined at once by inspection. But in any case of doubt it may readily be found without breaking the crystal by examining it in polarised light. There are good cleavages parallel to the two P planes and to the C plane. The crystals are very commonly twinned, the twin plane being C.

2. If one of the brilliantly coloured crystals be examined by reflection, and turned round in its own plane, without altering the angle of incidence, the colour disappears twice in a complete revolution. The vanishing positions are those in which the plane of incidence is the plane of symmetry. The colour is perhaps most vivid in a perpendicular plane; but for a very considerable change of azimuth from the perpendicular plane there is little variation in the intensity of the colour. There is no perceptible change of tint, but on approaching the plane of symmetry the colour gets more and more drowned in the white light reflected from the surface.

3. If instead of altering the azimuth of the plane of incidence a plane be chosen which gives vivid colour, and the angle of incidence be altered, the colour changes very materially. If we begin with a small angle the colour begins to appear while the angle of incidence is still quite moderate. What the initial colour is, varies from one crystal to another. As we increase the angle of incidence the colour becomes vivid, at the same time changing, and as we continue to increase the angle the change of colour goes on. The change is always in the order of increasing refrangibility;

for example, from red through green to blue. Not unfrequently, however, the initial tint may be green or blue, and on approaching a grazing incidence we may get red or even yellow mixed with the blue, as if a second order of colours were commencing.

4. The colours are not in any way due to absorption; the transmitted light is strictly complementary to the reflected, and whatever is missing in the reflected is found in the transmitted. As in the case of Newton's rings, the reflected tints are much more vivid than the transmitted, though, as will presently appear, for a very different reason.

5. As Dr Herapath remarked to me long ago, the coloured light is not polarised. It is produced indifferently whether the incident light be common light or light polarised in any plane, and is seen whether the reflected light be viewed directly or through a Nicol's prism turned in any way. The only difference appears to be that if the incident light be polarised, or the reflected light analysed, so as to furnish or retain light polarised perpendicularly to the plane of incidence, the white light reflected from the surface, which to a certain extent masks the coloured light, is more or less got rid of.

6. The character of the spectrum of the reflected light is most remarkable, and was wholly unexpected. A direct-vision hand spectroscope was used in the observations, and the crystal was generally examined in a direction roughly perpendicular to the plane of symmetry; but it is shown well through a wide range of azimuth of the plane of incidence. No two crystals, we may say, are alike as to the spectrum which they show, but there are certain features common to all. The remarkable feature is that there is a pretty narrow band, or it may be a limited portion of the spectrum, but still in general of no great extent, where the light suffers total or all but total reflection. As the angle of incidence is increased, these bands move rapidly in the direction of increasing refrangibility, at the same time increasing in width. The character of the spectrum gradually changes as the angle of incidence is increased; for example, a single band may divide into two or three bands.

The bands are most sharply defined at a moderate angle of incidence. When the angle of incidence is considerably increased, the bands usually get somewhat vague, at least towards the edges.

7. The commonest kind of spectrum, especially in crystals prepared on a small scale, which will be mentioned presently, is one showing only a single bright band; and I will describe at greater length the phenomena presented in this case.

When the angle of incidence is very small, the light reflected from the reflecting surfaces of the crystal shows only a continuous spectrum. As the angle of incidence is increased, while it is still quite moderate a very narrow bright band shows itself in some part of the spectrum. The particular part varies from one crystal to another; it may be anywhere from the extreme red to the extreme violet. It stands out by its greatly superior brightness on the general ground of the continuous spectrum, and when it is fully formed the reflection over the greater part of it appears to be total. The appearance recalls that of a bright band such as the green band seen when a calcium salt, or the orange band seen when a strontium salt, is put into a Bunsen flame. The bright band is frequently accompanied right and left by maxima and minima of illumination, forming bands of altogether subordinate importance as regards their illumination. Sometimes these seem to be absent, and I cannot say whether they are an essential feature of the phenomenon, which sometimes fail to be seen because the structure on which the bands depend is not quite regularly formed, or whether, on the other hand, they are something depending on a different cause.

Disregarding these altogether subordinate bands, and taking account of the mean illumination, it seems as if the brightness of the spectrum for a little way right and left of the bright band were somewhat less than that at a greater distance.

When the main band occurs at either of the faint ends of the spectrum, it is visible, by its superior brightness, in a region which, as regards the continuous spectrum, is too faint to be seen, and thus it appears separated from the continuous spectrum by a dark interval.

When the angle of incidence is increased, the band moves in the direction of increasing refrangibility, and at the same time increases rapidly in breadth. The increase of breadth is far too rapid to be accounted for merely as the result of a different law of separation of the colours, which in a diffraction spectrum would be separated approximately according to the squared reciprocal of the wave-length, while in bands depending on direct interference the phase of illumination would change according to the wave-length.

8. The transmitted light being complementary to the incident, we have a dark band in the transmitted answering to the bright band in the reflected. In those crystals in which the band is best formed, it appears as a narrow black band even in bright light. When the band first appears as we recede from a normal incidence it is extremely narrow, but it rapidly increases in breadth as the angle of incidence is increased.

9. Some of the general features of the phenomenon were prettily shown in the following experiment:—

Choosing a crystal in which the bright band in the reflected light began to appear, as the incidence was increased, on the red side of the line D, so that on continuing to increase the incidence it passed through the place of the line D before it had become of any great width, I viewed through the crystal a sheet of white paper illuminated by a soda flame. A dark ring was seen on the paper, which was circular, or nearly so, and was interrupted in two places at opposite extremities of a diameter, namely, the places where the ring was cut by the plane of symmetry. The light of the refrangibility of D was so nearly excluded from the greater part of the ring that it appeared nearly black, though slightly bluish, as it was illuminated by the feeble radiation from the flame belonging to refrangibilities other than those of the immediate neighbourhood of D. The ends of the two halves of the ring became feeble as they approached the plane of symmetry. A subordinate comparatively faint ring lay in this crystal immediately outside the main one.

10. Suspecting that the production of colour was in some way connected with twinning, I examined the cleft edge of some of the crystals which happened to have been broken across, and found that the bright reflection given by the exposed surface was interrupted by a line, much finer than a hair, running parallel to the C faces, which could be easily seen with a watchmaker's lens, if not with the naked eye. This line was dark on the illuminated bright surface exposed by cleavage, a surface which I suppose illuminated by a source of light not too large, such as a lamp, or a

window at some distance. The plane of incidence being supposed
normal to the intersection of the cleavage plane by the C faces,
on turning the crystal in a proper direction round a normal to
the plane of incidence, the light ceased to be reflected from the
cleavage surface, and after turning through a certain angle, the
narrow line, which previously had been dark, was seen to glisten,
indicating the existence of a reflecting surface, though it was much
too narrow to get a reflected image from off it. The direction of
rotation required to make the fine line glisten was what it ought
to be on the supposition that the fine line was the cleavage face of
an extremely narrow twin stratum.

11.   On examining the fine line under the microscope, it was
found to be of different thicknesses in different crystals, though in
those crystals which showed colour it did not vary very greatly.
On putting a little lycopodium on the cleavage face interrupted
by the fine line, it was seen that in those crystals which showed
colour the breadth of the twin stratum varied from a little greater
to a little less than the breadth of a spore. The thickness accordingly ranged somewhere about the thousandth of an inch,
such being the diameter of the spores. The stratum was visibly
thicker in those crystals which showed their bright band in the
red than in those which showed it in the blue.

12.   That the thin twin stratum was in fact the seat of the
colour, admitted of being proved by a very simple experiment. It
was sufficient to hold a needle, or the blade of a penknife (I will
suppose the latter), close to or touching the surface of the crystal
while it was illuminated by light coming approximately in one
direction, suppose from a lamp, or from a window a little way off,
and to examine the shadows with a watchmaker's lens. The light
reflected from the crystal comes partly from the upper surface,
partly from the twin stratum, partly from the under surface,
which, however, may be too irregular to give a good reflection.
The twin stratum is much too thin to allow of separating the
light reflected from its two surfaces in an observation like the
present, and it must therefore be spoken of as simply a reflecting
surface.

Corresponding to the three reflecting surfaces are three shadows,
where the incident light is cut off: (1) from the upper surface,
(2) from the twin stratum, (3) from the under surface. Where the

body casting the shadow is pretty broad in one part, as the blade of a penknife, the shadows in part overlap. The shadows are arranged as in the figure, where the numerals mark the streams of light reflected from the portions of the field on which they are respectively written, 1 denoting the stream reflected from the upper surface, 2, that reflected from the twin stratum, 3, that reflected from the under surface.

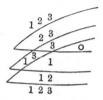

Let the crystal show by reflection, at the incidence at which the observation is made, say a green colour. Then this green colour is seen in the full field 123, though mixed with the white light reflected from the upper surface. The green is a good deal more vivid in the field 23, as the reflection from the upper surface is got rid of. The green is wholly absent from the fields 3, 0, 13, and 1. The field 3, and perhaps also the field 13, may show a little of the complementary red from transmitted light. The distinction between the fields 12 and 123 is not conspicuous, and often cannot be made out. The distinction, so far as it depends on the third shadow, is strongest between 3 and 0, and next to that between 13 and 1. We are not obliged, however, to have recourse to the third shadow, which is often difficult to see; the first two are amply sufficient.

Suppose we take a crystal which is broken at the edge so as to expose a cleavage surface interrupted by the cleavage of the narrow twin stratum. The stratum usually lies a good deal nearer to one C face than the other. Now when the two faces are turned uppermost alternately, and the distances between the first and second shadows are observed, they are found to be, as nearly as can be estimated, in the same proportion as the distances from the twin stratum to the two faces respectively.

Again, one of the crystals showed the exposed section of the twin stratum slightly inclined to one of the broad faces, which though smooth to the touch did not give a perfect reflection of objects viewed in it. On holding different parts of the blade of a penknife opposite to different parts of this face, the distance between the first and second shadows was found to vary, as nearly as could be guessed, in proportion to the thickness of crystal between the upper face and the twin stratum.

The conclusion was confirmed by observations made with sunlight; but the simple method of shadows is quite as good, and even by itself perfectly satisfactory.

13. Another useful method of observation, not so very simple as the last, is the following. A slit, suppose horizontal, not very narrow, is placed in front of the flame of a lamp at some distance, and an image of the slit is formed by a suitable lens, such as the compound achromatic objective of an opera-glass. The crystal is placed so as to receive in focus the image of the slit, being inclined at a suitable angle, usually in a plane perpendicular to the plane of symmetry. The eye is held in a position to catch the reflected light, and the images formed by the different reflections are viewed through a watchmaker's lens. If the slit be not too broad, the images formed by reflection from the upper surface, from the twin stratum, and from the under surface are seen distinct from each other, so that the light reflected from the twin stratum may be studied apart from that reflected from the upper and under surfaces.

In this mode of observation it can readily be seen, by turning the crystal in its own plane, and noticing the middle image, which is that reflected from the twin stratum, how very small a rotation out of the position in which the plane of incidence had been the plane of symmetry suffices to re-introduce the coloured light, which had vanished in that critical position, which appears to be a position not merely of absence of colour, but of absence of light altogether; at least if there be any it is too feeble to be seen in this mode of observation, though from theoretical considerations we should con-clude that there must be a very little reflected light, polarised perpendicularly to the plane of incidence.

14. On allowing a strong solution of chlorate of potash in hot water to crystallise rapidly, in which case excessively thin plates are formed in the bosom of the liquid, I noticed the play of colours by reflection mentioned by Professor Mills as belonging to the

crystals in general at an early stage of their growth. This, however, proved to be quite a different and no doubt a much simpler phenomenon. The difference was shown by the polarisation of the light, and above all by the character of the spectrum of the light so reflected, which resembled ordinary spectra of interference, and did not present the remarkable character of the spectra of the peculiar crystals.

15. When, however, the whole was left to itself for a day or so, among the mass of usually colourless crystals a few were found here and there which showed brilliant colours. These colours were commonly far more brilliant than those of the crystals mentioned in the preceding paragraph, and they showed to perfection the distinctive character of the spectrum of the peculiar crystals. It would have been very troublesome, if possible at all, to examine the twinning of such thin and tender plates as those thus obtained by working on a small scale; but the character of the spectrum, which is perhaps the most remarkable feature of the phenomenon, as well as the dependence of the colour on the orientation, may be examined very well; and thus anyone can study *these* features of the phenomenon, though he may not have access to such fine coloured crystals as those sent me by Professor Mills.

16. A certain amount of disturbance during the early stages of crystallisation, whether from natural currents of convection or from purposely stirring the solution somewhat gently so as not to break the crystals, seems favourable to the production of the peculiar crystals. When the salt crystallised slowly from a quiet solution I did not obtain them.

17. As it is easy in this way, by picking out the peculiar crystals from several crystallisations, to obtain a good number of them, the observer may satisfy himself as to the most usual character of the spectrum. It is best studied at a moderate incidence, as it is sharper than when the incidence is considerable. The spectrum most commonly shows a single intensely bright band, standing out on the general ground of a continuous spectrum of moderate intensity.

A few cases seem worthy of special mention. In one instance two bright bands were seen, one at each faint end of the spectrum, somewhat recalling the flame-spectrum of potassium salts. In another case a red, a green, and a blue band were seen, reminding one of the spectrum of incandescent hydrogen. This crystal in air was nearly colourless at moderate incidences, but showed red at rather high incidences. In another case the crystal was red of intense brilliancy in the mother-liquor, but was colourless when taken out, even at high incidences. Presumably the stratum in this case was so thick that a steeper incidence than could be obtained out of air was required to develop colour.

18. The number of coloured crystals obtained by crystallisations on a small scale, though very small, it is true, compared with the number of colourless ones, was still so much larger than Professor Mills' description of the rarity of the crystals had led me to expect, that I at one time doubted whether the simply twinned crystals which are so very common, if taken at a period of their growth when one component is still very thin, and of suitable thickness, might not possibly show the phenomenon, though the thin twin was in contact on one face only with the brother twin, the other face being in the mother-liquor or in air. The circumstances of reflection and transmission at the first surface of the twin plate must be very different according as it is in contact with the brother crystal, or else with the mother-liquor, or air, or some other fluid; and yet the peculiar spectrum was shown all the same whether the crystal was in air, or immersed in the mother-liquor, or in rock oil. However, to make sure of the matter I took a simply twinned crystal, and ground it at a slight inclination to the C face till the twin plane was partly ground away, thus leaving a very slender twin wedge forming part of the compound crystal, and polished the ground surface. On examining the reflected light with a lens, no colour was seen about the edge of the wedge, where the thickness of the wedge tapered away to nothing; and that, although the bands seen near the edge in polarised light, which was subsequently analysed, showed that had colours been producible in this way as they are by a thin twin stratum, they would not have been too narrow to escape observation.

In another experiment a simply twinned crystal was hollowed out till the twin plane was nearly reached. The hollowing was then continued with the wetted finger, so as to leave a concave smooth surface, the crystal being examined at short intervals in polarised light as the work went on, so as to know when the twin plane was pierced. But though in this case the twin plane formed a secant plane, nearly a tangent plane, to the worked surface, and near the section the twin portion of the crystal must have been very thin for a breadth by no means infinitesimal, as was shown by examination in polarised light, yet no colours were seen by reflection. I conclude therefore that the production of these colours requires the twin stratum to be in contact on *both* its faces with the brother crystal.

19. The fact that a single bright band is what most usually presents itself in the spectrum of the reflected light (though sometimes two or three such bands at regular intervals may be seen) seems to warrant us to regard that as the kind of spectrum belonging to the simplest form of twin stratum, namely, one in which there are just the two twin surfaces near together. The more complicated spectra seem to point to a compound interference, and to be referable to the existence of more than two twin planes very near together; and in fact in some of the crystals which showed the more complicated spectra, and which were broken across, I was able to make out under the microscope the existence of a system of more than two twin planes, close together. Restricting ourselves to what may be regarded as the normal case, we have then to inquire in what way the existence of two twin planes near together can account for the peculiar character of the spectrum of the reflected or transmitted light.

SECTION II.—*Of the Proximate Cause of the Phenomenon.*

20. Though I am not at present prepared to give a complete explanation of the very curious phenomenon I have described, I have thought it advisable to bring the subject before the Society, that the attention of others may be directed to it.

That the seat of the coloration is in a thin twin stratum, admits I think of no doubt whatsoever. A single twin plane does not show anything of the kind.

For the production of the colour the stratum must be neither too thick nor too thin. Twin strata a good deal thicker than those that show colour are common enough; and among the crystals sent to me I have found some twin strata which were a good deal thinner, in which case the crystal showed no colour.

The more complicated spectra which are frequently observed seem referable to the existence of more than two twin planes in close proximity. There is no reason to think that the explanation of these spectra would involve any new principle not already contained in the explanation of the appearance presented when there are only two twin planes, though the necessary formulæ would doubtless be more complicated.

Corresponding to a wave incident in any direction, in one component of a twin, on the twin plane, there are in general two refracted waves in the second component in planes slightly inclined to each other, and two reflected waves which also have their planes slightly inclined to each other, the angle of inclination, however, being by no means *very* small, as chlorate of potash is strongly double refracting. The planes of polarisation of the two refracted waves are approximately perpendicular to each other, as are also those of the two reflected waves; but on account of the different orientation of the two components of the twin, the planes of polarisation of the two refracted waves are in general altogether different from those of the incident wave and of its fellow, the trace of which on the twin plane would travel with the same velocity. In the plane of symmetry at any incidence, and for a small angle of incidence at any azimuth of the plane of incidence, the directions of the planes of polarisation of the two refracted waves agree accurately or nearly with those of the incident wave and its fellow. In these cases, therefore, an incident wave would produce hardly more than one refracted wave, namely, that one which nearly agrees with the incident wave in direction of polarisation. In these cases the colours are not produced. It appears, therefore, that their production demands that the incident wave shall be very determinately divided into two refracted waves, accompanied of course by reflected waves.

It seems evident that the thickness of the stratum affects the result through the difference of phase which it entails in the two refracted waves on arriving at the second twin plane. But whereas in the ordinary case of the production of colour by the interposition of a crystalline plate between a polariser and an analyser, we are concerned only with the difference of retardation of the differently polarised pencils which are transmitted across the plate, and not with the absolute retardation, it is possible that in this case we must take into account not only the difference of retardation for the differently polarised pencils which traverse the stratum, but also the absolute retardation; that is, the retardation of the light reflected from the second relatively to that reflected from the first twin plane.

21. I have not up to the present seen my way to going further. It is certainly very extraordinary and paradoxical that light should suffer total or all but total reflection at a transparent stratum of the very same substance, merely differing in orientation, in which the light had been travelling, and that, independently of its polarisation. It can have nothing to do with ordinary total internal reflection, since it is observed at quite moderate incidences, and *only within very narrow limits* of the angle of incidence*.

[POSTSCRIPT, from *Nature*, Vol. XXXII, p. 224, July 9, 1885.]

The appearance of Mr Madan's paper in *Nature*, Vol. XXXII, p. 102, induces me to offer some additional remarks on this subject.

In the discussion that followed the reading of my paper Mr Crookes referred to the closely analogous spectra exhibited by opals, as described in his paper (*Proc. Roy. Soc.*, Vol. XVII). This paper, though it came before me at the time when it was read, was not in my mind when I wrote my own. I called shortly afterwards at Mr Crookes' house, and saw the spectra of his opals. Supposing that there were sufficient grounds for the commonly received idea that the colours of the opal are due to fine tubes in the mineral, we did not at the time conceive that the phenomena could be the same; were it not for this, I should certainly have added to my paper a reference to that of Mr Crookes.

Mr Crookes was so good as to lend me his opals for more leisurely study. The further examination has so impressed me with the similarity of character of the spectra, that I am strongly

[* For further elucidation, see Lord Rayleigh, *Phil. Mag.* XXVI, 1888, pp. 256–265; reprinted in *Scientific Papers*, III, p. 204. The phenomenon is reviewed in Lord Rayleigh's Obituary Notice, reprinted at the beginning of the present volume.]

disposed to think that the colours of the opal and those of the chlorate crystals may be due to the same cause. This does not, however, lead me to attribute tubes or striæ to the chlorate crystals, the structure of which can comparatively easily be made out, but to doubt very greatly the theory which attributes the colours of opal to fine tubes.

Mr Madan does not profess to have actually seen in the chlorate crystals such tubes as he supposes to exist, nor could I see anything of the kind on examining some of the crystals I have got after the appearance of his paper. On the other hand, I notice that Brewster did not state that he had actually seen the supposed tubes, but merely inferred their existence from a comparison of the appearance under the microscope of the precious opal with that of hydrophane. And Mr Crookes tells me that an opal is not spoiled or affected by being immersed in water or even oil. The fact is that it is extremely difficult to make out what the actual structure is with which we have to deal in the case of the opal, whereas in the case of the chlorate crystals it is unmistakable. Moreover, in the case of the chlorate crystals there is a wonderful uniformity in the phenomena presented by the same crystal, extending, it may be, over nearly the whole of even a large crystal, whereas in the opal the colour extends over comparatively small patches; and even a single patch is seen under the microscope to present differences of structure in different parts. Hence if the colours in opal and those in the chlorate crystals are really due to a similar cause, it seems much more likely that a study of the phenomena of the chlorate crystals will throw light on those of the opal, than that the phenomena of the opal should furnish the key to the explanation of the colours of the chlorate crystals.

In truth, I do not see how the presence of tubes, if such there be in the opal, would account for the phenomena, and especially for the very peculiar spectrum exhibited. The supposition of the existence of rows of tubes leads one to look in the direction of diffraction. But I do not see how monochromatic light, or, at least, light almost monochromatic, can be obtained by diffraction. And even independently of this consideration there is one feature of the production of colour in the chlorate crystals which shows, at once and decisively, that at least in *their* case the colour cannot be due to diffraction. If an iridescent crystal be chosen with an even surface, and the flame of a candle in a dark room be viewed by reflection in it, it is found that the colour is seen in the direction of the regularly-reflected light. In fact, the coloured light forms a well-defined image of the flame of the candle, coinciding with, or overlapping, the colourless image due to reflection from the first surface. This differs altogether from what we get in the case of a grating, or in that of mother-of-pearl or Labrador spar. It agrees so far with the colours of thin plates, or the colours shown by reflection by certain quasi-metallic substances, such as several of the aniline dyes, though the production of colour in these three cases is due to three totally different causes.

It has been conclusively proved that the seat of the colour in the chlorate of potash crystals is in a very thin twin stratum; and I entertain myself little or no doubt that the colour depends in some way on the different orientation of the planes of polarisation in the two components of a twin, and on the difference of retardation of the two polarised pencils which traverse the thin stratum. But anything beyond this is at present only a matter of speculation. I see only two directions in one or other of which to look for a possible explanation; but as these could only be propounded at considerable length, and the matter has not at present advanced further, I refrained from saying anything about it in my former paper, nor will I further mention it here.

In conclusion, I would mention an interesting paper on "The Spectrum of the Noble Opal," by Prof. H. Behrens, a copy of which I have just received by the kindness of the author. In this paper, which is printed in the *Neues Jahrbuch für Mineralogie*, &c., 1873, the author, who was evidently unacquainted with Mr Crookes's paper when he wrote his own, has described and figured the peculiar spectra of several opals.

THE COEFFICIENT OF VISCOSITY OF AIR. By HERBERT TOM-
LINSON, B.A. Communicated, with the addition of two
Notes, by Professor G. G. STOKES, P.R.S.

[From the *Philosophical Transactions*, Vol. 177, Part II, 1886, pp. 767—799.
Received *Jan.* 6, read *Jan.* 14, 1886.]

Origin and Purpose of the Investigation.

THREE years ago I entered on a series of researches relating
to the internal friction of metals, little calculating, when I did so,
that the task which I had set myself would occupy almost the
whole of my spare time from that date to this. So, however, it
has been, and one of the many causes of delay has been the
necessity of making a re-determination of the coefficient of
viscosity of air; for the resistance of the air played far too
important a part in my investigations to permit of its being
either neglected or even roughly estimated. The coefficient of
viscosity of air may, according to Maxwell, be best defined by
considering a stratum of air between two parallel and horizontal
planes of indefinite extent, at a distance r from one another.
Suppose the upper plane to be set in motion in a horizontal
direction with a velocity of v centimetres per second, and to
continue in motion till the air in the different parts of the stratum
has taken up its final velocity, then the velocity of the air will
increase uniformly as we pass from the lower plane to the upper.
If the air in contact with the planes has the same velocity as the
planes themselves, then the velocity will increase v/r centimetres
per second for every centimetre we ascend. The friction between
any two contiguous strata of air will then be equal to that
between either surface and the air in contact with it. Suppose
that this friction is equal to a tangential force f on every square

centimetre, then $f = \mu v/r$, where μ is the coefficient of friction. If L, M, T represent the units of length, mass, and time, the dimensions of μ are $L^{-1}MT^{-1}$.

Several investigators have attempted to determine the co-efficient of viscosity of air, and the following table shows how very widely the results obtained differ among each other.

TABLE I.

Author*	Coefficient of viscosity of air in C.G.S. units	Temperature in degrees Centigrade
G. G. Stokes, from Baily's pendulum experiments	·000104	
Meyer, from Bessel's experiments.,.....	·000275	
Meyer, from Girault's experiments ...	·000384	
Meyer	·000360	18
Meyer (second Paper)+	·000333	8·3
,, ,, ...	·000323	21·5
,, ,, ...	·000366	34·4
Maxwell	·000200	18

Further, Maxwell finds the coefficient of viscosity of air to be independent of the pressure and to vary directly as the absolute temperature‡. The above author gives the following formula for finding μ, the coefficient of viscosity, at any temperature θ° C.:

$$\mu = ·0001878\,(1 + ·00365\theta).$$

Maxwell offers an explanation of the difference existing between his own results and those of Meyer, but states that "he has not found any means of explaining the difference between his own results and those of Professor Stokes." Professor Stokes

* For references see Maxwell's Bakerian Lecture, *Phil. Trans.* Vol. CLVI, 1866, p. 249.

† Meyer has more recently made other determinations of the coefficient, for which see the end of the Paper.

‡ This result does not seem to be confirmed by other experimenters. (See the end of the Paper.)

has, however, been good enough to inform me that, as at the time of making his deductions from Baily's experiments it was not known that the coefficient of viscosity of air was independent of the pressure, but, on the contrary, was assumed by him to vary directly as the pressure, the resistance offered by the residual air in Baily's partial vacua was underestimated, and, as a consequence, the deduced coefficient of viscosity was too small. It is to be hoped that Professor Stokes will at some future period apply the necessary corrections*, but as this has not yet been done, and as we have still no explanation of the discrepancies existing between the other values of μ given in Table I, I wished to make some independent observations on the viscosity of air for the purpose of ascertaining how far these would agree with those of Maxwell, in which I was inclined to place great confidence.

Maxwell employed the method of torsional vibrations of disks placed each between two parallel fixed disks at a small, but easily measurable distance, in which case, when the period of vibration is long, the mathematical difficulties of determining the motion of the air are greatly diminished. This method appeared to be a very good one, but, as I wished to make my determinations under conditions similar to those which held in my experiments on the internal friction of metals, I have employed the torsional vibrations of cylinders or spheres attached to a horizontal cylindrical bar and moving in a sufficiently unconfined space. The mathematical difficulties connected with the use of vibrating spheres are not so serious, but those in which cylinders are concerned are very considerable. They both, however, have been surmounted by Professor G. G. Stokes in his valuable paper "On the Effect of the Internal Friction of Fluids on the Motion of Pendulums†," and to this paper I am indebted for the mathematics essential to the purpose of the present inquiry.

Description of Apparatus and Mode of Experimenting.

A wire, *ab* (Fig. 1, p. 220), was suspended in the axis of an air-chamber, *W*, made of two concentric copper cylinders enclosing between them a layer of water. The outer diameter of the

[* See *ante*, Vol. III, p. 137.]
† *Camb. Phil. Soc. Trans.* Vol. IX, No. 10 (1850). [*Ante*, Vol. III, p. 1.]

air-chamber was 4 inches, the inner diameter 2 inches, and the length 4½ feet. Resting on the top of the air-chamber and wedged into it was a stout T-shaped piece of brass, C, to the lower extremity of which was clamped one end of the wire. The lower extremity of the wire was soldered or clamped at b to a vertical cylindrical copper bar bQ, which was in turn clamped at Q to the centre of a horizontal bar VV. The bar VV consisted of a piece of thin, hollow, drawn brass tubing, of which the length was 30·70 centims. and the diameter 1·420 centims. This bar was graduated into millimetres and carried two suspenders, S, S, which were clamped to it at equal distances from the centre (Fig. 3). The suspenders were each provided with an index such that their positions on the bar VV could be readily estimated to one-tenth of a millimetre. The mean diameter of the cylindrical portion, SK, of each suspender was 0·3366 centim., and the length of this portion 8·50 centims. To the ends, K, of the suspenders could be screwed (Fig. 3) hollow cylinders of stiff paper or metal, or spheres of wood; when the former were employed the suspenders were provided with disks, m, m, of the same diameter as the cylinders, and about 2 millims. in thickness. Two brass caps, D, D (Fig. 4), provided with screws about 8 centims. in length and 2 millims. in diameter, fit one into each end of the hollow bar VV, and can be easily removed from or placed in the latter.

To begin with, two cylinders or two spheres were screwed on to the ends of the suspenders (in the former case right up to the disks m, m), and the logarithmic decrement and the time of vibration determined from a very large number of vibrations. The cylinders or spheres were now unscrewed, and, the brass caps, D, D, having been temporarily removed for the purpose, two brass cylinders, h, h (Fig. 4), each of the same mass as either of the vertical cylinders or spheres which had just been removed, were, by means of companion-screws, cut along their axes, adjusted on to the screws attached to the caps D, D, and at such a distance from the latter as preliminary experiments had proved would give nearly the same vibration-period, when the caps should be replaced in the bar VV, as had existed before the vertical cylinders or spheres had been removed. The caps D, D, were now replaced in VV, and the logarithmic decrement, together with the time of vibration, was once more carefully determined. Observations such as these, when certain corrections presently to be mentioned had been applied, enabled one to calculate the effect of the resistance of the air on the vibrating vertical cylinders or spheres as far as the diminution of the amplitude of vibration was concerned.

The bar VV with its appendages was protected by a wooden box B of sufficient size to permit of vibrations, which, as regards the resistance of the air, were practically as free as in the open[*]. This box was provided with a window, EE, and two side-doors, lined with caoutchouc so as to fit air-tight; these side-doors were kept shut, except when it was necessary to make fresh adjustments. The torsional vibrations of the wire were observed by means of the usual mirror-and-scale arrangement, which is sufficiently shown in Fig. 1, where M is the light mirror reflecting an illuminated circle of light crossed by a vertical, fine, dark line on to a scale bent into an arc of a circle of 1 metre radius, and placed at a distance of 1 metre from the mirror.

My three years' experience of the internal friction of metals had taught me that this last is by no means constant unless the greatest care be taken to prevent slight fluctuations of temperature. The above-mentioned fact seems to have escaped the notice of Maxwell and Meyer, probably on account of the internal friction of the metal having a considerably less damping effect than the resistance of the air in their experiments. With me, however, especially in some cases, changes in the internal friction of the metal would have rendered it very difficult, nay, impossible, to attain the accuracy which I aimed at, and I deemed it advisable to protect the wire still further, as follows:—The top of the air-chamber W was well covered with baize, and surrounding W, and concentric with it, was a larger air-chamber X, made of tinned iron. This air-chamber was $11\frac{1}{2}$ inches in inner diameter, 15 inches in outer diameter, and 46 inches in height; the two concentric chambers of which it was composed enclosed between them a space 2 inches thick, stuffed with sawdust, whilst on the top of the chamber was placed a double cover A, also packed with sawdust. Passing through the outer air-chamber X, and through

---

[*] In Fig. 1 the cylinders appear to be closer to the sides of the box than they were in reality; the bar VV faced the window, but, for the sake of showing the arrangement of the cylinders better, it has been drawn facing the adjacent side of the box. The centres of the cylinders were at least six inches from the sides of the box.

the walls of W, were two metal tubes in which were placed two thermometers T_1, T_2, with their bulbs near the wire; these thermometers were made by indiarubber tubing to slide air-tight in the metal tubes. A section of the two chambers X and W passing through one of the thermometers is shown in plan in Fig. 2. The whole of this part of the apparatus rested on a stout wooden table, in which was pierced an aperture of a size just sufficient to allow the zinc tube Z, soldered to the air-chamber W, to pass through it and into the box beneath. A third thermometer T_3 served to give the temperature of the air in the box B, whilst the mean of the readings of T_1 and T_2 was used for the temperature of the wire. The thermometer T_3 was divided to one-tenth of a degree Centigrade, and had been tested at Kew; whilst the thermometers T_1 and T_2, which were graduated in degrees Centigrade, had been carefully compared by myself, degree by degree, with T_3.

The barometric pressure was registered by means of a delicate aneroid barometer, reading to $\frac{1}{100}$ of an inch, which has been in my possession for 15 years; this instrument I had recently compared with a standard mercury barometer*.

Before commencing the actual experiments on the viscosity of the air, it was found advisable to subject the wire to a preliminary training, in order not merely to diminish the internal friction of it, but also to make this last as constant as possible. In the first place, the wire was well annealed; this had the effect of reducing the internal friction of the hard-drawn metal to less than one-half of its previous amount†. In the next place, a load, equal to that of the cylinders or spheres to be used, having been suspended to VV, the wire was alternately heated to 100° C. and cooled again, this process being repeated for about a week, on each day of the week, until there was no further alteration of the internal

---

* In spite of the long period which has elapsed since this instrument was first made for me by the late Mr Becker, of Elliott Bros., the spring still shows a slight amount of permanent yielding, which during the last two years has altered the reading by ·015 inch.

† Either silver, platinum, or copper wires, well annealed, may be used with advantage. I should not recommend unannealed piano-steel wire as used by Maxwell; the last metal possesses, it is true, great elasticity, but the internal friction of silver, platinum, or copper can, by annealing, be made considerably less than that of the unannealed steel.

friction of the wire when cool. This treatment still further reduced very considerably the damping of the vibrations due to the wire. The manner in which the heating was effected will be shown in a future Paper, in which also will be recorded the results of experiments on the temporary effect of change of temperature on the torsional elasticity and internal friction of the metals used. When the wire had undergone this preliminary treatment, and all the arrangements were complete, the bar VV, with its appendant cylinders or spheres, as the case might be, was started by small impulses imparted by a worsted thread, until the arc of vibration, as reckoned from rest to rest, had reached about 400 divisions of the scale (about 10°, since 41·227 divisions represented 1°). After the arc of vibration from rest to rest had subsided to about 200 scale-divisions, the vibrator was again started, and this process was repeated until something like a thousand oscillations had been executed*. Finally the vibrator was re-started for the actual observations, through an arc of about 200 scale-divisions, and when about 50 oscillations had been executed after this last starting the readings were begun. Suppose that $a_1, b_1; a_2, b_2; a_3, b_3; a_4, b_4; a_5, b_5,$ and a_6 are eleven consecutive readings†, the ten corresponding arcs from rest to rest will be $a_1 + b_1, b_1 + a_2, a_2 + b_2, b_2 + a_3, a_3 + b_3, b_3 + b_4, a_4 + b_4, b_4 + a_5, a_5 + b_5, b_5 + a_6$. The means of $a_1 + b_1, b_5 + a_6; b_1 + a_2, a_5 + b_5; a_2 + b_2, b_4 + a_5; b_2 + a_3, a_4 + b_4,$ and of $a_3 + b_3, b_3 + a_4$ were written down, and if these agreed well with each other, which was almost invariably the case, the logarithmic decrement of the mean of the five means was taken. Now, say that n single vibrations have taken place between the end of this and the beginning of the next set of consecutive readings, the difference between the logarithms of the first and second total means will, when divided by $n + 10$, give the mean logarithmic decrement for a single vibration. The logarithmic decrement was found to be constant in each experiment within the limit of probable error; the deviations from uniformity were sometimes in one direction and sometimes in the opposite, and there was no evidence of any law of increase or diminution of the logarithmic decrement as the amplitudes decrease. In the

---

* The object of this treatment was to reduce the internal friction to its permanent condition, since long rest, or sometimes even a comparatively short rest, always raised sub-permanently the internal friction.

† This number was always taken.

intervals between one set and another of the readings, taken in the manner mentioned above, other readings were taken for the purpose of determining the vibration-period; the time of transit of the light across the centre of the scale, first in one direction and then in the opposite, was recorded for ten successive passages by means of a good watch provided with a seconds-hand, a similar series being recorded after every 200 vibrations. These last observations enabled the period of vibration to be determined with such exactness that we may completely disregard any error arising from want of precision in this respect. From time to time, at regular intervals, the readings of all three thermometers and of the aneroid barometer were taken, so that the mean pressure of the atmosphere, the temperature of the wire, and the temperature of the air in the box *B* could be calculated with the necessary accuracy. The greatest care was taken that the cylinders or spheres suspended from the horizontal bar *VV* should hang vertically; also that there should be no appreciable pendulous motion of the wire; if such motion existed it was checked by the hand before any of the readings were taken. Very great care was also taken in determining the moments of inertia of the vibrator in the various experiments, these being each obtained by several different methods*, which gave very concordant results. I shall have occasion in a future memoir to dwell on the various sources of error to which determinations of moments of inertia are liable; so it will suffice, perhaps, here to mention that this part of the work alone occupied my entire attention for nearly two weeks. The following five experiments, or rather sets of experiments, were made.

Experiment I.

The wire was of well-annealed copper, 97 centims. in length and 0·06272 centim. in diameter. Two cylinders, each having a mass of 70·19 grammes, were used. These cylinders were made of paper wrapped round a metal core a sufficient number of times to secure the requisite stiffness; the different layers of paper were pasted together, and when the whole was dry the metal core was

---

* The moments of inertia could be calculated with sufficient accuracy from the dimensions and mass of the vibrating system; they were, however, determined also indirectly by the two methods employed by Maxwell.

withdrawn; the outside of each of the cylinders was also coated with French polish to prevent the absorption of moisture. The mean diameter of each of the cylinders was measured by calipers reading to $\frac{1}{1000}$th of an inch, and estimated to $\frac{1}{10000}$th of an inch. In obtaining the value of the mean diameter of each cylinder, twenty measurements were made, ten at equal intervals along the whole length, and ten at the same intervals, but in a direction at right angles to the first. The measurements showed a very fair uniformity of diameter throughout the whole length, the mean being 1·0079 inches for one cylinder and 1·0108 inches for the other. In the calculations subsequently made it was assumed that the diameter of each cylinder was the mean of the two last given, *i.e.* was 1·0093 inches or 2·5636 centims. The lengths of the two cylinders were also very nearly the same, being 60·90 centims. and 60·85 centims. respectively; accordingly each cylinder was assumed to have a length of 60·875 centims. The ends of the cylinders consisted of wooden disks, into the centre of which was let a small brass disk provided with a screw, which was a companion to the screws at the ends of the suspenders, S, K, so that the cylinders could be screwed right up to the disks m, m (Fig. 3). The object of having the disks m, m, was to eliminate the effect of the friction of the air about the ends of the cylinders*, for Professor Stokes's mathematical investigations only apply strictly to cylinders of infinite length.

After the preliminary precautions previously mentioned had been taken the logarithmic decrement was determined from a great number of vibrations with the cylinders on; the cylinders were then each turned round their axes through a right angle, for the purpose of eliminating any error which might otherwise arise from the section of the cylinder being slightly elliptical instead of circular, and the logarithmic decrement was once more found. The cylinders were now unscrewed from the suspenders, and, the brass caps having been for the purpose removed from the hollow bar VV, the two brass cylinders h, h, were adjusted in the manner before mentioned, so that the vibration-period might remain very

---

* It would have been well to have had these disks much thicker. As it is, the disks would only imperfectly serve the purpose for which they were intended; the effect about the ends of the cylinders was, however, completely eliminated in Experiment IV. It would appear, moreover, from the results that with the long cylinders here used the effect mentioned above is neglectable.

nearly unaltered; the caps were then replaced. All the adjustments alluded to above were performed very carefully so as to avoid jarring the wire, for if this precaution be not taken the internal friction will be temporarily increased, and will not come back to its previous value until the wire has been vibrated for a considerable time. A period of more than an hour was now allowed to elapse, the wire during this time being kept more or less in a state of vibration, but not through a greater arc than that represented by 400 scale-divisions from rest to rest, when the logarithmic decrement was again determined. These processes were repeated during some eight or nine hours of each day through a period of three days, with the following *mean** results:—

Paper Cylinders on.

Temperature of the air in degrees Centigrade	Temperature of the wire	Barometric height in inches	Period of a single vibration in seconds	Logarithmic decrement for one vibration
12·02	12·43	29·872	6·8373	·0036476†

Paper Cylinders off.

12·25	12·31	29·817	6·8202	·0009103

The moment of inertia of the whole vibrator when the paper cylinders were on was 33773 in centimetre-gramme units.

Mathematical Formulæ necessary for the Investigation.

Before it can be shown how the results given above were made use of in finding the coefficient of viscosity of air, it will be necessary to point out how the requisite mathematical formulæ

---

* I have not thought it necessary to give here more than the mean values, as in a portion of a Paper on the internal friction of metals, which I hope shortly to be able to offer to the Royal Society, I have entered fully into the details of experiments very similar to these.

† Mean of eight trials, each of 200 vibrations, the numbers varying from ·0036300 to ·0036969.

can be obtained. I will first take the case of a cylinder vibrating horizontally under the influence of the torsional elasticity of a wire attached to its centre and hanging vertically.

Conceive the cylinder divided into elementary slices by planes perpendicular to its axis. Let r be the distance of any slice from the middle point, θ the angle between the actual and the mean positions of the axis, dF that part of the resistance experienced by the slice which varies as the first power of the velocity. Then, calculating the resistance as if the element belonged to an infinite cylinder moving with the same linear velocity, we have by Art. 31 of Prof. Stokes's paper,

$$dF = \frac{k'M'\pi}{\tau}\frac{d\xi}{dt},$$

where M' is the mass of fluid displaced by the slice, $d\xi/dt = r\,d\theta/dt$, τ is the vibration-period, and k' is a constant, provided the vibration-period, the diameter of the cylinder, and the nature of the fluid remain unchanged.

Let G be the moment of the resistance, l the whole length of the cylinder, a the radius of the cylinder, and ρ the density of the fluid; then

$$M' = \pi\rho a^2 dr, \quad \text{and} \quad G = \frac{\pi^2 k'\rho a^2 l^3}{12\tau}\frac{d\theta}{dt};$$

whence

$$\log_{10}\text{dec.} = \frac{\pi^2 k'\rho a^2 l^3}{24I}\log_{10}e \quad\dots\dots\dots\dots\dots(1),$$

I being the moment of inertia of the whole vibrator.

When we have a pair of cylinders of equal mass and dimensions suspended vertically from points equally distant from the axis of the wire, we can easily prove in a manner similar to the above that the logarithmic decrement due to the resistance of the air on the cylinders is expressed by the formula

$$\log_{10}\text{dec.} = \frac{\pi^2\rho\beta^2 l d^2 k'}{16I}\log_{10}e \quad\dots\dots\dots\dots(2)*.$$

---

* In this equation and in equation (4) the effect of the rotation of the cylinders about the axes is neglected. For the necessary correction see the end of the Paper. [The diameter of each cylinder is β, and the distance between their axes d. The exposition has been slightly condensed in the reprint.]

If the logarithmic decrement be known, we can determine k' from (2), and hence, by interpolation from the table given on p. 46* of Prof. Stokes's paper, $\mathfrak{m}$, this last being connected with μ, the coefficient of viscosity, by the formula

$$\mathfrak{m} = \frac{\beta}{4} \sqrt{\frac{\pi \rho}{\tau \mu}} \quad \dots\dots\dots\dots\dots\dots(3).$$

Since β, τ, and ρ are known, we can from (3) find μ.

In the case of two spheres of equal mass and dimensions there is no difficulty in obtaining the following formulæ from the data on p. 32* of Prof. Stokes's paper:—

$$\log_{10} \text{dec.} = \frac{\pi k' M' d^2}{4\,(I + 2kM')} \log_{10} e \quad \dots\dots\dots\dots(4),$$

where I is the moment of inertia of the whole vibrator, M' the mass of fluid displaced by each sphere, and k and k' are connected with μ by the equations

$$k = \frac{1}{2} + \frac{9}{4a} \sqrt{\frac{2\mu\tau}{\pi\rho}} \quad \dots\dots\dots\dots\dots\dots(5),$$

$$k' = \frac{9}{4a} \sqrt{\frac{2\mu\tau}{\pi\rho}} \left\{ 1 + \frac{1}{a} \sqrt{\frac{2\mu\tau}{\pi\rho}} \right\} \quad \dots\dots\dots\dots(6),$$

in which a is the radius of each sphere.

Application of the Mathematical Formulæ to the Results of Experiment I.

It will be seen that the logarithmic decrement with the paper cylinders on is ·0036476, whilst with the paper cylinders off it is ·0009103 ; therefore the logarithmic decrement due to the resistance of the air on the cylinders only is approximately ·0027373. I write 'approximately' because there are certain corrections to be applied which I will now proceed to describe. In the first place, the vibration-period, when the paper cylinders were on, though nearly the same as when the cylinders were off, was not quite the same. I therefore determined approximately the value of μ, without making this or the other small corrections

[* *Ante*, Vol. III, p. 52, and p. 34.]

to be mentioned presently, and used this value to obtain approximately the logarithmic decrement which would be due to the resistance of the air on the cylindrical bar VV and the cylindrical portions S, K, of the suspenders. The logarithmic decrement due to the resistance of the air on the other portions of the suspenders and on the disks m, m, was obtained by making independent observations, in which the bar was vibrated first with the suspenders on the bar, and then with the suspenders off, but with cylinders of equal mass placed inside the hollow bar VV, so that the time of vibration should remain unaltered.

Suppose that λ represents the logarithmic decrement due to the resistance of the air on the bar and the suspenders, and that t_1, t_2, are the vibration-periods with and without the paper cylinders respectively, then, with a sufficient degree of approximation, provided t_1 does not differ much from t_2, we have the amount to be added to the uncorrected logarithmic decrement equal to $\lambda\,(1 - t_2^{\frac{3}{2}}/t_1^{\frac{3}{2}})$.

Again, the temperature of the air and the pressure of the atmosphere were not quite the same with and without the paper cylinders. It can, however, be shown that for the small differences of temperature and pressure which we have here the logarithmic decrement will be independent of the temperature* and vary directly as the square root of the pressure; the amount to be added to the uncorrected logarithmic decrement, owing to the above causes, will therefore be $\lambda\,(1 - \sqrt{p_1/p_2})$, where p_1 and p_2 are the pressures with and without the paper cylinders respectively.

Further, when the cylinders were screwed on to the suspenders, about 4 mms. of the latter entered the former, so that the observed logarithmic decrement was less than it should be by an amount which would be nearly equal to the logarithmic decrement due to the resistance of the air on two vertical cylinders 4 mms. in length and 0·3366 cm. in diameter; this could be calculated to within

* The logarithmic decrement will not be independent of the temperature unless μ varies as the absolute temperature. If we adopt the results of recent experiments, the logarithmic decrement should approximately vary as $\sqrt{\dfrac{364+t}{273+t}}$, where t is the temperature in degrees Centigrade. The correction which this would entail I have neglected, as being inappreciable in these experiments.

a sufficient degree of approximation by using the approximate value of μ. The amount in this particular case was ·0000037.

Lastly, the temperature of the wire was not the same with and without the paper cylinders, but, as the effect of change of temperature had been determined previously, this difference could be allowed for.

No correction is required for any variation in the internal friction of the wire itself, arising from difference in the vibration-periods with and without the paper cylinders; for I had previously satisfied myself that the diminution of amplitude resulting from internal friction is nearly independent of the time of vibration.

Accordingly we have the following amounts to be added to the uncorrected logarithmic decrement:—

For difference of time of vibration with and without paper
    cylinders ....   + ·0000008

For difference of pressure of air ....   − ·0000005

For difference of temperature of the wire ....   − ·0000002

For portions of suspenders which enter the cylinders............   + ·0000037

                      Total...   + ·0000038

                      Corrected logarithmic decrement.........   ·0027411

In calculating ρ, the density of the air, I have assumed that the latter is half saturated with moisture, and that the mass of a cubic centimetre of dry air at 0° C., and under a pressure of 29·9217 inches of mercury, is ·0012930 gramme: thus, in the present instance,

$$\rho = \frac{29·872 - \frac{3}{8} \times ·206}{29·9217} \times \frac{273}{273 + 12·02} \times ·001293 = ·0012334.$$

The distance from each other of the axes of the two paper cylinders was 20·80 centims., and this distance was maintained in all the experiments which follow, except the last, where it was 20·78 centims.

From these and the previous data we can, by means of equation (2), get

$$k' = 1·6122;$$

and hence, by interpolation, we can obtain from the table on page 46 of Professor Stokes's paper

$$\mathfrak{m} = 1\cdot 1327.$$

Again, substituting this value of $\mathfrak{m}$ in equation (3), we obtain as the value of μ in C.G.S. units, at the temperature of 12°·02 C.,

$$\cdot 00018294.$$

Experiment II.

Two hollow cylinders, made of drawn brass tubing, and closed at both ends, were used instead of the paper cylinders. As measured by a gauge reading to $\frac{1}{100}$th of a millimetre, the mean diameter of one cylinder was 0·96446 centim., and of the other 0·96279 centim. These values were obtained by gauging each cylinder in ten different places, equidistant from each other, and in the calculations each cylinder was assumed to have a mean diameter of 0·96363 centim. The length of one cylinder was 60·92 centims., and of the other 60·85 centims., whilst the mean of these numbers, i.e. 60·885 centims., was assumed to be the length of each cylinder. The mass of each cylinder was 91·900 grammes, and when the cylinders were on the bar VV the moment of inertia of the whole vibrator, in centimetre-gramme units, was 36702. The value of the vibration-period was 7·0590 seconds. The temperature of the air was 14°·63 C., and the barometric height 29·707 inches. The uncorrected logarithmic decrement due to the resistance of the air on the cylinders was ·0012338, and the corrected logarithmic decrement was ·0012546. From these data was deduced a value of μ, at the temperature of 14°·63 C., of

$$\cdot 00017718.$$

Experiment III.

Everything else was arranged in the same manner as in Experiment I., but, instead of the annealed copper wire, an annealed silver wire, 97 centims. in length and 0·100863 centim. in diameter, was used. The paper cylinders employed in Experiment I. were used here, and when these cylinders were on the vibration-period was 3·0198 seconds. The temperature of the air was 11°·69 C., and the barometric height 30·207 inches. The uncorrected logarithmic decrement due to the resistance of the air against the cylinders was ·0016871, and the corrected logarithmic decrement ·0016905.   The value of μ at the temperature 11°·69 C. was calculated to be

$$·00018143.$$

Experiment IV.

Acting on the advice of Professor Stokes, I modified Experiment III. as follows. The logarithmic decrement was determined with the paper cylinders already used, and also with another pair of the same diameter, and made in the same manner, but having a length of 7·700 centims., the vibration-period being by the usual device maintained very nearly the same in both cases. The difference between the two logarithmic decrements, ·0024564 and ·0009933, will therefore equal the logarithmic decrement due to the resistance of the air on cylinders having each a length of (60·875 − 7·700) centims., i.e., 53·175 centims. When the longer paper cylinders were on the bar the vibration-period was 2·9994 seconds. The temperature of the air and the barometric height were 10°·64 C. and 30·057 inches respectively. The uncorrected logarithmic decrement was ·0014631, and the corrected logarithmic decrement ·0014638. The value of μ at the temperature of 10°·64 C., deduced from the above data, was

$$·00017955.$$

Experiment V.

The previous experiments had given such closely according values of μ that, though my investigations on the internal friction of metals only required that the formulæ for cylinders should give consistent results, I felt that it would be of interest to ascertain whether the use of spheres would be attended with the same satisfactory agreement. The main difficulty to be encountered with spheres is that the mass of a properly constructed spherical shell makes it rather unsuitable for experiments on the viscosity

13—2

of gases. After thinking over various plans of obtaining hollow spherical shells of sufficiently accurate make, and not feeling satisfied that I should be able to get, without much difficulty, what I wanted, I decided on using solid spheres made of fairly light wood. These spheres were specially turned for me, with instructions to make each as exactly as possible $2\frac{1}{2}$ inches in diameter. The turner executed his commission very fairly, for, on gauging each sphere at ten different places with calipers reading to $\frac{1}{1000}$th of an inch, I found that none of the readings differed from the mean by so much as ·3 per cent., and that the mean diameters of the two spheres were 2·5103 inches and 2·5007 inches respectively. In the calculations each sphere was reckoned as having a diameter of 2·5055 inches or 6·364 centims. The masses of the two spheres were not quite so equal as I could have wished, the apparent mass of one in air being 64·823 grammes, and of the other 63·761 grammes. No appreciable error will, however, be introduced by considering the apparent mass of each in air to be 64·292 grammes. The correction for the mass of air displaced by each sphere amounted to 0·168 gramme, so that in the calculations the mass of each sphere was taken as 64·460 grammes.

The spheres were attached to the suspenders S, K, in the same manner as the cylinders, but the disks were now dispensed with. The moment of inertia of the whole vibrator when the spheres were on was 30,927 in centimetre-gramme units, the vibration-period was 2·8791 seconds, and the temperature of the air and the barometric pressure were 9°·97 C. and 29·607 inches respectively. The uncorrected logarithmic decrement due to the friction of the air on the spheres was ·0003462, and the corrected logarithmic decrement was ·0003483.

In deducing the value of μ from the above data by the aid of equations (4), (5), and (6), I assumed, in finding $2kM'$, a value for μ equal to the mean of that got from the other experiments; this step is admissible, because $2kM'$ is very small compared with I^*. Having determined the value of k' by means of equation (4), I substituted it in equation (6), and thus obtained a quadratic equation for finding μ. The quadratic may, however,

---

* In fact, is quite neglectable in the case before us.

be converted into a simple equation by making use of the same value of μ as above in calculating the term $\dfrac{1}{a}\sqrt{\dfrac{2\mu\tau}{\pi\rho}}$, which was thus found to be 0·16085. The last number is not small compared with unity, and, had the final result proved to be as much as 10 per cent. greater or less than the mean of those got from the other experiments, the above conversion of the quadratic into the simple equation would not have been admissible. It will be seen eventually, however, that the conversion is legitimate, and the value of μ at a temperature of 9°·97 C. as determined from the simple equation is

$$·00019334.$$

Mathematical Formulæ required for the Effect of the Rotation of the Spheres or Cylinders about their own Axes*.

Professor G. G. Stokes has been good enough to furnish me with the following formulæ for the corrections not yet made for the effect of the rotation of the spheres or cylinders about their own axes.

Let λ_a be the logarithmic decrement due to the rotation, then for the spheres

$$\lambda_a = \frac{2\mu M'\tau}{I\rho}\,\frac{va+3+\dfrac{3}{va}+\dfrac{3}{2\,(va)^2}}{1+\dfrac{1}{va}+\dfrac{1}{2\,(va)^2}}\log_{10}e\ldots\ldots\ldots(7),$$

where I is the moment of inertia of the whole system, τ is the time of a vibration from rest to rest, M' is the mass of fluid displaced by each sphere, a is the radius of the sphere, and

$$\nu = \sqrt{\frac{\pi\rho}{2\mu\tau}}.$$

In the case of the cylinders, which were hollow, we have to take into account the effect of the air both inside and outside. For the air outside we may take

$$\lambda_a = \frac{2M'\mu\tau}{I\rho}\log_{10}eP \ \ldots\ldots\ldots\ldots\ldots(8),$$

* What follows was added Sept. 16, 1886. [See Prof. Stokes' second note, *infra*, p. 207.]

where P is the real part of the imaginary expression

$$ma\frac{1+\dfrac{3.5}{1(8ma)}+\dfrac{1.3.5.7}{1.2(8ma)^2}-\dfrac{1^2.3.5.7.9}{1.2.3(8ma)^3}+\dfrac{1^2.3^2.5.7.9.11}{1.2.3.4(8ma)^4}-\cdots}{1+\dfrac{1.3}{1(8ma)}-\dfrac{1^2.3.5}{1.2(8ma)^2}+\dfrac{1^2.3^2.5.7}{1.2.3(8ma)^3}-\cdots},$$

where

$$m=\sqrt{\frac{\pi\rho}{\mu\tau}}\,(\cos 45^\circ+\sqrt{-1}\,\sin 45^\circ).$$

On expanding P in descending powers of $f,=\sqrt{\dfrac{\pi\rho}{\mu\tau}}\,a$, we get

$$P=\frac{f}{\sqrt{2}}+1\cdot5+\frac{0\cdot375}{\sqrt{2}f}-\frac{0\cdot4922}{\sqrt{2}f^3}-\cdots\quad\ldots\ldots\ldots\ldots(9).$$

This series may be used with advantage in all the experiments relating to the cylinders to estimate approximately the effect of the air outside, but, unless the value of f is decidedly larger, the value of λ_a is best found from the formula

$$\lambda_a=\frac{4M'\mu\tau}{I\rho}\log_{10}e\cdot\frac{k^2+k'^2-1}{(k-1)^2+k'^2}\quad\ldots\ldots\ldots(10),$$

where k, k', are the quantities tabulated at p. 46 of Professor Stokes's Paper.

The corrections, as calculated from both formulæ, were found to agree satisfactorily.

For the air inside we may use, for such values of $\sqrt{\dfrac{\pi\rho}{\mu\tau}}\,a$ as we have here, the formula

$$\lambda_a=\frac{2M'\mu}{I\rho}(-Q)\log_{10}e\quad\ldots\ldots\ldots\ldots\ldots(11),$$

where M' is the mass of air inside, and Q is the real part of the imaginary expression

$$-\frac{\dfrac{2}{2.4}(ma)^2+\dfrac{4}{2.4^2.6}(ma)^4+\dfrac{6}{2.4^2.6^2.8}+\cdots}{1+\dfrac{1}{2.4}(ma)^2+\dfrac{1}{2.4^2.6}(ma)^4+\dfrac{1}{2.4^2.6^2.8}(ma)^6+\cdots},$$

which is of the form $\dfrac{A + \sqrt{-1}\,B}{C + \sqrt{-1}\,D}$, and of which the real part is

$$\frac{AC + BD}{C^2 + D^2}.$$

In the following table will be found the corrections necessary to be made for the rotations of the spheres and cylinders about their axes.

Cylinders.

Number of experiment	Log. dec. uncorrected for rotation about axes	Effect of air outside *	Effect of air inside *	Corrected log. dec.
I	·0027411	·0000313	·0000011	·0027087
II	·0012546	·0000030	·0000000	·0012516
III	·0016905	·0000173	·0000019	·0016713
IV	·0014638	·0000150	·0000017	·0014471

Spheres.

V	·0003483	·0000159	..	·0003324

There is still a further slight correction to make, inasmuch as the mercury of the barometer was not at 0° C. when the aneroid was compared with the mercury barometer, whereas the density of the air was calculated on the assumption of the mercury being at 0° C. The correction is very slight, but the closeness of agreement of the different experiments justifies us in making it. It will be sufficient for this purpose to multiply each value of μ as determined from the above table by $(1 + ·00018t)$, where t is the temperature at which the aneroid was compared with the mercury barometer. Applying all the corrections, the final results are as follows :—

---

* In making these corrections an approximate value of μ was used.

TABLE II.—Cylinders.

Number of experiment	Length in centims.	Diameter in centims.	Distance between the centres in centims.	Vibration-period in seconds	Tempera-ture in degrees Centigrade	Coefficient of viscosity of air in c.g.s. units
I	60·875	2·5636	20·80	6·8373	12·02	·00017900
II	60·885	0·9636	20·80	7·0590	14·63	·00017680
III	60·875	2·5636	20·80	3·0198	11·69	·00017767
IV	53·175	2·5636	20·80	2·9994	10·64	·00017581

Spheres.

V	..	6·364	20·78	2·8811	9·97	·00017626

Taking the means of the numbers in the sixth and seventh columns, we find that the value of μ at a temperature of 11°·79 C. is

$$·00017711.$$

The Effect of the Presence of Aqueous Vapour on the Viscosity of Air.

The above experiments extended over a period of some months, during which the air was in various conditions with respect to being saturated with aqueous vapour, so that for a rough ap-proximation we may assume that the mean value for μ just given will apply to air half saturated with vapour at a temperature of 12° C., and it would appear that the presence of the small quantity of aqueous vapour which this implies would not affect the value of μ to an extent equal to that of the probable error in experimenting. From the careful investigations of Mr Crookes[*] we learn that at a temperature of 15° C., and under pressures of from 760 to 350 millims., the presence of aqueous vapour has little or no influence on the logarithmic decrement. By the aid of Professor Stokes's note[†], I have estimated that at 15° C., and

---

[*] *Phil. Trans.*, Part II, 1881, p. 427.

[†] See p. 440 of the above Paper. [*Supra*, p. 100.]

under a pressure of 760 millims., the air when *saturated* with aqueous vapour would be *more* viscous than perfectly dry air[*] to the extent of only ·2 per cent. It is not until the air is under a less pressure than 350 millims. that the aqueous vapour begins to show appreciable effect, but when the rarefaction is great the moist air becomes considerably *less* viscous than dry air.

According to Maxwell[†] damp air over water at a temperature of 21°·11 C., and under a pressure of 101 millims., is *less* viscous than dry air by about $\frac{1}{80}$th part.

On the whole it would seem that the aqueous vapour in the air used in my experiments would hardly influence the value of μ to the extent of ·1 per cent.

The presence of carbon dioxide in the air would still less affect the result, as not only is the viscosity of carbon dioxide not very remote from that of air, but the amount of the gas present is also very minute.

Comparison of the Results of Recent Investigations of the Coefficient of Viscosity of Air.

In the beginning of this memoir I pointed out the very large discrepancies which existed between the results of different experimenters, but, since I entered on my task, not only have I acquired fresh information respecting what had already been done, but also quite recently fresh investigations have been made. Table III contains the required information.

TABLE III.

Authority	Coefficient of viscosity of air at 0° C.	Method
O. E. Meyer ‡.........	·0001875	Oscillating plates
„	·0001727	Transpiration
Puluj ‡	·0001798	„
Schneebeli §	·0001707	„
Obermayer §	·0001705	„

[*] Mr Crookes adopted great precautions to render the air dry.
[†] *Phil. Trans.*, Vol. CLVI, 1866.
[‡] *Phil. Mag.*, Vol. XXI, 1886, p. 220.
[§] *Archives Sci. Phys. Nat.*, Vol. XIV, 1885.

In order to reduce my own observations to 0° C., I have made use of the investigations of Professor Silas W. Holman on the effect of temperature on the viscosity of air*. According to the exceedingly careful and elaborate observations of this experimenter, the coefficient of viscosity of dry air is not proportional to the absolute temperature, but

$$\mu_t = \mu_0 (1 + 0\cdot002751t - 0\cdot00000034t^2) \quad\ldots\ldots\ldots(12),$$

where t is the temperature in degrees Centigrade, and μ_t, μ_0, are the coefficients of viscosity at $t°$ C. and 0° C. respectively.

My own observations were made with too small ranges of temperature to show the relation between the value of μ and the temperature, but the above formula expresses more nearly this relation as deduced from my experiments than [Maxwell's] formula

$$\mu_t = \mu_0.(1 + 0\cdot00366t).$$

Adopting, therefore, formula (12), we have the following equation for determining the value of μ at any temperature:—

$$\mu_t = \cdot00017155\,(1 + \cdot002751t - \cdot00000034t^2)\ \ldots\ldots(13).$$

The differences between the observed and calculated values of μ_t for the five different sets of experiments are given below:—

Experiment	Observed value of μ_t	Calculated value of μ_t	Difference
I	·00017900	·00017760	+ ·00000140
II	·00017680	·00017850	− ·00000170
III	·00017767	·00017704	+ ·00000063
IV	·00017581	·00017653	− ·00000072
V	·00017626	·00017622	+ ·00000004

The probable error is about ·2 per cent., and, considering the manner in which the five sets of experiments varied as regards their conditions, it would seem that, even when all allowance has been made for aqueous vapour, &c., the number ·00017155 must represent the value of μ_0 for *dry* air within at least ½ per cent. Now, this number agrees fairly with the values of μ_0 obtained by other observers with the transpiration method; it is, however, more than 9 per cent. less than that obtained by Meyer with oscillating plates, and by Maxwell. The mathematical difficulties

* *Phil. Mag.*, Vol. XXI, 1886.

attending Professor Meyer's method of oscillating plates have been already mentioned, but the method of Professor Maxwell does not seem open to these objections, and indeed appeared to me to be so good that I for some time attempted, though in vain, to account for the difference between Maxwell's result and my own. Professor G. G. Stokes has, however, kindly interested himself in the matter, and has shown in the accompanying note the possibility of Maxwell's result being too high. I may perhaps be allowed to add that, if we only take the first two of the five sets of Maxwell's experiments, in which two the distances of the fixed from the oscillating plates are so great as to render any error such as suggested by Professor Stokes very small, we obtain a value for the coefficient which is nearly identical with that obtained by myself*.

Addendum.

Note on the preceding Paper, by Professor G. G. Stokes, P.R.S.

[Received *January* 14, 1886.]

The consistency of Mr Tomlinson's different determinations of the coefficient of viscosity of air, notwithstanding the great variation in the circumstances of the experiments, and the consistency with one another of the numbers got by a different process by Maxwell, led me to endeavour to make out the real cause of the difference, and I think the main part, at any rate, of it can be explained by a very natural supposition.

The fact that Mr Tomlinson worked with air in its ordinary state, whereas Maxwell's air was dry, even if it tends in the right

[* The uncertainty arising from viscosity of the suspension is however greater in these sets.

Prof. A. H. Leahy has pointed out (Maxwell's *Scientific Papers, errata* to Vol. ii), that in finding the value of the moment *A* employed by Maxwell in his equations (23) and (24), a value of the radius of the disk is used different from the one previously recorded in the memoir, and that if this value is changed the result for the viscosity will be reduced so as to approximate to the values obtained by more recent experimenters.

In reply to an inquiry, Prof. Leahy states that the result of his examination of Maxwell's apparatus in 1886 was that the source of error suspected by Prof. Stokes (*infra*) was not sensible, and that the numerical error above mentioned, along with minor ones in calculating the effect of the edges of the disks, cleared up the discrepancy.]

direction, would evidently not go nearly far enough. But it occurred to me that the effect of any error of level in the movable disks employed by Maxwell must have been much greater than might at first sight appear. For suppose a very small error δ to exist, and suppose the fixed disks adjusted to be parallel to the movable ones in the position of equilibrium of the latter. Then the two systems must be, very nearly indeed, parallel throughout the motion, since the angle of oscillation of the movable disks to one side or other of the position of equilibrium is very small. If 2α be the whole amplitude, the greatest error of parallelism will be of the order $\delta\alpha$, and it would naturally appear at first sight that the effect of so small an error of parallelism must be insignificant for any such error of level as we can reasonably suppose to have existed. But a little consideration will show that this need not be the case when the distance between the fixed and movable disks is very small compared with the diameter of the latter. For suppose the disk to have been rotated through a small angle ρ round a vertical axis; the rotation ρ may be decomposed into a rotation $\rho \cos \delta$ round the axis of figure, and a rotation $\rho \sin \delta$ round a horizontal axis in the plane of the disk. As regards the former, the motion takes place as supposed in the investigation. But as regards the latter the disk oscillates about a horizontal axis in its own plane. Now, when the disks are very near one another this oscillation entails a squeezing thinner of the stratum of air opposite to one half of the disk, and a widening of the stratum opposite the other half, the two halves being alternately squeezed thinner and widened; and, since for such slow motions the air is practically incompressible, this transfer of air cannot be effected without a motion of the air along the surface of the disk far larger than what would be produced by an equal rotation about the axis of figure. Accordingly a very slight error of horizontality in the movable disk might produce a sensible error in the result, though an error of direction of similar amount in the orientation of the fixed disk would be quite insignificant in its influence on the final result.

This conclusion is fully borne out by the result of mathematical calculation founded on the equations of motion of a viscous uncompressed fluid. The calculation becomes very simple if we treat the distance between the disks as very small compared with the radius, neglect the special actions about the edge, and further neglect the inertia of the air, as we safely may, since it was small in Maxwell's experiments, especially those in which the disks were at a small distance apart, and therefore the influence of viscosity the greatest; or those again in which the air was rarefied.

Let the plane of a movable disk in its position of equilibrium be taken for the plane of xy, the axis of figure for the axis of z, and the intersection of a horizontal plane with the plane of the disk for the axis of y; and let the opposed fixed plane be parallel to the plane of xy, and at a distance h from it. Let a be the radius of the disk.

First, as regards motion round the axis of figure. Let ω be the angular velocity of the disk. Then, according to the simplifications adopted, the motion of the fluid will be a motion of simple shearing, such that the velocity at a point whose semi-polar coordinates are r, θ, z, will be $\omega r (h - z)/h$ in a direction perpendicular to the radius vector. It will suffice to write down the moment of the force which this calls into play, which is

$$\frac{\pi \mu a^4 \omega}{2h} \quad \dots\dots\dots\dots\dots\dots\dots(A).$$

Next, for motion round the axis of y. Let ω' be the angular velocity; u, v, w, the components of the velocity; U, V, the mean values of u, v, from 0 to h. Consider the prism of fluid standing on the base $dx\,dy$, and extending between the planes. As the volume of the prism is diminished at the base by $\omega' x\,dx\,dy\,dt$ in the time dt, the excess of the volume of the fluid which flows out across the face $h\,dy$, whose abscissa is $x + dx$, over that which flows in across the face $h\,dy$, whose abscissa is x, plus the similar difference for the pair of faces $h\,dx$, must equal $\omega' x\,dx\,dy\,dt$. This leads to the equation

$$h \frac{dU}{dx} + h \frac{dV}{dy} = \omega' x \dots\dots\dots\dots\dots(1).$$

But, for motion between two close parallel planes, the velocity parallel to the plane, and its components in two fixed directions in that plane, vary as $z (h - z)$, and therefore

$$u = \frac{6z (h - z)}{h^2} U, \qquad v = \frac{6z (h - z)}{h^2} V \dots\dots(2).$$

The first equation of motion is

$$\frac{dp}{dx} = \mu \left(\frac{d^2u}{dx^2} + \frac{d^2u}{dy^2} + \frac{d^2u}{dz^2} \right) \quad \dots\dots\dots\dots(3).$$

Now, on account of the smallness of h, the space-variations of the components u, v, of the velocity are much greater for z than for x or y. Hence in (3), and the corresponding equation for v, the first two terms in the right-hand members may be omitted, giving, by (2),

$$\frac{dp}{dx} = -\frac{12\mu}{h^2} U, \qquad \frac{dp}{dy} = -\frac{12\mu}{h^2} V,$$

and then, from (1),

$$\frac{d^2p}{dx^2} + \frac{d^2p}{dy^2} = -\frac{12\mu\omega'}{h^3} x,$$

or, in polar coordinates,

$$\frac{d^2p}{dr^2} + \frac{1}{r}\frac{dp}{dr} + \frac{1}{r^2}\frac{d^2p}{d\theta^2} = -\frac{12\mu\omega'}{h^3} r \cos\theta \dots\dots\dots(4);$$

and if we take, as we may, p to mean the excess of pressure over the pressure in equilibrium, we have the conditions that p shall vanish when $r = a$, and that p shall not become infinite at the centre.

The equation (4) and the conditions at the mouth and centre may be satisfied by taking

$$p = f(r) \cos\theta,$$

which gives, from (4),

$$f''(r) + \frac{1}{r} f'(r) - \frac{1}{r^2} f(r) = -\frac{12\mu\omega'}{h^3} r.$$

The integral of this equation is

$$f(r) = -\frac{3\mu\omega'}{2h^3} r^3 + Ar + \frac{B}{r},$$

where A, B, are arbitrary constants. The conditions at the centre and mouth give

$$B = 0, \qquad A = \frac{3\mu\omega'a^2}{2h^3},$$

whence

$$p = \frac{3\mu\omega'}{2h^3}(a^2r - r^3) \cos\theta.$$

The moment of this pressure about the axis of y is

$$\iint pr \cos \theta . r \, dr \, d\theta,$$

or $\qquad \dfrac{\pi\mu\omega' a^6}{8h^3}$ ....(B).

The moments (B) and (A) are as $a^2\omega'$ to $4h^2\omega$, and the works of these moments in the time dt are as $a^2\omega'^2$ to $4h^2\omega^2$. If this ratio be denoted by n to e, and ω, ω', are the components of an angular velocity round an axis in the plane of xz, inclined at an angle δ to the axis of z,

$$\tan^2 \delta = \frac{4h^2}{a^2}\, n.$$

In Maxwell's experiments a was 5·28 inches, and when the fixed and movable disks were closest h was 0·18475. If we suppose the whole loss of energy 8 per cent. greater than that due to rotation round the axis of figure, to which it was deemed to be due, we have $n = 0·08$, giving $\delta = 1° 8'$. Now, no special adjustment was made to secure the strict horizontality of the movable disks, or at least none is mentioned; the final adjustment is stated to have been that of the fixed disks, which were presumably adjusted to be parallel to the movable ones, and at the desired distance. Hence such small errors of level as that just mentioned may very well have occurred.

SECOND NOTE.—*On the Effect of the Rotations of the Cylinders or Spheres round their own Axes in increasing the Logarithmic Decrement of the Arc of Vibration.*

[Received *October* 22, 1886.]

In Art. 9 of my paper on Pendulums* I pointed out that in the case of a ball pendulum the resistance due to the rotation of the sphere round its axis need not be regarded, on account of the large ratio which the distance of the centre from the axis of suspension bears to the radius of the sphere. In Mr Tomlinson's experiments the corresponding ratio is not near so great, and its

---

[* *Camb. Phil. Trans.*, Vol. IX, 1850: *ante*, Vol. III, p. 22.]

squared reciprocal is not small enough to allow us to neglect the correction altogether, especially in the case of the spheres, the radius of which is much larger than that of the cylinders. In both cases the problem admits of solution.

In both cases the motion of the suspended body may be regarded as compounded of a motion of translation, in which the centre oscillates in an arc of a circle, and a motion of rotation about its axis of figure, supposed fixed; and, the motion being small, the effects of the two may be considered separately. It is the latter with which we have at present to deal. As regards the motion of translation, the spheres or cylinders were sufficiently far apart to allow us to regard each as out of the influence of the other, and accordingly as oscillating in an infinite mass of fluid; and this is still more nearly true as regards the motion of rotation. The problem, then, is reduced to this: a sphere or cylinder performs small oscillations of rotation about its axis of figure, which is vertical and regarded as fixed, in an infinite mass of viscous fluid; it is required to determine the motion, and thereby to find the effect of the fluid in damping the motion of the system of which the suspended body forms a part.

In the case of the sphere the problem of determining the motion of the fluid is identical with that solved by Professor von Helmholtz in a paper published in the 40th volume of the 'Sitzungsberichte' of the Vienna Academy, p. 607, and reprinted in the first volume of his collected works, p. 172, with the exception that the arbitrary constants which occur in the integral of the fundamental ordinary differential equation are differently determined, since the condition that the motion shall not become infinite at the centre is replaced by the condition that it shall not be infinite at an infinite distance.

In the present case the motion is necessarily symmetrical about the axis, so that it is alike all round any circle that has the axis for its axis; it is, moreover, tangential to the circle. Let the fluid be referred to polar coordinates r, θ, ϖ; r being the distance from the centre, θ the inclination of the radius vector to the axis, and ϖ as usual. Then, taking ρ, μ, to denote the density and coefficient of viscosity, and observing that $v = q \cos \varpi$, where q is the velocity, we easily get from the second equation of motion, by putting, as we may, $\varpi = 0$ after differentiation,

$$\frac{d^2q}{dr^2} + \frac{2}{r}\frac{dq}{dr} + \frac{1}{r^2 \sin\theta}\frac{d}{d\theta}\left(\sin\theta\frac{dq}{d\theta}\right) - \frac{q}{r^2 \sin^2\theta} - \frac{\rho}{\mu}\frac{dq}{dt} = 0 \;...(1);$$

and we have the condition at the surface

$$q = \omega a \sin\theta \quad \text{when} \quad r = a \;.................(2),$$

where ω is the angular velocity of the sphere, and a its radius.

The motion with which we have to deal is periodic, subject to a secular diminution. The latter being actually very slow, it will suffice, in calculating the force of the air on the sphere, to take the motion as periodic, and expressed, so far as the time is concerned, by the sine or cosine of nt. It will be more convenient, however, to use the symbolical expression e^{int}, where $i = \sqrt{(-1)}$. The general equation (1) and the equation of condition (2) can both be satisfied by taking q to be expressed, so far as θ is concerned, by $\sin\theta$. Assuming, then,

$$q = e^{int}\sin\theta f(r) \;.......................(3),$$

and writing

$$\frac{i\rho n}{\mu} = \frac{i\pi\rho}{\mu\tau} = m^2 \;.......................(4),$$

we have

$$f''(r) + \frac{2}{r}f'(r) - \frac{2}{r^2}f(r) - m^2 f(r) = 0...........(5).$$

Taking $+m$ for that root of the imaginary m^2 which has its real part positive, we have for the integral of (5), subject to the condition of not becoming infinite at an infinite distance

$$f(r) = A e^{-mr}\left(\frac{1}{r} + \frac{1}{mr^2}\right) \;.................(6).$$

Omitting the pressure in equilibrium, we shall have for the force of the fluid on an element of the sphere a tangential pressure (say T, referred to a unit of surface) acting perpendicularly to the plane passing through the axis and the element, the expression for which, reckoned positive when it acts in the direction of ϖ positive, is

$$T = \mu\left(\frac{dq}{dr} - \frac{q}{r}\right)_{r=a} = \mu e^{int}\sin\theta\left(f'(a) - \frac{f(a)}{a}\right);$$

and the moment of the force taken all over the sphere is

$$\int_0^\pi T . 2\pi a^2 \sin\theta . a \sin\theta \, d\theta = \tfrac{8}{3}\pi\mu a^3 e^{int}\left(f'(a) - \frac{f(a)}{a}\right)$$

$$= 2M'\mu' e^{int}\left(f'(a) - \frac{f(a)}{a}\right)$$

if $\mu' = \mu/\rho$, and M' is the mass of the fluid displaced by the sphere.

Now we have, by (2), (3),

$$\omega a = e^{int} f(a),$$

whence the expression for the moment becomes

$$2M'\mu'\left(\frac{af'(a)}{f(a)} - 1\right)\omega.$$

To get the whole moment, the above must be doubled, as there are two spheres. If θ be the angular distance of the vibrating system from the position of equilibrium, we may write $d\theta/dt$ for ω; and if the mixed imaginary within parentheses, with sign changed, be denoted by $P + iQ$, the real part, P, will be that which affects the arc of vibration, the imaginary part falling upon the time, which we do not want. The Napierian logarithmic decrement in one vibration will be got by dividing half the real part of the expression for the moment of the forces by the moment of inertia, or, say, MK^2. It will therefore be $2M'\mu'P/MK^2$.

Now we get, from (6),

$$1 - \frac{af'(a)}{f(a)} = \frac{ma + 3 + \dfrac{3}{ma}}{1 + \dfrac{1}{ma}};$$

and, taking the real part of this, we get finally, after reduction,

$$\text{Nap. log. dec.} = \frac{2\mu'M'}{MK^2}\,\frac{\nu a + 3 + \dfrac{3}{\nu a} + \dfrac{3}{2\nu^2 a^2}}{1 + \dfrac{1}{\nu a} + \dfrac{1}{2\nu^2 a^2}}\,\tau \quad \ldots\ldots(7),$$

where

$$\nu = \sqrt{\left(\frac{\pi}{2\mu'\tau}\right)}.$$

In the case of the cylinder the motion is in two dimensions, and is most conveniently referred to polar coordinates r, θ, the origin being in the axis. The radius of the cylinder will be denoted by a, the outer or inner radius according as we are dealing with the air outside or inside.

The mode of proceeding is precisely analogous to that in the case of the sphere, and, q being the whole velocity, we have

$$q = e^{int} f_1(r) \dots\dots\dots\dots\dots(8),$$

where

$$f_1''(r) + \frac{1}{r} f_1'(r) - \frac{1}{r^2} f_1(r) - m^2 f_1(r) = 0 \dots\dots(9);$$

and the condition at the surface gives

$$e^{int} f_1(a) = \omega a \dots\dots\dots\dots(10).$$

If T be the tangential pressure on the cylinder,

$$T = \pm \mu \left(\frac{dq}{dr} - \frac{q}{r} \right)_{r=a} \dots\dots\dots\dots(11),$$

the sign being $+$ or $-$ according as we are dealing with the air outside or inside. The moment of this pressure on a length, l, of the cylinder is

$$\pm 2\pi \mu a^2 l e^{int} \left(f_1'(a) - \frac{1}{a} f_1(a) \right) = \pm 2M' \mu' \left(\frac{a f_1'(a)}{f_1(a)} - 1 \right) \omega \dots(12).$$

The equation (9) cannot be integrated in finite terms. Nevertheless, in the case of the air outside, the expression (12) for the moment may be obtained in a finite form in terms of two functions, k, k' which I had occasion to tabulate for the purpose of finding the resistance of a viscous fluid to a pendulum of the form of a cylindrical rod.

Putting, as in my former paper,

$$f_1(r) = f_0'(r) \dots\dots\dots\dots(13),$$

(f_1, f_0, are the functions there denoted by F_2, F_3), we have

$$f_0''(r) + \frac{1}{r} f_0'(r) - m^2 f_0(r) = 0 \dots\dots\dots\dots(14).$$

Now in both problems (that of my former paper and that of the present note) the function $f_0(r)$ satisfies the same differential equation (14) and the same condition of vanishing at infinity. Hence the function $f_0(r)$ is the same in the two cases, save as to

14—2

the value of the arbitrary constant, which is a factor of the whole, and which disappears from the expression (12) as well as from those of k and k'.

The definition of k and k' is given by equation (99) of my former paper, viz.,

$$1 - \frac{4f_0'(a)}{m^2 a f_0(a)} = k - ik' \quad \text{..................(15).}$$

Now by (13), (14), and (15),

$$1 - \frac{af_1'(a)}{f_1(a)} = 1 - \frac{af_0''(a)}{f_0'(a)} = 2 - \frac{m^2 a f_0(a)}{f_0'(a)}$$

$$= 2\frac{k+1-ik'}{k-1-ik'} = 2\frac{k^2-1+k'^2+2ik'}{(k-1)^2+k'^2} \, ;$$

whence we get, as before, for the part of the logarithmic decrement due to the external air, in consequence of the rotations of the two cylinders round their own axes, M' denoting the mass of air which would be displaced by one if solid and of radius a,

$$\text{Nap. log. dec.} = \frac{4M'\mu'\tau}{MK^2} \cdot \frac{k^2-1+k'^2}{(k-1)^2+k'^2} \quad \text{.........(16).}$$

In the table given in Art. 37 of my paper, $\mathfrak{m}$ denotes half the modulus of ma, or

$$\frac{a}{2}\sqrt{\frac{\pi}{\mu'\tau}}.$$

This table is not available for calculating the effect of the internal air, for which we must have recourse to the differential equation (9). The integral of this equation, expressed in ascending series, subject to the condition of not becoming infinite at the origin, is

$$f_1(r) = A\left\{r + \frac{m^2 r^3}{2\cdot 4} + \frac{m^4 r^5}{2\cdot 4^2\cdot 6} + \frac{m^6 r^7}{2\cdot 4^2\cdot 6^2\cdot 7} + \cdots\right\},$$

which gives

$$\frac{af_1'(a)}{f_1(a)} - 1 = \frac{\dfrac{m^2 a^2}{4} + \dfrac{m^4 a^4}{2\cdot 4\cdot 6} + \dfrac{m^6 a^6}{2\cdot 4^2\cdot 6\cdot 8} + \cdots}{1 + \dfrac{m^2 a^2}{2\cdot 4} + \dfrac{m^4 a^4}{2\cdot 4^2\cdot 6} + \dfrac{m^6 a^6}{2\cdot 4^2\cdot 6^2\cdot 8} + \cdots} \quad \text{...(17).}$$

Let the numerator of this fraction be denoted by $E + iF$, and the denominator by $G + iH$, where E, F, G, H are real; then the real part will be $EG + FH$ divided by $G^2 + H^2$, and we shall have for the correction due to the internal air

$$\text{Nap. log. dec.} = \frac{2M'\mu'}{MK^2} \cdot \frac{EG+FH}{G^2+H^2} \quad \dots\dots\dots(18).$$

When the modulus of ma is small, it is rather more convenient to expand (17) according to ascending powers of ma. This may be done by actual division, or more conveniently by assuming a series with indeterminate coefficients, and using the non-linear differential equation of the first order in z obtained from (9) by putting $f_1'(r) = zf_1(r)$. Carried as far as to a^{12}, the development is

$$\frac{m^2a^2}{4} - \frac{m^4a^4}{96} + \frac{m^6a^6}{1536} - \frac{m^8a^8}{23040} + \frac{13m^{10}a^{10}}{4423680} - \frac{11m^{12}a^{12}}{55050240};$$

and, denoting the modulus of ma by f, and taking the real part, we have

$$\text{Nap. log. dec.} = \frac{2M'\mu'}{MK^2}\left\{\frac{f^4}{96} - \frac{f^8}{23040} + \frac{11f^{12}}{55050240} - \dots\right\}\dots(19).$$

This series must not be used when f is at all large, as the convergence is too slow, and, as appears by a theorem due to Cauchy, it becomes actually divergent when $f = 3\cdot340$* nearly, whereas the series in (17) are always convergent, and when f has the above value converge rapidly.

When f is decidedly large the series in (17), though ultimately convergent, begin by diverging, so that the calculation is troublesome, and moreover my table giving k and k' is not carried beyond $f = 8$, as the calculation by a different method then becomes very easy. In this case we should employ the integral of (9), which is of the form e^{-mr} or e^{mr} multiplied by a descending series. The former exponential only will come in when we are treating of the external air, and the latter only when of the internal.

* The square root of the smallest real root of the equation

$$1 - \frac{x}{2.4} + \frac{x^2}{2.4^2.6} - \dots = 0.$$

The series would have become divergent still earlier if the equation just written had had an imaginary root with a modulus smaller than $3\cdot340\dots^2$.

For the external air the integral is of the form

$$f_1(r) = Be^{-mr} r^{-\frac{1}{2}} \left\{ 1 + \frac{1.3}{1.(8mr)} - \frac{1^2.3.5}{1.2\,(8mr)^2} \right.$$

$$\left. + \frac{1^2.3^2.5.7}{1.2.3\,(8mr)^3} - \ldots \right\} \ldots\ldots\ldots\ldots\ldots\ldots(20),$$

the signs being alternately + and −, and the new factors in the numerator being two less and two greater than the last factor in the term before. We get from (12), (20), and the expression for the logarithmic decrement in terms of T and the moment of inertia,

Nap. log. dec. $= \dfrac{2M'\mu'\tau}{MK^2} \times$ real part of

$$ma \frac{1 + \dfrac{3.5}{1.8ma} + \dfrac{1.3.5.7}{1.2\,(8ma)^2} - \dfrac{1^2.3.5.7.9}{1.2.3\,(8ma)^3} + \ldots}{1 + \dfrac{1.3}{1.8ma} - \dfrac{1^2.3.5}{1.2\,(8ma)^2} + \dfrac{1^2.3^2.5.7}{1.2.3\,(8ma)^3} - \ldots}\ldots(21).$$

Instead of the latter part of (21), in which, however, the law of either series is manifest, we may use its development according to descending powers of a, which is

$$ma + \frac{3}{2} + \frac{3}{8ma} - \frac{24}{(8ma)^2} + \frac{252}{(8ma)^3} - \frac{3456}{(8ma)^4}$$

$$+ \frac{60768}{(8ma)^5} - \frac{1327104}{(8ma)^6} + \ldots \ldots\ldots\ldots\ldots(22).$$

The expression for the correction for the internal air will be got from the above by changing the sign of ma and of the whole, or, in other words, by changing the signs of the 2nd, 4th, 6th ... terms in the series in (21) or (22). It will be remembered that ma is $f(\cos 45° + i \sin 45°)$.

APPENDIX (*received November 15th*, 1886).

In the previous experiments the *main* loss of energy arising from the friction of the air may be characterised as being due to the fact that the air is *pushed**. A small portion, however, of the loss is occasioned by the rotation of the cylinders or spheres about their own axes, and in this case the air may be said to be *dragged*. Professor G. G. Stokes has, in the preceding note, deduced formulæ by means of which this last portion of the whole loss of energy can be calculated, and it seemed of interest to determine whether the coefficient of viscosity of air would prove to be the same as before, when the air was *entirely dragged*. This will occur when only one sphere or one cylinder is used, whose axis is made to coincide with the axis of rotation. Accordingly I followed out a suggestion of Professor Stokes in the manner detailed in the following experiments.

Experiment VI.

A paper cylinder was made by wrapping drawing-paper several times round a metal cylinder, which had been turned true throughout its whole length, the different layers being pasted together. When dry, the paper cylinder was removed from its metal core, and its external diameter very carefully gauged by calipers reading to $\frac{1}{1000}$th of an inch at six different places equidistant from each other. It was then gauged at the same distances from the ends, but in directions at right angles to the first. The following were the two sets of gauges:—

	Set I Diameter in inches	Set II Diameter in inches
	6·026	6·073
	6·083	6·010
	6·106	6·051
	6·106	6·020
	6·090	6·030
	6·010	6·006
Mean	6·0701	6·0323

[* A balancing pair of cylinders or spheres were suspended from the ends of a vibrating horizontal bar, *supra*, p. 178.]

The circumference of the cylinder was next measured by a steel tape at five different equidistant places:

Circumference in centims.

48·64
48·66
48·60
48·56
48·35

Mean   48·562

Allowing for the thickness of the steel tape, the circumference is 48·485.

From the measurements made with the calipers and tape, the mean diameter of the cylinder was 15·370 and 15·433 centims. respectively, and the total mean 15·4015 centims.

It will be observed that the external diameter is nearly, but not quite, uniform throughout; this no doubt arises from the fact that the paper was not quite uniform in thickness. Inside, as far as could be judged by inserting a straight-edge, the bore of the cylinder was perfectly uniform throughout.

The inside diameter was determined by the calipers at the top and bottom, at eight different places in all. It was also determined by gauging the thickness of the walls of the cylinder at the top and bottom by means of a wire gauge, and subtracting twice the thickness from the external diameter as measured by the tape. The internal diameter, measured in the two different ways mentioned above, was exactly the same for both, namely, 14·872 centims. The mean of the internal and external diameters is 15·1395 centims., and the mean radius 7·5698 centims.

The length of the paper cylinder was 60·80 centims., and the mass, allowing for the air displaced, was 543·6 grammes.

The wire was inserted into a hole bored in the centre of one end of a vertical brass rod 2 millims. thick and 15 centims. long, and there soldered: the other extremity of the rod was soldered into the centre of a horizontal, hollow, brass tube, of length 17·85 centims., of diameter 1·25 centims., and of mass 29·20 grammes.

From the hollow brass bar the paper cylinder was suspended; two holes, whose centres were 2½ centims. from the top, being cut in the walls of the paper cylinder for this purpose.

Great care was taken in arranging the cylinder, so that the axis of rotation might coincide in direction as accurately as possible with the axis of the suspended system. As the paper cylinder did not quite hang truly, it was made to do so by placing small strips of tinfoil, as riders, on the top of the cylinder, and these strips were carefully padded down by hand to the walls of the cylinder. The usual previous precautions having been taken, the logarithmic decrement was determined from seven sets of observations, each involving 100 vibrations, as follows :—

Number of observation	Logarithmic decrement
1	·0026307
2	·0025856
3	·0025837
4	·0025810
5	·0025700
6	·0025849
7	·0025550

The observations were consecutive, and the mean of them is ·0025844.

The paper cylinder was now removed, and in its place was substituted a much shorter cylinder, made partly of paper and partly of tinfoil, and having nearly the same mass and mean radius. The dimensions of this cylinder were as carefully measured as those of the longer cylinder, with both steel tape and calipers. The mean of the inside and outside radius was 7·5132 centims., and its real length was 12·80 centims. Since the radius is, however, not quite the same as that of the longer cylinder, we must assume its length to be $12·80 \times \left(\dfrac{7·5132}{7·5698}\right)^3$ centims., or 12·52 centims. if we are to use it for the purpose mentioned below.

The same pieces of tinfoil as had been used with the long cylinder were used here, and for the same purpose. The logarithmic decrement was then determined by six sets of experiments, each involving three times the number of vibrations employed with the long cylinder.

Number of observation	Logarithmic decrement
1	·0009162
2	·0009015
3	·0008743
4	·0008871
5	·0009019
6	·0008993

These, like the others, are consecutive observations, and the mean of them is ·0008967.

Applying the corrections, mentioned in the paper, for small differences in the vibration-periods, temperature, etc., when the two cylinders were used, we have for the logarithmic decrement due to a cylinder $60·80 - 12·32$ centims. or $48·28$ centims. in length the value

$$·0017029.$$

It follows, from Professor Stokes's formulæ, that the logarithmic decrement arising from the friction of the air against the inner and outer walls taken together will be

$$\frac{M}{I}\frac{\mu\tau}{\rho} \log_{10} e \, (\sqrt{2}f + \sqrt{2} \times 0·375f^{-1} - \sqrt{2} \times 0·4922f^{-3} + \text{etc.}),$$

$$f \text{ being equal to } \sqrt{\frac{\pi\rho}{\mu\tau}} \cdot a,$$

where a is the mean radius of the cylinder, τ the vibration-period, μ the coefficient of viscosity, ρ the density of the air, M the mass of air which would be contained in a cylinder of the same length, and having an internal radius equal to a, and I the moment of inertia.

The values of I and τ were 36966 centimetre-grammes and 3·6038 seconds respectively. The corrected height of the barometer was 29·354 inches, and the temperature 12°·225 C. The value of ρ was calculated, as usual, on the supposition that the air is half saturated with moisture.

The terms $0·375f^{-1}$ and $0·4922f^{-3}$ are so small that we may calculate them by using an approximate value of μ, and the series converges so rapidly that it is quite unnecessary to include any more terms in it*.

The value of μ, determined from the data given above, was found to be

$$·00017580.$$

---

* Indeed, the third term might have been dispensed with in this case, but not in the next experiment.

Experiment VII.

The copper wire used in the last experiment was about $4\frac{1}{2}$ feet in length and 0·1 centim. in diameter. This was now changed for one of the same length, but of 0·063 centim. diameter, so that the vibration-period became 8·930 seconds. The rest of the arrangements were the same as in Experiment VI. The corrected logarithmic decrement was ·0027040, and the value of μ deduced as above was found to be ·00017902 at a temperature of 13°·100 C.

The mean of the two last experiments is ·00017741 at a temperature of 12°·663 C. This result agrees so well with the mean of those deduced from the previous experiments that it is unnecessary to make any alteration in the formula already given for finding the viscosity at any temperature.

I have entered more into the details of these last experiments, as I think the present method can be more advantageously employed than any of the others. Indeed, by spending sufficient time over the experiments, whereby the errors likely to arise from the somewhat unstable nature of the internal friction of the metal may be more perfectly eliminated, it seems likely that very considerable accuracy can be attained by it.

[*Note* added Dec. 8th, 1886:—A much greater number of observations were afterwards made with the same cylinders and wires, and resulted as follows:—With the wire used in Experiment VI. the value of μ obtained was ·00017708 at a temperature of 12°·225 C., and with the finer wire of Experiment VII. the value was ·00017783 at a temperature of 13°·075 C. The mean of these values is ·00017746 at 12°·650 C., as compared with ·00017711 at 11°·79 C., the mean of the other five sets of experiments. If we allow for the difference of temperature by using the previously given formula*, the agreement between these two means is perfect.]

In conclusion, my warmest thanks are due to Professor Stokes for his valuable suggestions and advice throughout the investigation. To myself the experimental verification of Professor Stokes's formulæ has been a source of great pleasure.

[* *Supra*, p. 202.]

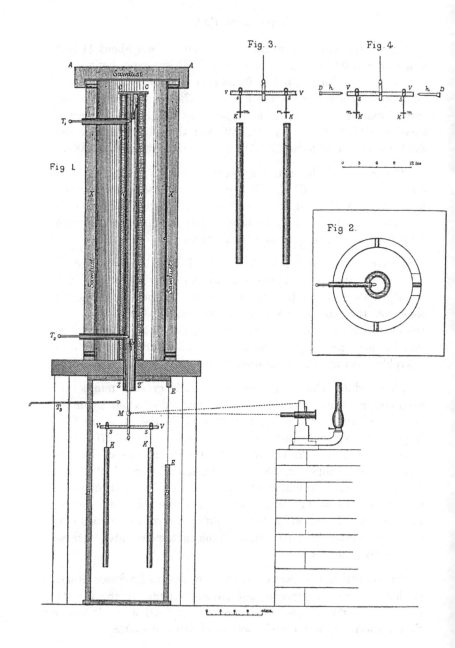

Fig. 3.

Fig. 4.

Fig 1.

Fig 2.

NOTE ON THE DETERMINATION OF ARBITRARY CONSTANTS WHICH
APPEAR AS MULTIPLIERS OF SEMI-CONVERGENT SERIES.

[From the *Proceedings of the Cambridge Philosophical Society*,
Vol. VI, Pt. VI. Read *June* 3, 1889.]

IN three papers communicated at different times to the
Society*, I have considered the application of divergent series to
the actual and easy calculation, to an amply sufficient degree of
accuracy, of certain functions which occur in physical investigations,
but which can of course be considered quite apart from their
applications. These functions present themselves as the complete integrals of certain linear differential equations, or it may
be as definite integrals which lead to such differential equations,
of which they form particular integrals; and as of course the
theory of the complete integrals includes that of any particular
integrals, the subject is best regarded from the former, and more
general, point of view.

The independent variable was taken as in general a mixed
imaginary, and the complete integral was expressed in two ways,
either by ascending series which were always convergent, or by
exponentials multiplied by descending series which were always
divergent (except in very special cases in which they might
terminate), though when the divergent series were practically
useful they were of the kind that has been called semi-convergent.
In either form of the complete integral, the arbitrary constants
appeared as multipliers of the infinite series (of the ascending or
descending as the case might be), or it might be, in part, of a
function in finite terms. The determination of the arbitrary
constants, a thing in general so easy, formed here one of the chief

* *Camb. Phil. Trans.*, Vol. IX, p. 166 [*ante*, Vol. II, p. 329]; Vol. X, p. 105
[*ante*, Vol. IV, p. 77]; Vol. XI, p. 412 [*ante*, Vol. IV, p. 283].

difficulties; and the capital problem may be stated to be, to find the relations (linear relations of course) between the arbitrary constants in the one and those in the other of these two forms of the complete integral.

In the papers referred to, this was always effected by means of a third form of the complete integral, in which it was expressed by definite integrals, their coefficients forming a third set of arbitrary constants. The first two forms of integral were useful for numerical calculation, the one or the other being preferred according as the modulus of the independent variable was small or large; the second form indeed could be used *only* when the modulus was sufficiently large, so that the adoption of the first form in that case was not *merely* a matter of preference; the first form could theoretically be used in any case, but the numerical calculation would become inconveniently or even impracticably long if the modulus were large. The third form was not convenient for numerical calculation, and was used only as a journeyman solution, for connecting the arbitrary constants in the first and second forms of integral with one another, by connecting them each in the first instance with the set in the third form of solution. I remarked that in the event of our not being able to obtain a solution of the differential equation in the form of definite integrals, the use of the first two forms of integral would not therefore fall to the ground; the linear relations between the arbitrary constants in the first and those in the second form could still be obtained numerically, though in an inelegant and more laborious manner, by calculating numerically from the ascending and descending series for the same value of the variable, and equating the results.

My attention has recently been recalled to the subject, and I have been led to perceive that the constants in the first two forms of the integral may readily be connected without going behind the series themselves, so that the expression of the integral of the differential equation by means of definite integrals may be dispensed with altogether; and even if we failed to obtain a solution in this form the two sets of arbitrary constants could be connected exactly by means of known transcendents, and not merely approximately by numerical calculation.

The ascending, and always convergent, series treated of in the three papers already referred to were particular cases of one which, on dividing the whole by a certain power of the variable, has for general or $(m + 1)$th term

$$u_m = \frac{\Gamma(m + a)\,\Gamma(m + b)\ldots}{\Gamma(m + h)\,\Gamma(m + k)\ldots}\,x^m \ \ldots\ldots\ldots\ldots(A),$$

there being at least one more Γ-function in the denominator than in the numerator, so that the series is always convergent. The connexion of the constants in the ascending and descending series was made to depend on two things; one, the determination of the critical amplitudes of the imaginary variable x, or $\rho\,(\cos\theta + i \sin\theta)$, in crossing which the arbitrary constant multiplying one of the divergent series was liable to change, and the mode of that change; the other, the determination, for some one value of θ lying within those limits, of that function of ρ to which the whole expression by ascending series was ultimately equal when ρ became infinite. The value of θ always chosen was such as to make all the terms in one of the ascending series regularly positive; accordingly in the series whose general term is written above it would be $\theta = 0$, giving $x = \rho$. Now when ρ is very large the series diverges for a great number of terms, but at last we arrive at the greatest term, u_{m_1}, suppose, after which the series begins to converge. For a great number of terms in the neighbourhood of u_{m_1}, the ratio of consecutive terms is very nearly a ratio of equality, but the product of those ratios presently begins to tell. Let α and β be two positive quantities as small as we please; then the number of integers lying between $(1 - \alpha)\,m_1$ and $(1 + \rho)\,m_1$ will increase indefinitely as ρ and consequently m_1 increases indefinitely, and moreover the ratio of Σu_m taken for values of m lying between the limits $(1 - \alpha)m_1$ and $(1 + \beta)m_1$ will ultimately bear to the whole series from 0 to ∞ a ratio of equality. Hence in considering the ultimate value of the series we may restrict ourselves to the portion of it mentioned above.

Now when m is very great we have ultimately by a known theorem

$$\Gamma(m + a) = \sqrt{2\pi(m + a - 1)}\,\{(m + a - 1)/e\}^{m+a-1}$$

or
$$\sqrt{2\pi m}\,.\,m^{m+a-1}\left(1 + \frac{a - 1}{m}\right)^m e^{-(m+a-1)},$$

or
$$\sqrt{2\pi m}\,.\,m^{a-1} m^m e^{-m}.$$

Let $h + k + \ldots - a - b - \ldots = s$, and let t be the excess of the number of Γ-functions in the denominator of u_m over the number in the numerator; then the expression for u_m becomes ultimately

$$(2\pi m)^{-t/2} \cdot m^{-s+t} \cdot (e/m)^{tm} \rho^m \ldots\ldots\ldots\ldots(B).$$

The ratio of consecutive terms, which may be obtained from this expression, or more readily directly from (A), is since m is supposed very large $m^{-t}\rho$, and hence for the greatest term we may take

$$m_1{}^t = \rho \ldots\ldots\ldots\ldots\ldots\ldots(C).$$

Strictly speaking m_1 would be the integer next over the (in general) fractional value of m which satisfies the above equation, but it is easy to see that in passing to the limit we may suppose the equation satisfied exactly. Within the specified limits of that portion of our series which it suffices to consider, we see at once that m_1 may be written for m when we are dealing with any finite power of m, since α and β may be supposed to vanish *after* ρ has been made infinite. We need therefore only attend to the last portion (v) of u_m where

$$v = (e/m)^{tm}\rho^m = e^{tm}m^{-tm}m_1{}^{tm}.$$

Now treating m as continuous, *i.e.* not necessarily integral, and putting w for $\log v$, we have

$$w = t(1 + \log m_1 - \log m)\, m, \quad = tm_1 \text{ when } m = m_1,$$

$$\frac{dw}{dm} = t(\log m_1 - \log m), \quad = 0 \text{ when } m = m_1,$$

$$\frac{d^2 w}{dm^2} = -\frac{t}{m}, \quad = -\frac{t}{m_1} \text{ when } m = m_1,$$

whence putting $m = m_1 + \mu$ we have by Taylor's theorem

$$w = tm_1 - \frac{t\mu^2}{2m_1} + \ldots$$

This series proceeds according to powers of μ/m_1, which lies between $-\alpha$ and β, and therefore vanishes in the limit, and therefore ultimately

$$w = tm_1 - \frac{t\mu^2}{2m_1}, \quad v = e^{t\left(m_1 - \frac{\mu^2}{2m_1}\right)}.$$

Now between the limits $(1 - \alpha) m_1$ and $(1 + \beta) m_1$ of m consecutive terms of the series Σv are ultimately equal, and we may replace Σv by $\int v dm$ or $\int v d\mu$; and the limits of μ are $- \alpha m_1$ and $+ \beta m_1$, that is in the limit $- \infty$ and $+ \infty$. Hence in the limit

$$\Sigma v = \int_{-\infty}^{\infty} v d\mu = \sqrt{\frac{2\pi m_1}{t}} e^{t m_1},$$

$$\Sigma u_m = t^{-\frac{1}{2}} (2\pi m_1)^{\frac{1-t}{2}} m_1^{-s+t} e^{t m_1} \quad \dots\dots\dots\dots\dots(D),$$

which may be expressed by (C) in terms of ρ.

The function of ρ to which the complete series Σu_m bears ultimately a ratio of equality when ρ is infinite is thus found, without the necessity of expressing the series in the first instance for general values of ρ by means of a definite integral.

The same method will evidently apply to the series whose general term is formed from (A) by integrations or differentiations with or without intervening multiplications by powers of x, since this process will merely introduce factors of the form $m + c$ into the numerator or denominator or both, and in passing to the limit for $\rho = \infty$ these factors may be put outside after writing m_1 for m.

On a Graphical Representation of the Results
of Dr Alder Wright's Experiments on Ternary Alloys*

[From the *Proceedings of the Royal Society*, XLIX, Jan. 29, 1891.]

SUPPOSE three liquids, such as water, ether, and alcohol, of
which the third is miscible in all proportions with either of the
others, are mixed together, the temperature being kept constant.
According to circumstances, the mixture forms a single liquid
mass, or separates into two. In the latter case, if we suppose
that the liquids had been merely gently poured together, and
imagine the upper and under portions separately to be homogeneous
to start with, this state of things would not remain; an alteration
of composition would take place close to the surface of separation
on both sides, depending on the relative solubilities, &c., of the
ingredients. If now the two altered strata were mixed up with
the rest of the portions to which they respectively belong, the
same thing would go on again, and so on till a condition was
reached in which what we may call an equilibrium of composition
on the two sides of the surface of separation had been attained.
As this equilibrium depends only on the molecular forces, which
are insensible at sensible distances, it is evident that the equilibrium would not be disturbed by removing a part of either the
upper or the under liquid, or by adding to it liquid of exactly the

[* The following preface to this note is taken from the Paper, "On Certain
Ternary Alloys, Part IV," by Dr C. R. Alder Wright, F.R.S. and others, in which
the note occurs.

"A method of graphically representing the results of the experiments described
in the previous portions of these researches has been kindly suggested to one of us
by Sir G. G. Stokes, founded on a principle which he regards as self-evident. We
subjoin a note which he has been so good as to draw up for us, explaining the
application of this method, and then describe some further experiments which we
have instituted with a view to test the correctness of the assumed principle."]

same composition as itself. This final state would take place only very slowly in the manner conceived above; but if the mixture be well agitated the total surface of separation, where alone the change of composition can go on, is greatly increased, and, moreover, the altered strata are mixed up with the rest of the liquids to which they respectively belong, so that the final state is reached comparatively quickly. I think I have seen an experimental verification of this anticipation, namely, that equilibrium depends only on the compositions of the upper and lower mixtures, and not on their quantities, in a French serial, but I have not the reference.

The same principles would apply to ternary alloys, which form a homogeneous mass, or separate into two, as the case may be; but of course the difficulty of preserving a constant temperature is much greater, as well as that of giving sufficient agitation to bring about the final condition.

It seemed to me that, for giving an insight into the results of experiments with ternary alloys, a mode of graphical representation might be usefully employed which is already well known. It is the same as that which Maxwell used for the composition of colours, at least with one slight addition. In this way the whole of the circumstances of the experiment, so far as they are material, would be exhibited to the eye*.

Let A, B, C be three liquids, such as water, ether, alcohol, or else lead, zinc, tin, in fusion, of which the third (which for distinction may be called the solvent) may be mixed in all proportions with either the first or the second. Take a triangle, ABC (Fig. 1), which may be of any form, but is most conveniently chosen equilateral; and, to represent the composition of any mixture of the three, imagine weights equal to those of the substances A, B, C placed at the points A, B, C, and find their centre of gravity, P. To each different set of proportions $A : B : C$ (the letters here denoting weights) will correspond a different position of P, which

[* The theory of the equilibrium of coexistent phases had been fully developed by Willard Gibbs, *Trans. Connecticut Acad.* 1875–8, who also introduced this triangular diagram for a triple system: but his work was just beginning to attract general attention at this date. For the type of system here discussed, defined by the next sentence in the text, cf. Ostwald, *Allgemeine Chemie*, 2 ed. 1896–1902, pp. 984 and 1026 *seq.*]

point will serve to represent to the eye the composition of an
actual or ideal alloy (supposing the substances to be metals) formed
of the three metals in the given proportions. If the quantity of
the solvent be sufficient, P will represent on the diagram the composition of an actual alloy. If it be insufficient, the alloy represented as to composition by P will be ideal only; and on attempting
to form it the mass will separate into two layers. If we suppose
the agitation to have been sufficient, there will be equilibrium of
solution at the surface of junction, and the mass will have reached
its final state. Supposing this condition to have been attained, let
the two portions be analysed, and the points Q, R representing
their compositions be laid down on the diagram, and joined by a

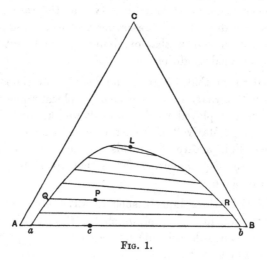

Fig. 1.

straight line. From the construction, this line must pass through
the point P if there has been no loss by volatilisation or oxidation.
Let the same thing be done for several other proportions of the
ingredients. Then the points Q, R will lie in a curve $aQLRb$,
cutting AB in two points a, b, which represent, the first, a saturated
solution of B in A, the second, a saturated solution of A in B.
Call this curve the *critical curve*, and the lines such as QR *tie-lines*,
or simply *ties*. Then the critical curve and the system of ties will
represent the complete result of the experiments, supposing them
to have been exactly made. Alloys of a pair may conveniently be
called *conjugate*. Intermediate tie-lines may be interpolated by
eye; or if we prefer we may substitute for the system of ties

their envelope, on which plan the result of the experiments would be completely represented by two curves, the critical curve and the envelope.

The critical curve separates mixtures of which alloys can actually be formed from those on attempting to form an alloy of which the mass separates into two layers. In the latter case, if through P we draw a tangent to the envelope, cutting the critical curve in Q, R, the points Q, R will represent the compositions of the portions into which the mass separates, while their weights will be as PR to PQ.

If L be the limiting position of the chord QR, or, in other words, the point of contact with the critical curve of a common tangent to it and the envelope, as P tends to coincide with L, the two strata into which the mass separates tend to become identical in nature. If we take a mixture of A and B, represented by a point c in ab, and continually increase the quantity of C from 0, the point P will ascend from c towards C until it reaches the critical curve. At this stage the quantity of the second alloy has just dwindled away to nothing, its nature, so long as there was any of it left, differing from that of the other alloy. If, however, the point c lies in the line CL, on increasing the quantity of C the two alloys merge into one.

On communicating to Dr Alder Wright this mode of graphical representation, he tried it on a large scale on the results of two pairs of series from the former experiments. In one pair the temperature was 650°, and the proportion by weight of zinc to lead was 2 to 1 in the first case, and 1 to 2 in the second. In the other pair the weights of zinc and lead were equal, and the temperature 650° in one case and 800° in the other. In the first pair the agreement of the critical curves was very good, but the agreement in the direction of the ties was not by any means equally good. In the upper part of the figure, corresponding to the case in which there was a considerable quantity of tin, though not enough by any means to prevent the formation of two layers in the entire mass, the difference of inclination ranged to about 5°, the ties in the first case being inclined to those in the second as if they had been turned round in the direction of a line passing through the lead corner of the triangle, and turning round in the direction from lead-zinc to lead-tin. In the second pair of series in which the weights of lead and zinc were equal, and the temperature was 650° in the first case and 800° in the second, the critical curve for 800° was of the same general character as that for 650°, but lay a little inside it, which is just what was to be expected, on account of the increase of solubility attending the higher temperature. Moreover, the critical curve for 650° agreed very fairly with those for the same temperature in the first pair, notwithstanding the difference in the proportion of lead to zinc in the three cases.

I had not anticipated the greater accordance existing between the critical curves in different cases for the same temperature than that shown in the direction of the ties. But, when the plottings revealed it, it seemed to me that the cause was not far to seek. When the molten mass has as yet been but slightly stirred, the superposed alloys, supposed to be severally homogeneous, will most likely be represented on the diagram by points, one or both of which lie outside the critical curve. In this condition an alloy represented by an external point, having the metal C to spare, will be capable of dissolving bodily a portion of the other. This process accordingly, being something analogous to the solution of a salt till saturation is obtained, will go on as the stirring proceeds, and be sensibly complete in a moderate time. The two alloys will then be represented by two points lying on the critical curve. Such alloys may be said to be *associated*. But the passage from merely associated to truly conjugate alloys, as the stirring proceeds, is likely to be decidedly slower. For now neither alloy can bodily dissolve any portion, however small, of the other; there can only be an interchange of constituents across the surface of separation.

The critical curve may be otherwise defined as the curve expressing the saturation of the solvent C with a mixture in given variable proportion of the remaining substances A, B. That it is really such, a little consideration suffices to show. The determination accordingly of the critical curve furnishes us with definite information, even though we do not go into the ulterior question of the condition of conjugation.

Perhaps the attainment of true conjugation might involve more stirring than would be practically feasible with molten metallic mixtures. The most hopeful way would seem to be to fuse the mass at a higher temperature than that intended for the experiment, stirring it well, and then let down the temperature to that intended, stirring all the time, and avoiding too rapid a fall of the temperature.

If truly conjugate alloys were obtained, and portions of each were taken and fused together at the temperature at which the alloys were made, the compositions ought to be the same as before. But if the alloys were merely associated, then, even if the stirring in the second part of the experiment were sufficient to ensure conjugation, the compositions would not be the same as the original, nor would they be independent of the proportion of the two alloys which the operator took for fusing together.

[Note by C. R. A. W., Feb. 25, 1891.—Sir G. G. Stokes has pointed out to me that the diagram, Fig. 1, shows at once that, inasmuch as the difference between the percentages of the solvent in two conjugate alloys vanishes for the pair, a, b, being *nil* for each, and again for the pair which merge into one, represented by the point L, it must necessarily be a maximum for some intermediate pair; and also that, in order to preserve the continuity of conditions, we must, in crossing L, pass from the upper alloy to the lower, and *vice versâ*. Hence, if the entire system of ties could be determined, so as to obtain every possible pair of conjugate points lying, one on one side, the other on the other side, of L, and if these values were plotted on the abscissa and ordinate system, the curve representing the difference between the percentages of the solvent, after having ascended and attained a maximum elevation, must descend again to the base line at a point corresponding with L. If we wish to continue the curve beyond that point, we must now take the ordinates negative instead of positive, the same in magnitude as before, and the curve having crossed the base line, and attained a minimum elevation, will ultimately ascend again to the final point on the base line.]

ON AN OPTICAL PROOF OF THE EXISTENCE OF SUSPENDED
MATTER IN FLAMES.   By Sir G. G. STOKES, Bart., F.R.S.
(*In a letter to Professor Tait.*)

[From the *Proceedings of the Royal Society of Edinburgh.*
Read *June* 15, 1891.]

8 BELGRAVE CRESCENT, EDINBURGH,
June 13, 1891.

DEAR PROFESSOR TAIT,—I write to put on paper an account
of the observation I mentioned to you to-night, in case you
should think it worth communicating to the Royal Society of
Edinburgh.

In the course of last summer I was led, in connection with
some questions about lighthouses, to pass a beam of sunlight,
condensed by a lens, through the flame of a candle. I noticed
that where the cone of rays cut the luminous envelope there were
two patches of light brighter than the general flame, which were
evidently due to sunlight scattered by matter in the envelope
which was in a state of suspension. The patches corresponded
in area to the intersection of the double cone by the envelope,
and their thickness was, I may say, insensibly small. Within the
envelope, as well as outside, there was none of this scattering.
The patches were made more conspicuous by viewing the whole
through a cell with an ammoniacal solution of a salt of copper,
or through a blue glass coloured by cobalt. In the former case
the light from the flame was more weakened than the scattered
light, which was richer in rays of high refrangibility; in the latter
case the patches were distinguished by a difference of colour, the
patches being blue, while the flame (with a suitable thickness
of blue glass) was purplish. The light of the patches exhibited

the polarisation of light scattered by fine particles—that is to say, when viewed in a direction perpendicular to the incident light it was polarised in a plane passing through the beam and the line of sight.

When the beam was passed through the blue base of the flame there was no scattered light. A luminous gas flame showed the patches indicating scattered light like the flame of a candle, but less copiously. They were not seen in a Bunsen flame or in the flame of alcohol, but were well seen in the luminous flame of ether. When a glass jar was inverted over burning ether, the blue part, which does not show scattered light, extended higher till, just before the flame went out, the luminous part disappeared altogether. A Bunsen flame, fed with chloride of sodium, did not show the phenomenon, though the flame was fairly luminous.

The phenomenon shows very prettily the separation of carbon (associated, it may be, with some hydrogen) in the flame, and at the same time the extreme thinness of the layer which this forms. It shows, too, the mode of separation of the carbon, namely, that it is due to the action of heat on the volatile hydrocarbon or vapour of ether, as the case may be. At the base, where there is a plentiful supply of oxygen, the molecules are burned at once. Higher up the heated products of combustion have time to decompose the combustible vapour before it gets oxygen enough to burn it. In the ether just going out, for want of fresh air, the previous decomposition does not take place, probably because the heat arising from the combustion is divided between a large quantity of inert gas (nitrogen and products of combustion) and the combustible vapour, so that the portion which goes to the latter is not sufficient to decompose it prior to combustion.

In the Bunsen flame fed with chloride of sodium, the absence of scattered light tallies with the testimony of the prism, that the sodium is in the state of vapour, though I would not insist on this proof, as it is possible that the test of scattering sunlight is not sufficiently delicate to show the presence of so small a quantity of matter in a solid or liquid state.—Yours, sincerely,

G. G. STOKES.

P.S.—I fancy the thinness of the stratum of glowing carbon is due to its being attacked on both sides—on the outside by oxygen, on the inside by carbonic acid, which with the glowing carbon would form carbonic oxide.

[*When the above was written I was not acquainted with the previous paper by Mr (G. J.) Burch, published in Vol. XXXI of *Nature* (p. 272), nor did any of the scientific friends to whom I had mentioned the observation seem to be aware of it. Had I known of it, I should not have thought my paper worthy of being presented to the Royal Society of Edinburgh, as Mr Burch has anticipated me in the fundamental method of observation.

The reaction mentioned in the postscript is to be taken merely as a specimen of the reactions, on the inside of the carbon stratum, by which the carbon may be re-engaged in a gaseous combination. Carbonic oxide is only one of the combustible gases, not originally present, which are formed during the process of combustion, and are found inside the envelope in which the combustion is going on.—G.G.S., *November* 20, 1891.]

[* This postscript occurs in the reprint in *Nature*, Vol. XLV, p. 133.]

ON THE REACTIONS OCCURRING IN FLAMES.

[From the *Proceedings of the Chemical Society*, Feb. 4, 1892.]

"4, WINDSOR TERRACE, MALAHIDE, IRELAND,
"23rd September, 1891.

"DEAR DR ARMSTRONG,

"I enclose a little optico-chemical paper*, that is to say, one in which the method is optical, but the results are of interest, such as they have, rather from a chemical point of view. I use, to express it in short terms, a flame as a screen on which to receive an image of the sun.

"The reaction mentioned in the P.S. is to be taken as a specimen of reactions of the kind, for though it probably takes place, there are doubtless others also, as there are a lot of compounds found in the interior of the flame.

"I read the other day your address to the Junior Engineering Society, in which you speak of oxygen as combining with hydrogen in preference to carbon; I should have supposed it would have been the other way. Not only does the facility with which steam is decomposed by glowing carbon favour this view, but it seems to me to fit better with the phenomena of flames. According to my notions, we must carefully distinguish between the changes which take place in the partial combustion of a molecule and those which are produced in neighbouring molecules as a result of the heat thus produced. We may, for the sake of a name, call the former pure-chemical, and the latter thermo-chemical. The action of the heated walls of a tube is of the thermo-chemical kind; it involves a regrouping of the existing molecules under the molecular agitation of a hot body, without bringing a fresh reagent (suppose oxygen) into play from outside the molecule. I think that in the blue base of the flame of a candle, where oxygen is plentiful, we have pure-chemical changes. The blue shell invests for a little way the highly luminous shell, like a calyx investing a corolla, and

[* *Supra*, p. 232.]

I think the thin shell of glowing carbon, to which the bulk of the light is due, owes its origin to a thermo-chemical change, the heat being derived from the combinations with oxygen which take place just outside it.

"I imagine that the hydrocarbon spectrum is due to a gas formed by a pure-chemical as distinguished from a thermo-chemical change. But what gas is it? It is commonly supposed to be acetylene. To me it seems more probable that it is marsh gas, formed by a pure-chemical, not a thermo-chemical, change. According to my notion, this unknown gas (x, say) is a hydro-carbon, which when burnt without admixture of other hydro-carbons would show but feebly if at all the hydrocarbon spectrum. More especially might this be expected to take place if it were burnt at a reduced pressure, or considerably diluted with, say, nitrogen. For in order that x should show its spectrum its molecule must be in a state of violent agitation, which it might be expected to be if it had been born as a result of partial com-bustion, but would not be merely because it was going to be slain by union with oxygen.

"I have not seen a statement as to the spectrum of marsh gas, as pure as may be, when burnt. Perhaps you know about it. As hydrocarbons in general (I don't know how it is as to marsh gas) show the same spectrum, x must be some gas of a simple kind formed in the process of partial combustion, though probably (at least under ordinary circumstances) itself burnt almost im-mediately afterwards.

"Yours very truly,

"G. G. STOKES."

"21st January, 1892.

"DEAR PROFESSOR ARMSTRONG,

"Perhaps I may be allowed to add a few words in explanation of what I meant by thermo-chemical change. I had not in view thermic measurements. I will endeavour to explain my ideas by an example. Let us contrast (a) the formation of water from mixed oxygen and hydrogen, (b) the formation of acetylene and hydrogen from marsh gas at a high temperature. In both cases alike molecular agitation is required to bring about the change; in (a) if the change be brought about at one point, the consequent agitation supplies the requisite disturbance to the neighbouring molecules, and the change is propagated with explosions; but in (b) I picture to my own mind the change as taking place in this way. When sufficient heat is supplied to the gas from without, the collisions of the molecules of the marsh gas become sufficiently violent to allow the carbon atoms in a pair to get into a condition in which their tendency to self-combination comes into play, and in the coalescing a portion of the total hydrogen in the pair is thrown off. But the continuance of this change is dependent on a continuous supply of heat from without, under which it is gradually effected. I should call (a) a pure-chemical change, even though heat at one point is necessary to start it, and I should call (b) thermo-chemical, even though chemical affinities are concerned in it.

"The results of Professor Smithells seem to me to make it probable that x may be carbonic oxide.

"Yours very truly,

"G. G. Stokes."

THE INFLUENCE OF SURFACE-LOADING ON THE FLEXURE OF BEAMS. By Prof. C. A. CARUS WILSON. (Extract.)

[From the *Philosophical Magazine* (5), XXXII, Dec. 1891, pp. 500—503.]

SINCE communicating the above*, Sir George Stokes has gone very fully into this problem, and has kindly allowed me to quote

[* The problem considered is that of a slip of glass supported on parallel rollers at B and C and subject to pressure P per unit length applied through a roller on the other surface at A. The glass, when thus stressed, is examined in

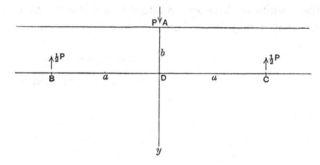

a direction parallel to the rollers, between crossed Nicols, along the line AD, so that the colours of polarization may reveal the nature of the strain. A measurable feature is the neutral points which appear as black dots on the coloured field; they are at the places at which the two principal stresses, transverse and longitudinal, are equal, where therefore the strain is a uniform dilatation and the refraction is not double; and it is their theoretical determination, for comparison with the results of observation, that is under consideration.

The author had obtained from M. Boussinesq the value of the stress produced throughout the glass by the application of the surface-force P along the roller A, when the depth of the glass is unlimited. In fact, in his book *Application des Potentials...*, 1885, M. Boussinesq had found by methods like those employed by Stokes on the theory of diffraction in 1849 (*ante*, Vol. II, p. 257), that due to a point-force Q at A the traction across any element of interface parallel to the boundary face, in the glass, is a simple radial push directed from A and of intensity $\dfrac{3Q}{2\pi r^2} \cos^2\theta$, where θ is the angle made by the radius vector with the

the following extracts from letters I have received from him on the subject.

" Let A be the point in the upper surface where the pressure (P) is applied; B, C the points of support below, which I suppose to be equidistant from A; D the middle point of BC. Let y be measured downwards from A; denote BD or DC by a, and AD by b. You have the expressions for the stresses produced by P in an infinite solid $\left(x = \dfrac{2P}{\pi}.\dfrac{1}{y}\right)$, and the question is, What system must we superpose on this to pass to the actual case ? This, as I showed you, is the system of stresses produced by a system of forces applied to the surface. The forces consist—(1) of the two pressures $\frac{1}{2}P$ at B and C; (2) of a continuous oblique tension below, represented in drawing by a fan of tensions directed at every point of the lower surface from the point A.

" Imagine now the beam cut into two by a plane along AD. Consider one half only, say that on the B side. Everything will remain the same as before, provided we supply to the surface AD forces representing the pressures or tensions which existed in the undivided beam. On account of the symmetry, the direction of these must be normal.

" At D the vertical pressure on a horizontal plane in the infinite solid is compounded with an equal vertical tension due to the fan. Hence, of the vertical pressure in AD which must be superposed on the vertical pressure in the infinite solid, we know thus much without obtaining a complete solution of the problem, namely, that it must equal *minus* $2P/\pi b$ at D and 0 at A. If we suppose it to vary uniformly between, we are not likely to be far wrong.

normal to the face : from this it is readily derived by a line-integration that the effect of force P per unit length applied along the roller A is $2P/\pi r \cos \theta$, or $2P/\pi y$, which is the expression indicated above in brackets near the beginning. The author verified optically the formula kP/y along AD, for a slip of glass supported underneath on a flat surface; and proceeded to a result for the original problem by superposing on this vertical traction that arising from the pure bending of the slip under the opposing applied torques. It was found that the position of the neutral points along AD indicated for k the value ·726 instead of $2/\pi$ or ·64. The object of Sir George Stokes' investigation is to remove this discrepancy. It is to be observed that the rollers B and C are considered to be so far from AD that the local effects of the surface-loadings arising from them are immaterial.]

"This leads to the following expression for the vertical pressure in AD:—

$$\frac{2P}{\pi}\left(\frac{1}{y}-\frac{y}{b^2}\right).$$

"Now for the horizontal. We know that the complete system of external forces must satisfy the conditions of equilibrium of a rigid body. The direction in each element of the fan passes through A, about which therefore the fan has no moment. Hence the moment of the horizontal forces along AD taken about A must equal $\frac{1}{2}Pa$. Again, the resultant of the semi-fan is a force passing through A, and its vertical component is $\frac{1}{2}P$. Its horizontal component is the integral of

$$\frac{2Pb^2}{\pi}\cdot\frac{xdx}{(b^2+x^2)^2},$$

taken from 0 to infinity,. or P/π.

"Hence of the horizontal forces along AD we know these two things:—

(1) The sum must equal P/π,

(2) The moment round A must equal $\frac{1}{2}Pa$.

"In default of a knowledge of the law according to which the force varies with y, it is natural to take it, for a more or less close approximation, to be expressed by the linear function $A + By$, or say Y. To determine the arbitrary constants A, B, we have only to equate the integral of Ydy to P/π, and that of $Yydy$ to $\frac{1}{2}Pa$, the limits being 0 to b. We thus get for the expression for the tension at any point of AD,

$$\frac{P}{b}\left(\frac{4}{\pi}-\frac{3a}{b}\right)+\frac{6P}{b}\left(\frac{a}{b}-\frac{1}{\pi}\right)\frac{y}{b}.$$

"At neutral points the vertical pressure equals *minus* the horizontal tension, giving

$$\left(\frac{6\pi a}{b}-8\right)\frac{y^2}{b^2}+\left(4-\frac{3\pi a}{b}\right)\frac{y}{b}+2=0;$$

or, putting for shortness $\dfrac{3\pi a}{b}-4=m$,

$$2m\left(\frac{y}{b}\right)^2-m\frac{y}{b}+2=0,\quad\text{therefore}\quad\frac{y}{b}=\frac{1}{4}\pm\sqrt{\frac{1}{16}-\frac{1}{m}}.$$

For the neutral points to be real and different, we must have

$$m > 16, \quad \frac{2a}{b} > \frac{40}{3\pi}.$$

When the neutral points coalesce into one, we have m equal to 16, y equal to $\frac{1}{4}b$; and for the ratio of the span to the depth, $2a/b$ equal to $40/3\pi$, or $4\cdot245$, or, say, the span is $4\frac{1}{4}$ times the depth.

"As regards the horizontal tension at points along AD, you take a linear function of y as I do, and your condition of moments is the same as my (2), but in lieu of my (1) you do what is equivalent to taking the total tension nil. You further omit the correction to the vertical pressure when we pass from a solid of infinite depth to one terminated by a plane below. You further take the coefficient of P/y as k, a constant to be determined by the observations, instead of $2/\pi$.

"Taking the place of the neutral point (at one-fourth of the depth) and the ratio of span to depth as given by my formulæ, and then treating them as if they had been the results of experiment, and substituting in your formulæ for the determination of k, I got $0\cdot7947$ instead of $0\cdot64$. The largeness of your coefficient is I think fully accounted for by the employment of the formulæ which you used.

"In your method you take the stress belonging to the solid supposed infinitely deep, and superpose it on the stress corresponding to a pure bend.

"This comes to the same thing as retaining three terms only in the equation I gave in my letter for determining the y of the neutral points.

"The equation thus becomes

$$\frac{6\pi a}{b}\frac{y^2}{b^2} - \frac{3\pi a}{b}\frac{y}{b} + 2 = 0,$$

or

$$2m\frac{y^2}{b^2} - m\frac{y}{b} + 2 = 0,$$

where

$$m = \frac{3\pi a}{b} \text{ instead of } \frac{3\pi a}{b} - 4.$$

"When the two neutral points merge into one, we have in

both cases alike y equal to $\frac{1}{4}b$, and the only difference is that $3\pi\, a/b$ equals m instead of m *plus* 4.

"If you had supposed the coefficient for the infinite solid to be an unknown quantity k, and had applied your observations to determine it, using my formulæ instead of your own, you would have got something very close indeed to 0·64.

"It is noteworthy that in your problem, taken as one in two dimensions, the theoretical stresses in the planes of displacement are independent of the ratio between the two elastic constants; in other words, independent of the value of Poisson's ratio."

I have calculated the positions of the neutral points from Sir George Stokes's formula

$$\frac{y}{b} = \frac{1}{4} \pm \sqrt{\frac{1}{16} - \frac{1}{m}}$$

for spans of 88, 100, and 120 millim. in a beam 128 millim. long × 5·5 millim. wide × 19 millim. deep. These are given in the following Table in the 2nd and 3rd columns. The results of actual observations are given in columns 4 and 5; while columns 6 and 7 give the same points as found by plotting the intersection of the curves of pure bending and loading (infinite solid assumed):—

Span	Distance of Neutral Points from top edge, by					
	Sir George Stokes's formula		Observation		Intersection of curves	
88.........	6·3	3·2	6·4	3·3	6·9	2·7
100.........	7·0	2·5	7·2	2·5	7·3	2·3
120.........	7·7	1·8	7·8	1·8	7·8	1·75

The error by the intersection method is greater in proportion as the span is smaller, as might have been expected.

If the observed positions of the neutral points are inserted in Sir George Stokes's formula, the value 0·64 is obtained for the constant k in the equation $x = \dfrac{2P}{\pi} \cdot \dfrac{1}{y}$.

On the best Methods of Recording the direct Intensity of Solar Radiation.

[From the *Report of the British Association*, September, 1892. Drawn up by Sir G. G. Stokes, Chairman of the Committee*.]

THE work of the Committee during the past year has been confined to an examination, both experimental and theoretical, of Balfour Stewart's second actinometer when used as a dynamical instrument. Actinometers may be divided into two classes, which may conveniently be denominated *dynamical* and *statical* respectively. In those of the dynamical class the mercury or other fluid employed is examined while the head of the column is in motion, in consequence of the exposure being varied by suitable manipulation, and readings of the column are taken at chosen times, or else the times are noted when the top of the column reaches chosen readings. In the statical class the instrument is allowed to attain its permanent state, subject of course to a secular change, such as that due to the varying altitude of the sun, and the results are deduced from the stationary readings of two or more thermometers. Herschel's, Hodgkinson's, and Stewart's first actinometers are examples of the dynamical kind; the black bulb thermometer, Violle's actinometer, and Stewart's second actinometer, when used as he intended, are examples of the statical class.

Stewart's second actinometer has been already described (see Reports of the Association for 1886, p. 63, and 1887, p. 32), but to save the trouble of reference it may be well briefly to mention that it consists of an envelope of thick copper, closed on all sides except as regards a small hole to allow the sun's rays to enter, and one to allow the stem of the central thermometer to pass through. In the actual instrument the envelope is cubical, and its temperature is determined by three thermometers, with their bulbs sunk in

[* Cf. *supra*, p. 137.]

hollow chambers in the thick metal, two (A, B) in the front face of the cube, or that turned towards the sun, the third (C) in the back face. The internal thermometer (D) has a lenticular bulb, which is mounted so that it lies at the centre of the cube, or nearly so, and has its plane perpendicular to the incident rays. The copper cube is surrounded by a thick coating of felt, and this by a covering of thin brass. The object of this arrangement is to make the temperature of the copper cube sensibly the same all round, and at the same time to prevent it from changing more than very slowly when the instrument is exposed. The thermometers A, B, C were graduated to degrees, D to half-degrees, all Fahrenheit.

With a view to increasing the effect of the radiation from the sun on the thermometer D its bulb was ordered to be of green glass of a particular kind. This, as we ascertained from Mr Casella, who made the instrument, occasioned a great deal of trouble, as not only had a pot of green glass to be made specially for the purpose, but many thermometers broke in the process of construction, the fracture taking place at the junction of the green glass of the bulb with the colourless glass of the stem. In a future instrument we should not think of encountering these difficulties, since, as will presently appear, our researches led us to the conclusion that little advantage, if any, was gained by the substitution of green for colourless glass in the construction of the bulb.

The principle of the instrument was to make any point of the radiation thermometer (D) look, so to speak, in all directions outwards at an envelope of uniform temperature, except as to directions lying within a very small solid angle (that subtended by the hole), within which the sun's rays were admitted. If the direct rays of the sun had been used the solid angle in question could only be made small on condition of admitting only a very small amount of the sun's rays, which would not have sufficed to raise the temperature of D sufficiently above that of A, B, C. To reconcile the two conditions of allowing the bulb of D to be almost wholly surrounded by the copper envelope, and at the same time permitting a sufficient amount of solar radiation to fall upon it, a lens was introduced, mounted on a stem perpendicular to the front face, to which face the plane of the lens was parallel, and from which the lens was distant by its focal length. In this way the necessary hole in the envelope need hardly be wider than the image of the sun, though it was convenient to allow some margin in order to provide for the contingency of the pointing of the instrument not being very exact. The lens was provided with two diaphragms for optional use, one having twice the area of the other.

The observations which have been taken at intervals during the past year with a view to test the practical working of Stewart's second actinometer have been made by Professor McLeod; the reduction of the observations has been mostly done by the Chairman, with whom also Professor McLeod has been in frequent communication as to the lines of inquiry. In consequence of other engagements, the observations have not yet been subjected to so complete a reduction as the care with which they have been made deserves; but enough has been done to serve as a guide to the inquiry, and to permit of some general conclusions as to the behaviour of the instrument.

As has been already stated, the instrument was intended for use as a statical actinometer with permanent exposure. But it seemed desirable in the first instance to study the march of the thermometers when the instrument was first exposed to radiation from the sun, or the sun's rays were cut off after it had been exposed for some time. This seemed to hold out a better prospect of obtaining a thorough insight into the working of the instrument than if it had been at once used as a statical actinometer; besides which the latter use would have involved some outlay in the way of providing some sort of equatorial mounting and clock movement, and it did not seem desirable to go to the expense of this unless preliminary testing showed that the instrument was likely to be successful when used as a statical actinometer.

The temperature of the case was determined from the readings of the thermometers A, B, C by taking first the mean of A and B, and then the mean of that and C. It was found, however, that A and B always read almost exactly alike, and C was not usually more than one or two tenths of a degree lower. In any future instrument it would doubtless be sufficient to determine the temperature of the copper case by a single thermometer sunk in one of the side faces, midway between the front and back face.

In spite of the felt packing the temperature of the case was found to change more rapidly when the instrument was exposed to the sun than was to be desired, and Professor McLeod found it an improvement to introduce a screen of tinned iron placed a little distance in front of the front side of the cube, and of course provided with a hole for letting the sun's rays through that were to fall upon the thermometer D. In most of the observations the case thermometers were merely sunk in their holes, the sides of which the bulbs might or might not touch in one or two places. It was feared that in spite of the slowness of the change of temperature of the case, the lagging of the case thermometers might possibly introduce a sensible error. Accordingly the effect was tried of introducing a packing of reduced silver between the bulb of the thermometer and the wall of the cavity in which it was inserted. By packing in this manner one of the thermometers A, B, and leaving the other unpacked, it was possible to judge whether any sensible error was to be apprehended from lagging. It was found that the packed thermometer was a little more prompt, but the difference of temperatures read off was very small, little more than emerging from errors of observation.

In the first regular observation on the march of the thermometers under insolation, the four thermometers were read before exposure, then the instrument was exposed, and the thermometers read at intervals of a minute for a quarter of an hour, by which time thermometer D had become sensibly stationary, having risen $51°\cdot3$, while the case thermometers rose about $2°$, the excess of D over the temperature of the case rising to $49°\cdot7$. The sun was then screened off, and the reading of all the thermometers at intervals of one minute continued for about half an hour. During this time the case thermometers continued slowly to rise, the total rise in the half-hour amounting to $1°\cdot5$; the central thermometer fell, pretty rapidly towards the beginning, slowly near the end, till it stood only $0°\cdot4$ or $0°\cdot5$ above the case thermometers. The sky was very clear, and there were no clouds near the sun; and as the insolation began at XII, 26, the decrease of the sun's altitude during insolation was but small.

It remains to be shown whether, and if so in what way, a measure of the radiation can be obtained from the results.

Let θ be the temperature of the insolated thermometer, T that of the case as measured by the case thermometers, q the coefficient of cooling, the rate of cooling being taken as following Newton's law, r the rate of heating of D due to solar radiation. Then in the time dt the increment $d\theta$ of D's temperature is made up of the gain, rdt, due to radiation and the loss, $q(\theta - T)$, due partly to convection, partly to the excess of the radiation from D to the case over that from the case to D. We have therefore.

$$\frac{d\theta}{dt} + q(\theta - T) = r \quad \dotsb(1).$$

If we suppose r and T constant, or subject only to slow secular changes, so that they may be deemed constant in the integration, we have

$$\theta = T + \frac{r}{q} + ce^{-qt} \dotsb(2).$$

Hence if u denotes the excess of temperature of the central thermometer over that of the case, we see from (2), or directly from (1), that u tends to the limit

$$u = \frac{r}{q} = \lambda r, \text{ say} \dotsb(3),$$

when the time t which has elapsed since exposure, or whatever other change it may have been in the disposition of the instrument, is large enough to permit of our neglecting the last term in (2). The constant λ, the reciprocal of q, in (3) denotes a time, which may conveniently be called the *lagging time* of the thermometer D.

Were the actinometer used as a statical instrument the simple expression (3) is all that we should be concerned with. The quantity r varies as the radiation, but involves a coefficient depending on the particular instrument and for a given instrument on the area of the diaphragm used, and on the presence or absence of the quartz plate which is furnished for covering the aperture. The constant q need not be determined, as it is associated with a coefficient depending on the instrument. By itself alone the actinometer gives only the ratio of variation of the radiation. To obtain an absolute measure the actinometer would have to be compared, once for all, with some actinometer which gives absolute results. We believe that the main object which Stewart had in view was to furnish an instrument which might supply a means of detecting possible variations in the intrinsic intensity of radiation from the sun, corresponding, suppose, to the sun-spot period; and for this object the same instrument would be employed throughout, so that we should not be concerned with absolute measures.

In studying, however, the march of thermometer D when the instrument is exposed, or else the sun's rays cut off, we must have recourse to equation (2), and now we can no longer dispense with a knowledge of the value of the constant q. The easiest way of determining it seemed to be to make use in the first instance of the readings in the latter portion of the observation, when thermometer D, after having been heated by exposure, was cooling in consequence of the sun's rays having been intercepted by a screen. In this case $r = 0$, and we have simply from (2)

$$u = ce^{-qt} \dots\dots\dots\dots\dots\dots\dots\dots\dots(4).$$

Hence, if we plot the observations, taking the time for abscissa and the logarithm of the excess u for ordinate, we ought to get a series of points lying in a straight line.

On laying down the observations on paper it was found that, after a slight initial irregularity, the dots representing the observations lay extremely closely in a straight line until the excess u, which began at $49°\!\cdot\!7$, was reduced to about $3°$. They then began to fall a little too high, and the height above the straight line representing the previous observations kept increasing as we proceeded. We have not investigated the cause of this variation, but it seems possible that it may have been due to a slight lagging of the case thermometers. As these were still rising, though D was falling, the lagging would make the temperature of the case appear a little too low, and therefore the excess u a little too great, and therefore the actual reduction of u would be somewhat less than the calculated. The difference between the real and apparent temperature of the case would be too small sensibly to affect the result until the absolute excess u became comparatively small. We are not, however, concerned with such small excesses u in the actual use of the instrument. The lagging time deduced came out $5\!\cdot\!6$ minutes.

The reciprocal of this was then introduced into equation (2), which was then applied to the reduction of the first portion of the observation, that portion, namely, which was concerned with the rise of D consequent on exposure. As θ and T are observed and q deemed to be known, the equation contains only two unknown quantities, namely, r, which depends on the radiation, and the arbitrary constant c. These might be determined by any two not unfavourably selected observations of the series, and then the observed and calculated values of u ought to agree for the rest. This, however, was found to be by no means the case, and the differences between theory and observation were far too methodical to be attributable to errors of observation. Equation (2) was then tried as a mere formula of interpolation, q being taken as a disposable constant as well as r and c. Any three observations would of course theoretically suffice for the determination of the three constants, and then the formula would give the calculated final excess, to which r is theoretically proportional, or the calculated value of u for any other observation of the series.

The numerical calculation is much facilitated by choosing for the determination of the constants three observations equidistant in time. If t_0 be the time of the first of the three and Δt the chosen interval, we have from (2)

$$u_0 = \frac{r}{q} + ce^{-qt_0};$$

$$\Delta u_0 = -ce^{-qt_0}(1 - e^{-q\Delta t});$$

$$\Delta^2 u_0 = ce^{-qt_0}(1 - e^{-q\Delta t})^2.$$

These equations give

$$1 - e^{-q\Delta t} = -\frac{\Delta^2 u_0}{\Delta u_0},$$

which determines q, and then

$$\frac{r}{q} = u_0 - \frac{(\Delta u_0)^2}{\Delta^2 u_0},$$

which gives the calculated final excess.

A rough calculation showed that four minutes was a very suitable interval Δt to choose, which also agreed with the result of actual trials. When various trios were taken from different parts of the series not too near the end, as there the differences became small, and consequently errors of observation would be telling, the calculated final excesses came out remarkably accordant. It thus appeared that equation (2) was no mere formula of interpolation, but that it was very well satisfied, provided, at least, the higher part of the series were not included. The limit to which the excess tended when it had become nearly stationary was evidently a little, though only a little, lower than the calculated limit. This is not to be wondered at, because in the calculation it was assumed that the cooling followed Newton's law, which it is known is not sufficiently accurate when the excess of temperature is as great as 40° or 50°, the cooling in such a case being more rapid than if Newton's law had been followed exactly, the constant involved in it being determined by observations taken with more moderate excesses of temperature.

The values of q as determined by different trios did not come out so closely accordant as the calculated final excesses, as might indeed have been expected from the nature of the equations. Still they agreed in showing that to satisfy the insolation observations the coefficient of cooling q must be taken distinctly larger, in the ratio of about 5 to 4, than when the thermometer cooled after exposure. When a beam of the rays of the sun falls on the front face of the thermometer a portion of heat is absorbed directly by the mercury under the place where the rays strike. As mercury is opaque the portion thus warmed would in the first instance form a thin stratum adjacent to the surface by which the rays entered. Of course currents of convection would arise in the mercury, and also the fluid metal would conduct the heat. But if the heat thus tends to get diffused, on the other hand there is a constant renewal of the superficial heating. Now this specially heated stratum, however thin, helps to raise the mean temperature of the surface, but contributes comparatively little to the mean temperature of the mass; in fact, if it were infinitely thin it would contribute infinitely little. Now the rate of cooling is determined by the average temperature of the surface taken all round, whereas the indication of the thermometer is determined by the average temperature of the whole mass of mercury. Hence the mean temperature of the surface is greater than the mean temperature of the mass; and therefore, if the rate of cooling is supposed to be determined by the temperature indicated by the thermometer, in

other words, to be what it would have been if there had been no
such inequality of temperature in different parts of the mass, we
must to make up for it take a larger coefficient of cooling.

Hitherto a single series only of observations has been mentioned.
In fact, a considerable number of series were taken, but as the
general mode of treatment and the general character of the results
are pretty nearly the same throughout, it does not seem necessary
to mention them except when they were made for the special
elucidation of particular points.

In the first series the diaphragm used with the lens was the
larger one, of $\sqrt{2}$-inch diameter. It seemed desirable to compare
the results obtained with this and with the smaller diaphragm
of 1-inch diameter. Accordingly, on a day when the sky was
clear, series were taken with the two diaphragms in succession.
On reducing the results it was found that the effect of radiation
through the larger diaphragm was as nearly as possible double that
through the smaller.

The object of the quartz plate was to prevent possible irregularities arising from the action of the wind, which, it was thought,
might cause some interchange between the air inside and outside
the cube. It seemed desirable to try the instrument with and
without the quartz plate. Comparative series were accordingly
taken on a clear and not windy day with and without the quartz
plate. The effect was in round numbers about 10 per cent. less
with plate on than with plate off. When the plate is used there
is loss by reflection from the two surfaces, besides which there
may also conceivably be loss by absorption. The loss by reflection
can easily be calculated by Fresnel's formula for the intensity
of reflected light. If we disregard the double refraction, and take
μ for the refractive index answering to the mean of the heat rays
incident, and take account of the rays reflected an even number of
times, as well as of those which are not reflected at all, we have
for the intensity of the transmitted light, that of the incident
being 1,

$$\frac{2\mu}{\mu^2 + 1}.$$

On multiplying the calculated final excess got from the observations without any plate by the above factor, it came, within the
limits of errors of observation, the same as the calculated final

excess obtained from the observations with the quartz plate on. It follows that there is no sensible loss due to absorption in the quartz plate. It is to be remembered that the rays that fell upon the quartz had already passed through the glass lens, and also that in the radiation from the sun it is a comparatively small proportion of heat rays that are absorbed by glass and similar substances.

It remains to be explained in what way we were led to the conclusion that the employment of green instead of colourless glass for the bulb of the insolation thermometer must have made but little difference in the results obtained.

Imagine a thermometer to be suddenly exposed to solar radiation, as in Stewart's second actinometer, and consider what its behaviour ought to be on the two extreme suppositions: (1) that the mercury in contact with the glass reflects perfectly all the rays that fall upon it, but that the shell is partially opaque; (2) that the mercury reflects only partially, but that the shell is perfectly diathermanous.

On the first supposition the mercury would not be warmed at all by the rays which fell upon it, but only by conduction from the shell, which itself would be heated by absorption of a portion of the rays that fell upon it, either as they came from the sun or as they were on their way back after reflection at the surface of the mercury. The rise of temperature of the shell would ultimately vary as the time elapsed. But if the shell were at a given temperature the total heat received by the mercury from the shell would vary ultimately as the time during which it has been passing in. But as the temperature of the shell is not constant, but its rise varies ultimately as the time since exposure, the total heat received by the mercury will vary ultimately as the integral of a quantity which varies as the time, and will therefore vary ultimately as the square of the time.

On the second supposition the mercury receives its heat directly from the sun, and the total heat received varies ultimately as the time during which it has been receiving it.

Now in the actual observation the gain of heat was found to be ultimately sensibly proportional to the time elapsed, not to the square of the time, as may be inferred from the fact that the rate of increase was decreasing from the first. We may conclude therefore that the gain of heat was due almost entirely to the imperfection of the reflection from the mercury, which entails direct absorption by the mercury of the portion which failed to be reflected, and only in a comparatively insignificant degree due to absorption of heat by the shell in the passage of the heat through it. We may therefore infer that the substitution of green for colourless glass in the shell of the bulb would make but little difference in the results obtained. This agrees with the experience of Captain Abney, who was led by his experiments on the diathermancy of various kinds of glass to suppose that a thermometer with a bulb of green glass would rise decidedly higher in sunshine than one with a shell of colourless glass, but found on trial that the substitution of green for colourless glass made only a slight difference.

That the rise should be due chiefly to absorption of radiant heat by the mercury is not to be wondered at. We do not know whether actual experiments have been made on the reflecting power of mercury in contact with glass, but we should probably not be far wrong in estimating it at about 65 per cent., which is about the reflecting power of speculum metal in air. This would leave as much as 35 per cent. of the incident rays to be absorbed by the mercury.

In some of the experiments the change of temparature of the case was barely slow enough to allow of regarding T as constant in the integration of (1). But it is easy to prove that if T vary slowly, though not infinitely slowly, in order to correct for the finiteness of the rate of change, we have merely to add the term $-dT/qdt$ to the right-hand member of (2).

ON THE NATURE OF THE RÖNTGEN RAYS.

[From the *Proceedings of the Cambridge Philosophical Society*, Vol. ix, 1896, pp. 215-6. Read *Nov.* 9, 1896.]

IN this communication the author explained the views he had been led to entertain as to the nature of the Röntgen rays, and to a certain extent the considerations which had led him to those conclusions. As Röntgen himself pointed out, the X rays have their origin in the portion of the wall of the Crookes' tube on which the so-called cathodic rays fall, and it is natural that our notions as to the nature of the X rays should be intimately bound up with those we entertain as to the nature of the cathodic rays. Two different views have been adopted on this question. Several eminent German physicists hold that the cathodic rays are essentially a process going on in the ether, the nature of which nobody has been able to explain; and that if any propulsion of molecules from the cathode accompanies them, it is merely a secondary phenomenon. The other view is that the cathodic rays are not proper rays at all, but that they are essentially streams of molecules. The latter view is that which, so far as the author knows, is universally adopted in this country. The author expressed the fullest conviction that the cathodic rays are no mere process going on in the ether, but that the propulsion of molecules is of the very essence of the phenomenon; only it is to be remembered that the molecules are not to be thought of as acting merely dynamically, by virtue of their mass and velocity; they are carriers of electricity; and it would seem to be mainly to this circumstance that some at least of their effects are due. He indicated what he believed to be the true answers to the objections of those who regard the cathodic rays as processes in the ether; and adopting the theory that they are streams of molecules, explained how, in his opinion, this theory, taken in connection with the more salient features of the X rays to which the cathodic rays give birth, leads us to a theory of the nature of the X rays.

Everything leads us to regard the X rays as being, like rays of light, some process going on in the ether; and sufficient indications of their polarization appear to have been obtained to lead us, at least when those indications are taken along with the undoubted polarization of the Becquerel rays with which they have so many properties in common*, to refer the Röntgen as well as the Becquerel rays to a disturbance transverse to the direction of propagation.   The absence of refraction, which is so remarkable a feature of the X rays, leads us to regard their progress through ponderable matter as taking place by vibrations in the ether existing in the interstices between the ponderable molecules; a view which if correct leads us incidentally to a somewhat novel view as to the mechanism of the refraction of light.   The absence, or almost complete absence, of diffraction and interference of the X rays leads to one of two alternatives:—either that they are of excessively short wave-length, or that they are non-periodic or only very slightly periodic, the X light being on the latter supposition regarded as a vast succession of independent pulses analogous to the "hedge-fire" of a regiment of soldiers†.   According to the author's view, each electrically charged molecule on arrival at the target gives rise to an independent pulse, and the vastness of the number of pulses depends on the vastness of the number of molecules in even a minute portion of ponderable matter.

[* Now however traced to a different cause : cf. *infra*, p. 274.]

[† On the question whether these alternatives are essentially different, see, on the other side, Lord Rayleigh, *Nature*, LVII, 1898, p. 607, reprinted in *Scientific Papers*, IV, p. 353.   According to the view expressed in the text there must be some kind of statistical regularity in the succession of such pulses, in order that they may reinforce each other, and thus constitute ordinary light capable of regular refraction.   Cf. *infra*, p. 272.]

On the Nature of the Röntgen Rays.

[The Wilde Lecture; delivered *July* 2, 1897. From *Memoirs and Proceedings of the Manchester Literary and Philosophical Society*, Vol. XLI.]

EVER since the remarkable discovery of Professor Röntgen was published, the subject has attracted a great deal of attention in all civilised countries, and numbers of physicists have worked experimentally, endeavouring to make out the laws of these rays, to determine their nature if possible, and to arrange for their application. I am sorry to say that I have not myself worked experimentally at the subject; and that being the case, there is a certain amount of presumption perhaps in my venturing to lecture on it. Still, I have followed pretty well what has been done by others, and the subject borders very closely on one to which I have paid considerable attention; that is, the subject of light.

In Röntgen's original paper he stated that it was shown experimentally that the seat of these remarkable rays was the place where the so-called cathodic rays fall on the opposite wall of the highly-exhausted tube in which they are produced. I will not stop to describe what is meant by cathodic rays. It would take me too much away from my subject, and I may assume, I think, that the audience I am now addressing know what is meant by that term. This statement of Röntgen's was not, I think, universally accepted. Some experimentalists set themselves to investigate the point by observing the positions of the shadows cast by bodies subjected to the discharge of the Röntgen rays—to investigate, I say, the place within the tube from which the rays appeared to come. Now, when the shadows were received on a photographic plate, and the shadow was joined to the substance casting the shadow, and the joining lines were produced backwards, as a rule they tended more or less nearly to meet somewhere within the tube—Crookes' tube, I will now call it—and some people seem to have had the idea that at that point of meeting or approximate meeting there was something going on which was the source of these rays. I have in my hands a paper published in St Petersburg by Prince B. Galitzin and A. v. Karnojitzky, which contains some very elaborate photographs obtained in this way. A board was taken and ruled with cross lines at equal intervals, and at the points of intersection nails were stuck in in an upright position. The board was placed on top of the photographic plate, with an opaque substance between—a substance which these strange Röntgen rays are capable of passing through, though it is impervious to light. The shadows cast by the nails were obtained on the photograph, and this paper contains a number of the photographs. It is remarkable, considering the somewhat large space in the tube over which the discharge from the cathode is spread, that the shadows are as sharp as they actually are; and the same thing may be affirmed of the ordinary shadows of the bones of the hand, for instance, which one so frequently sees now. Another remarkable point in these photographs is that in some cases it appears as if there were two shadows of the same nail, as though there were two different sources from which these strange rays come, both situated within the Crookes' tube. Now, have we a right to suppose that the place of meeting of the lines by which the shadows are formed, prolonged backwards into the tube, is the place which is the seat of action of these rays? I think we have not. If a portion of the Crookes' tube which is influenced by the cathode discharge be isolated by, we will say, a lead screen containing a small hole, you get a portion of the cathodic rays which come out through that small hole, and you can trace what becomes of them beyond. It is found that the influence is decidedly stronger in a normal direction than in oblique directions. Professor J. J. Thomson, of Cambridge, who has worked a great deal experimentally at this subject, mentioned that to me as a striking thing. You might imagine that the fact that the shadows appear to be cast approximately from a source within the tube could be accounted for in this way. Supposing, as Röntgen believed, that the seat of the rays is in the place where the cathode discharge falls on the surface of the glass, those which come in an oblique direction have to pass through a greater thickness of glass than those which come in a normal direction. Now, glass is only partially transparent to the Röntgen rays; therefore the oblique rays would be more absorbed in passing through the glass than the rays which come in a normal direction. I mentioned that to Professor Thomson, but he said he thought the difference between the intensity of the rays which come out obliquely and those which come out in a normal direction was much too great to be accounted for in that way*. I will take it as a fact, without entering at present into any speculation as to the reason for it, that the Röntgen rays do come out from the glass wall more copiously in a normal direction than in an oblique direction. Assuming this, we can rightly say that the results obtained by Prince Galitzin and M. Karnojitzky, and similar results obtained by others, do not by any means prove that the seat of the rays is within the tube. Suppose, for example, that the tube were spherical, and a portion of this spherical surface were reached by the cathodic rays; if the Röntgen rays which passed outside came wholly, we will say, in a normal direction, produce the directions backwards and you will get the centre of the tube. But we have no right to say from that there is anything particular going on in the centre of the spherical tube. The result is perfectly compatible with Röntgen's original assertion, which I believe to be true, as to the seat of the rays.

Everything tends to show that these Röntgen rays are something which, like rays of light, are propagated in the ether. What, then, is the nature of this process going on in the ether? Some of the properties of the Röntgen rays are very surprising, and very unlike what we are in the habit of considering with regard to rays of light. One of the most striking things is the

---

* I have found by subsequent inquiry that the experiment referred to was not made by Professor Thomson himself, but by Mr J. C. McClelland, in the Cavendish Laboratory, and that on being recently repeated with the same tube the effect of the X rays was found to be by no means so much concentrated towards the normal to the wall of the tube as in the former experiment. It seems likely that the difference may have been due to use of the tube in the interval, which would have made the exhaustion higher, and caused the X rays given out to be of higher penetrative power, so as to render the increased thickness of glass which the rays emerging obliquely had to pass through to be of less consequence. But the subject is still under examination. In consequence of the result obtained in the second experiment, the statement in the text should be less absolute; but it may very well have happened that in the experiments of others the conditions may more nearly have agreed with those of the first experiment, causing what we may call the resultant activity of the X rays to have had a direction leaning towards the normal drawn from the point casting the shadow to the wall of the tube.

facility with which they go through bodies which are utterly opaque to light, such, for example, as black paper, board, and so forth. If that stood alone it would, perhaps, not constitute a very important difference between them and light. A red glass will stop green rays and let red rays through ; and just in the same way if the Röntgen rays were of the nature of the ordinary rays of light, it is possible that a substance, although opaque to light, might be transparent to them. So, as I say, that remarkable property, if it stood alone, would not necessarily constitute any great difference of nature between them and ordinary light. But there are other properties which are far more difficult to reconcile with the idea that the Röntgen rays are of the nature of light. There is the absence, or almost complete absence, of refraction and reflection. Another remarkable property of these rays is the extreme sharpness of the shadows which they cast when the source of the rays is made sufficiently narrow. The shadows are far sharper than those produced under similar circumstances by light, because in the case of light the shadows are enlarged as the effect of diffraction. This absence, or almost complete absence, of diffraction is then another circumstance distinguishing these rays from ordinary rays of light. In face of these remarkable differences, those who speculated with regard to the nature of the rays were naturally disposed to look in a direction in which there was some distinct difference from the process which we conceive to go on in the propagation and production of ordinary rays of light. Those who have speculated on the dynamical theory of double refraction have been led to imagine the possible existence in the ether of longitudinal vibrations, as well as those transversal vibrations which we know to constitute light. If we were to suppose that the Röntgen rays are due to longitudinal vibrations, that would constitute such a very great difference of nature between them and rays of light that a very great difference in properties might reasonably be expected. But assuming that the Röntgen rays are a process which goes on in the ether, are the vibrations belonging to them normal or transversal ? If we could obtain evidence of the polarisation of those rays, that would prove that the vibrations were not normal but transversal. But if we fail to obtain evidence of polarisation, that does not at once prove that the vibrations may not after all be transversal, because the properties of these rays are such as to lead us *à priori* to expect great

17—2

difficulties in the way of putting in evidence their polarisation, if, indeed, they are capable of polarisation at all. Several experimentalists have attempted, by means of tourmalines, to obtain evidence of polarisation, but the result in general has been negative. Of the two photographic markings that ought to be of unequal intensity on the supposition of polarisation, one could not say with certainty that one was darker than the other. Another way of obtaining polarised light is by reflection at the proper angle from glass or other substance; but, unfortunately for the success of such a method, the Röntgen rays refuse to be regularly reflected, except to a very small extent indeed. The authors of the paper to which I have already referred appear to have had some success with the tourmaline. Like others who have worked at the same experiment, they took a tourmaline cut parallel to the axis and put on top of it two others, also cut parallel to the axis, and of equal thickness, which were placed with their axes parallel and perpendicular respectively to that of the under tourmaline. But they supplemented this method by a device which is not explained in the paper itself, although a memoir is referred to in which the explanation is to be found—at least by those who can read the Russian language, which unfortunately I cannot. I can, therefore, only guess what the method was. It is something depending on the superposition of sensitive photographic films. I suspect they had several photographic films superposed, took the photographs on these, and then took them asunder for development, and after development put them together again as they had been originally. They consider that they have succeeded in obtaining evidence of a certain amount of polarisation. If we assume that evidence to be undoubted, it decides the question at once. But as the experiment, as made in this way, is rather a delicate one, it is important for the evidence that we should consider as well what we may call the Becquerel rays. If time permits, I shall have something to say about these towards the close of my lecture, but, for the present, I shall say merely that they appear to be intermediate in their properties between the Röntgen rays and rays of ordinary light*. The Becquerel rays undoubtedly admit of polarisation, and the evidence appears on the whole pretty conclusive that the Röntgen rays, like rays of

[* Of their actual complex constitution much exact knowledge has of course since been gained.]

ordinary light, are due to transversal, and not to longitudinal, vibrations. It remains to be explained, if we can explain it, wherein lies the difference between the nature of the Röntgen rays and rays of ordinary light which accounts for the strange and remarkable difference in the properties of the two. I may mention that, although Cauchy and Neumann, and some others who have written on the dynamical theory of double refraction, have been led to the contemplation of normal vibrations, Green has put forward what seems to me a very strong argument against the existence of normal vibrations in the case of light. The argument Green used always weighed strongly with me against the supposition that the Röntgen rays were due to longitudinal vibrations; and the experiments by which, as I conceive, the possibility of their polarisation has now been established, go completely in the same direction, showing that they are due, assuming them to be some process going on in the ether, to a transversal disturbance of some kind.

Now, the so-called cathodic rays are, as we may say, the parents of the Röntgen rays. Consequently, if we are to explain the nature of the Röntgen rays, it is very important that we should have as clear ideas as may be permissible of the nature of the cathodic rays. Now, two views have been entertained as to the nature of the cathodic rays. According to one view, they are not rays of light at all, but streams of molecules which are projected from the cathode, and, if the exhaustion within the tube be sufficient, reach the opposite wall. That was the idea under which Crookes worked in his well-known experiments, and, so far as I know, it is the view held by all physicists in this country. Another opinion, however, has been published, and there are some eminent physicists who favour it, especially, I think, in Germany. According to this latter opinion, the cathodic rays are, like rays of light, some process going on in the ether. The cathodic ray, coming from the cathode towards the opposite wall of the tube, is invisible as such if you look across it. There is in reality a faint blue light ordinarily, but not necessarily, seen when you look across it. Lenard, in his most elaborate and remarkable experiments, succeeded in producing the cathodic rays within a space from which the gas was so very nearly completely taken away that, although the cathodic rays passed freely through the space, there was no appearance of the blue light when you viewed their path transversely. They produced, however, the ordinary effect of phosphorescence at the other end of the tube. The appearance, then, may be analogous to that of a sunbeam coming from a hole in the clouds. If it were not for the slight amount of dust and suspended matter in the air, the sunbeam would be invisible if you looked across it. But as the air is never free from motes, you see the path of the sunbeam when you look across it by the light reflected from these motes. Something of the same kind may be conceived to take place with regard to the cathodic rays if they are some process going on in the ether. But there are very great difficulties in the way of this second hypothesis, and especially as regards certain properties of the cathodic rays. In the first place, they act mechanically. In Crookes' experiments he succeeded in causing a light windmill, if I may so describe it, to spin rapidly under the action of the rays. And when they were received on a very thin film of blown glass, the glass was actually bent under them as they fell upon it. But that is not all. These cathodic rays appear to proceed in a normal direction from the cathode, and ordinarily proceed in straight lines. But—and this is the important point—they are capable of being deflected in their path both by electro-static force and by magnetic or electro-dynamic force. Nothing whatever of the kind occurs with rays of light, and there are enormous, almost insuperable difficulties in the supposition of any such deflection occurring if the cathodic rays are a process going on in the ether. I will not go into all the arguments for and against the two views, especially as the cathodic rays only enter incidentally into the subject I have chosen to bring before you. I will confine myself to one or two chief difficulties in the way of the supposition that the cathodic rays are streams of molecules. In his admirable experiments Lenard produced the cathodic rays in a tube which was highly exhausted, but not exhausted to the very highest degree that art can obtain. When you get to such tremendous exhaustions as that you cannot get the discharge to pass through the tube. What did he do? Previous experiments had shown that certain metals—aluminium especially—are, or appear to be, to a certain extent transparent to these rays. Working on the supposition that an aluminium plate is, to a certain extent, transparent to these rays, Lenard constructed a tube, highly exhausted, but not to the very last degree. Then a window of aluminium foil—a very small aperture for mechanical reasons—was fastened in an air-tight manner at the end of the tube, to lead into a second tube provided with a phosphorescent screen. The cathodic rays produced in the first tube fell upon the aluminium plate, and, as Lenard supposed, passed through it as rays of light would pass through glass. And so he got them into the second tube, and it not being necessary to make an electric discharge pass through the second tube, he could exhaust it to the very highest power of skill that he had. It was a work of days and days. The cathodic rays behaved in this very highly exhausted tube like ordinary cathodic rays. We are asked to assume that we are dealing here with a vacuum, and according to Lenard that shows—and no doubt it would if we grant the assumption—that it is no longer a question of matter, but of some process going on in the ether*. And, apparently on the strength of that very elaborate experiment, Röntgen in his first paper seems to have been of the opinion that the cathodic rays were something going on in the ether. But are we justified in assuming that we are here dealing with a perfect vacuum? I do not think we are. I believe it passes the power of art to produce a perfect vacuum. You always have a little residue of which you cannot absolutely get rid, and some of Lenard's own figures show the effect of the residual gas. He isolated by screens a small part of the cathodic discharge in the second tube, and received it on a phosphorescent screen. He represents the phosphorescent light in the tube as consisting of a bright nucleus surrounded by a less bright halo. The bright nucleus was such as would be produced if the cathodic rays were rays of light, provided that that light were incapable of diffraction. But, then, how do you account for the halo? The blue light by which the cathodic rays are seen under ordinary circumstances is due, I believe, to an interference of the projected molecules with the molecules of the gas. In some of Lenard's experiments he received the cathodic rays in the first tube into the air, and a considerable amount of this blue light was seen. The appearance was much as if you had admitted a beam of light into a mixture of milk and water. To my mind this fainter halo in the most

* Even if the vacuum were perfect, and the result were still the same, that would not disprove the theory that the cathodic rays are streams of molecules, for the molecules might have been obtained from the aluminium window itself.

refined of Lenard's experiments, lying outside this well-defined nucleus, was evidence that the vacuum, in spite of all the skill and time expended upon it, was not perfect.   And for aught we know to the contrary—I believe, indeed, it is the case—the cathodic rays in the second highly-exhausted tube were really streams of molecules coming from the residual gas in the tube. But now comes a difficulty with regard to the passage of the cathodic rays through an aluminium plate.   If the cathodic rays were something going on in the ether we might very well understand that an aluminium plate might be transparent to them although opaque to ordinary rays of light.   But if the cathodic rays are really streams of molecules, how can we imagine that they get through the plate ?   Do they get through the plate ? I do not believe they do.   Do they riddle the plate like a bullet going through a thin piece of board ?   I do not think it.   Suppose you have a trough containing a solution of sulphate of copper, and at the ends of it you have two copper plates ; if you send an electric current through the trough, copper is eaten away at the anode and deposited at the cathode.   Now, suppose you divide this trough into two by a plate of copper, you still have copper eaten away at the original anode and copper deposited at the original cathode.   The interposed plate really divides the cell into two, in each of which electrolysis goes on, so that you have not only copper eaten away at one end of the trough and deposited at the other, but in your interposed plate you have copper eaten away at one side and deposited at the other.   So it may be that the second surface of the aluminium foil becomes, as it were, a new cathode, and starts cathodic rays.   This, perhaps, is not what we should have anticipated before.   Still, there is nothing unnatural in it, and nothing, it seems to me, in consequence of which you would be obliged to reject the theory which makes the cathodic rays to be streams of molecules.   There are one or two other difficulties mentioned by Wiedemann, but I do not think they are at all serious ; they are certainly not so serious as the one I have just referred to.   I will therefore pass on.   The possibility of deflecting the cathodic rays by electrostatic and magnetic forces seems to be an insuperable difficulty in the way of the theory which. makes them to be a process going on in the ether ; but both of these are perfectly in accordance with what was to be expected on the supposition that they are streams of molecules, provided you remember that these molecules are highly charged with electricity. A moving charged body behaves as regards deflection like an electric current. Again, if you have highly-charged molecules in the neighbourhood of a positively or negatively statically-charged body, they will be attracted or repelled, and the deflections of the rays are precisely what was to be expected according to that theory. I think we may assume that the cathodic rays are really streams of electrified molecules which strike against the opposite wall of the tube, or, as I will now call it, the target. Now, when a molecule, coming in this way from the cathode, strikes the target, how does the molecule act? It may act in two ways. It may act as a mass of matter, infinitesimal though it be, by virtue of its momentum —by virtue of its mass and velocity—and it may act also as a charged body, a statically-charged body. What the appropriate physical idea is of a statically-charged body is more than I can tell you. I was talking not long ago to Lord Kelvin about it—and he is a far higher authority in electrical matters than I am—and he considers that the physical idea of a statically-charged body is still a mystery to us. Well, if these charged molecules strike the target we may think it exceedingly probable that by virtue of their charge they produce some sort of disturbance in the ether. This disturbance in the ether would spread in all directions from the place of disturbance, so that each projected molecule would on that supposition become, on reaching the target, a source of ethereal disturbance spreading in all directions*. Well, what is the character of such a disturbance? The problem of diffraction, dynamically considered, may be supposed to reduce itself to this. Suppose you have an infinite mass of an elastic medium, and suppose a small portion is disturbed in the most general way possible, what will take place? A wave of disturbance will spread out spherically from the place of disturbance†. You might at first sight suppose that you could have a wave, in any limited region of which you might have a transversal disturbance in some one direction, the same all through the thickness of the shell

[* This view was expressed independently by E. Wiechert, *Koenigsberg Memoirs*, 1896.]

† If the medium be compressible there will be two waves, that which travels the more swiftly consisting of normal vibrations; but the opinion has already been expressed that it is transversal vibrations with which we are concerned.

occupied by the wave, though naturally the direction of disturbance might vary from one region to another more or less distant region. But the dynamical theory shows that that is not possible. In any limited region, or elementary area, as we may regard it, of the wave, as you pass in a direction perpendicular to the front, the disturbance in one direction must be exchanged for a disturbance in the opposite direction, in such a manner that ultimately—that is, when the radius of the wave is very large compared with its thickness—the integral of the disturbance in one direction, which we may designate as positive, must be balanced by the integral of the disturbance in the opposite, or negative, direction. The simplest sort of "pulse," as I will call it, in order to distinguish it from a periodic undulation, would be one consisting of two halves in which the disturbances were in opposite directions. The positive and negative parts are not necessarily alike, as one may make up by a greater width, measured in the direction of propagation, for a smaller amplitude; but it will be simplest to think of them as alike, except as to sign. The following figure represents this conception, the positive and negative halves being distinguished by a difference of shading*.

According to the view here put forward, the Röntgen emanation consists of a vast succession of independent pulses, starting respectively from the points and at the times at which the individual charged molecules projected from the cathode impinge on the target. At first sight it might appear as if mere pulses would be inadequate to account for the effects produced, seeing that in the case of light we have to deal with series consisting each of a very great number of consecutive undulations. But we must bear in mind how vast, according to our theoretical views, must be the number of molecules contained in the smallest quantity of ponderable matter of which we can take cognisance by our senses. Hence, small as is the quantity of matter projected in a given short time from the cathode, it may yet be sufficient to give rise to pulses the number of which is inconceivably great. It remains to consider in what way this conception may enable us to explain

[* This holds good also on the electric theory.]

the most striking properties of the Röntgen rays in relation to the contrasts which they offer to rays of light.

The most elementary difference, as being one which has relation only to propagation in the ether, consists in the absence, or, at any rate, almost complete absence, of diffraction. As the different pulses are by hypothesis quite independent of one another, we have to explain this phenomenon for a single pulse.

In the figure let CB be a portion of a spherical pulse spreading outwards from the centre of disturbance (which I will call O) from

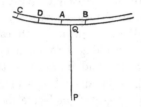

which it came, P a point in front of the wave, where the disturbance which will arrive there is sought. From P let fall a normal PQ on the front of the wave, and let AB, taken around Q, be a small portion of the spherical shell which at the present moment is the seat of the pulse, and suppose the breadth of AB to be small compared with PQ and with the radius of the shell, but large compared with the shell's thickness. Let CD be an element of the shell of similar size to AB, but situated in a direction from P distinctly inclined to PQ; and supposing all the disturbance in the shell stopped except what occupies one or other of the elements AB, CD, let us inquire what will be the disturbance subsequently produced at P in the two cases respectively.

I have shown elsewhere* that in our present problem the disturbance at P is expressed by a double integral taken over such portion of the surface of a sphere with P for centre and bt for radius (b being the velocity of propagation) as lies within the disturbed region, which in this case is the spherical shell or a part of it. It will be convenient to think of a series of spheres drawn round P with radii bt for increasing values of t. When t is such that the sphere just touches the shell at Q, and then goes on

* "On the Dynamical Theory of Diffraction." *Cambridge Philosophical Transactions*, Vol. IX, p. 1, or *Collected Papers*, Vol. II, p. 243, Arts. 19—22.

increasing, the disturbance is nearly the same all over that portion of the surface of the sphere which lies within the small region AB, and that, whether we take the portion of the expression for the disturbance at P which depends on the disturbance (displacement or velocity) at the surface of the sphere whose radius is bt, or the portion which depends on the differential coefficient of the displacement or velocity with respect to a radius vector drawn from O. Consequently the positive and negative parts of the disturbance will reach P in succession. But if instead of the small portion AB of the shell we take CD, lying in a direction from P not very near the normal, it is easy to see that the positive and negative parts of the disturbance expressed by our double integral, reaching as they do P simultaneously, almost completely cancel each other. And this cancelling is so much more nearly complete as the obliquity is greater, and likewise as the thickness of the shell is smaller. If, then, the disturbance in the ether consequent on the arrival of any projected molecule at the target is very prompt, lasting it may be only a very small fraction of the period of a single vibration of the ether in the case of light, our shell will be so thin that a small isolated portion of the Röntgen discharge is propagated so nearly wholly in the direction of a normal to the wave that the almost complete absence of diffraction is thus accounted for*.

The explanation which has just been given of the apparent absence of diffraction in the case of the Röntgen rays is closely analogous to the ordinary explanation of the existence of rays and shadows. It differs, however, in this respect, that here we are dealing with a single pulse, whereas in the case of light we are dealing with an indefinite succession of disturbances. In order to understand the sharpness of the shadows produced by the Röntgen

---

* It is known that there is a difference of quality in Röntgen rays, and that the Röntgen discharge may be filtered by absorption. It is known also that the increased exhaustion in a Crookes' tube, which is accompanied by increasing difficulty in sending a discharge through it, has the effect of giving rise to increasing penetrative power in the Röntgen rays which it gives out. It seems to me probable that this difference of quality corresponds to a more or less close approach to perfect abruptness in the production of disturbance in the ether when a molecule propelled from the cathode reaches the target, and accordingly to a less or a greater thickness in the outward-travelling shell of disturbance in the ether; and that at relatively high exhaustions the molecules are propelled with a higher velocity, and so give rise to a more prompt disturbance when they reach the target.

rays, we are not obliged to suppose that the disturbance is periodic at all. It must be partly negative and partly positive, and that being the case, if the thickness of the shell is very small, the amount of diffraction will be very small, too. Those who have attempted to obtain evidence of the diffraction of the Röntgen rays have been led to the conclusion that if the rays are periodic at all the period is something enormously small—perhaps thirty times, perhaps a hundred times, as small as the wave-length of green light. It seems difficult to imagine by what process you could get such very small vibrations, if vibrations there be. It is easier to understand how the arrival of charged molecules at the cathode might produce disturbances which are almost abrupt*.

Well, then, this is what I conceive to constitute the Röntgen rays. You have a rain of molecules coming from the electrically-charged cathode, which you may think of as the rain-drops in a shower. They strike successively on the target, each molecule on striking the target producing a pulse, as I have called it, in the ether, which is essentially partly positive and partly negative; and you have a vast succession of these pulses coming from the various points of the target which are not protected by some screen interposed for the purpose of experiment.

This explains the absence, or almost complete absence, of diffraction. But that is not all we have to explain; we have still a very serious thing behind. What is it that constitutes the difference between the Röntgen rays and rays of ordinary light in consequence of which the one are not refracted, or only in an infinitesimal degree, while the other are freely refracted? This difficulty led me to conceive of a theory, which I believe to be new, as to the nature of refraction itself—as to the nature of what takes place, for example, when light is refracted through a prism. Suppose we have light of a definite refrangibility, and a prism on which it may be made to fall. When the light is admitted we commonly imagine—at least, I believe so—that the light is immediately refracted, and with proper appliances you get the spectrum. Immediately? I do not think so. How is it that light travels

[* For the experimental evidence as regards diffraction, cf. Haga, Tiddens, and Wind, 1897—1901, *Archives Néerlandaises*, VIII, p. 412.

Recently (*Phil. Mag.* Aug. 1903), R. J. Strutt has inferred the absence of polarisation in the rays from the absence of dichroism in the fluorescence excited by them in crystals of magnesium platino-cyanide; cf. *ante*, Vol. IV, p. 17.]

more slowly through a refracting medium than through vacuum?
There are different conjectures which have been advanced. One
is that the ether within refracting media is more dense than the
ether in free space. Another is that while the density is the
same the elasticity is less. Then, there have been speculations as
to the ether being loaded with particles of matter.

Take a piano. If you strike a note a string is set in vibration. You would hardly hear any sound at all if it were rigidly
supported. But it rests on a bridge communicating with a
sounding-board, and the sounding-board presents a broad surface
to the air, and is set in motion by the string. The sounding-
board and the string form a compound vibrating system. In the
same way it may be that the molecules of the glass, or other
refracting medium, and the ether form between them a compound
vibrating system, and *when the motion is fully established,* the two
vibrate harmoniously together. But how does it get to be established? We can hardly imagine otherwise than that the ether
is excessively rare compared with ponderable matter*. Well,

---

* The views as to the nature of refraction, which I have endeavoured to explain,
lead me incidentally to make a remark on another subject not, indeed, very closely
connected with it. From the first, Röntgen recognised as the seat of the X rays
which he had discovered the place where the cathodic rays fall on the wall of the
Crookes' tube. This place is indicated to the eye by the fluorescence of the glass.
But we are not on that account to regard the fluorescence as the cause of the
Röntgen rays, or even to regard the Röntgen emission as a sort of fluorescence.
I have seen it remarked, as indicating no very close connection between the two,
that with a metallic target we have a copious emission of Röntgen rays though
there is no fluorescence, and that when a spot on the glass wall of a Crookes' tube
has for some time been exposed to a rather concentrated cathodic discharge, though
the fluorescence which it exhibits under the action of the cathodic discharge becomes
comparatively dull, as if the glass were in some way fatigued for fluorescence, it
emits the Röntgen rays as well as before.

Fluorescence is undoubtedly indicative of a molecular disturbance; but in what
precise way this disturbance is brought about by the cathodic discharge, is a matter
on which I refrain from speculating. But whatever be the precise nature of the
process, it seems pretty evident that it can only be by repeated impacts of molecules
from the cathode that a sufficient molecular disturbance can be got up to show
itself as a visible fluorescence.

Suppose a shower of molecules from the cathode to be allowed suddenly to fall
on the anti-cathode, and after raining on it for a little to be as suddenly cut off.
According to the views I entertain as to the nature of the Röntgen rays, the
moment the shower is let on the emission of Röntgen rays begins, it lasts as long
as the shower, and ceases the moment the shower is cut off. But the fluorescence
only gradually, quickly though it may be, comes on when the shower is allowed to
fall, and gradually fades away when the shower is cut off. So far from the

supposing the ethereal vibrations start and reach a set of molecules, they are somewhat impeded by the molecules, and they tend also to move the molecules. But as the molecules are relatively very heavy, it may be that it takes some considerable time for the molecules to be set sensibly in motion. Now, if the system of molecules is exceedingly complex, a mode of motion of the molecules, or it may be of the constituent parts of the molecules, may be found such that the system tends to vibrate in practically any periodic time that you may choose; only as you choose one time or another the mode of vibration will be different; and, again, according to the direction in which the molecules are successively made to vibrate the actual mode of vibration will be different. Well, I conceive that the difference between the propagation of the Röntgen rays and rays of ordinary light with reference to passing through a prism depends upon that. When you let a ray of light fall upon a refracting medium such as glass, motions begin to take place in the molecules forming the medium. The motion is at first more or less irregular; but the vibrations ultimately settle down into a system of such a kind that the regular joint vibrations of the molecules and of the ether are such as correspond to a given periodic time, namely, that of the light before incidence on the medium. That particular kind of vibration among the molecules is kept up, while the others die away, so that after a prolonged time—the time occupied by, we will say, ten thousand vibrations, which is only about the forty thousand millionth part of a second—the motion of the molecules of the glass has gradually

fluorescence being in any way the cause of the Röntgen emission, there seems reason to think that if it exercises any effect upon it at all, it is rather adverse than favourable. For it has been found that when the target is metallic, and gets heated, the Röntgen discharge falls off; and fluorescence, like a rise of temperature, involves a molecular disturbance, though the kind of disturbance is different in the two cases.

As the fluorescence of the glass wall and the emission of X rays are two totally different effects of the same cause, namely, the molecular bombardment from the cathode, the intensity of the one must by no means be taken as a measure of the intensity of the other, even with the same tube. The former effect would appear to be the more easily produced. This consideration removes a difficulty mentioned at p. 10 of the paper by Prince Galitzin and M. v. Karnojitzky, as attending the supposition that the X rays originate in the points in which the cathodic rays fall on the wall of the tube or other target. Nor need it surprise us that in some cases the shadows seem to indicate more than one source of action, when we remember that from a given point more than one normal can be drawn to a given closed surface.

got up until you have the molecules of the glass and the ether vibrating harmoniously together. But in the case of the Röntgen rays, if the nature of them be what I have explained, you have a constant succession of pulses independent of one another. Consequently there is no chance to get up harmony between the vibrations of the ether and the vibrations of the body*.

Go back to the case of light passing through glass. When the regular combined vibration is established you have a kinetic energy, due partly to the motion of the ether and partly to the motion of the molecules. If you make abstraction of the loss of energy by reflection, the rate at which the energy passes within the glass must be the same as it has outside, and consequently there must be the same energy for one wave-length, which corresponds to one period of the vibration, inside as outside. But if the kinetic energy of the ether is the same for the same volume inside and outside, and you have in addition inside a certain amount of kinetic energy due to the motion of the molecules, the two taken together can only make the energy for a wave inside the same as for a wave outside on the condition that the velocity of propagation inside is less than the velocity of propagation outside. That is the theory† I have been forced to adopt as to the nature of refraction in conse-

[* If there were any statistical regularity in the succession of the impulses, this conclusion would not apply. The essential distinction between the Röntgen radiation and ordinary light would thus be the almost total absence of any orderly features in the impulses constituting the former. The Röntgen impulses are independent of each other, being separately excited, like the traffic of passengers in a street, or the 'hedge-fire' of a company of soldiers. On the other hand, the molecular shocks, transmitted through the ether, which constitute the 'white light' from an incandescent body, would be regarded as having some sort of statistical order in their distribution, as have the velocities of the molecules of a gas, and for a similar reason.

A simple isolated impulse, travelling in free ether, could be drawn out by reflexion from an *ideal perfectly reflecting* grating, into a periodic train of undulations, in number equal to that of the elements of the grating, as explained by Young and enforced by Lord Rayleigh, this being the physical parallel to a Fourier mathematical analysis of the impulse. But the argument in the text insists that the single pulse cannot be refracted by matter, because refraction implies that the molecules of the matter are vibrating fully in concert with the ether, and this requires the acquisition by them of a store of elastic energy which could only have been supplied by previous pulses related in some way to the one under consideration. Cf. footnote, p. 255, *supra*.]

[† Rather it is perhaps the aspect as regards energy, of the theory of dispersion, including anomalous dispersion, gradually evolved from the ideas of Stokes, Maxwell, Rayleigh, Sellmeier, Ketteler, Helmholtz, Kelvin, and others.]

quence of the ideas I hold as to the nature of the Röntgen rays ; and if you adopt that theory I think everything falls into its place. When you have the Röntgen rays falling on a body, the motion of the ether due to them is interfered with by the molecules of the body, more or less. No body is perfectly transparent to these rays, and on the other hand perhaps we may say no body is perfectly opaque. That all falls into its place on this supposition as to the nature of the action of the ether on the molecules. Now, why is it that the Röntgen rays do not care whether you present them with black paper or white paper ? What is the cause of blackness ? The light falling upon the paper produces motion in the ultimate molecules. In the case of a transparent substance you have a compound vibrating system going on, vibrating without change. But in the case of an absorbing medium the vibrations which after a time are produced in the molecules spread out into adjoining molecules, by virtue of the communication of the molecules with one another, and are carried away ; so that in the case of an absorbing medium there is a constant beginning to set the molecules in vibration ; but they never get to the permanent state, because the vibration is carried away by communication from one molecule to another. But in the case of the Röntgen rays you have done with the pulse altogether long before any harmonious vibration between the ether and the molecules can be established ; so that a state of things is not brought about in which you get a, comparatively speaking, large vibration of the molecules. Consequently, the Röntgen rays do not care whether you give them black paper or not.

I must not keep you more than a minute or two longer; but I do not like to close this lecture without saying a word or two regarding the Becquerel rays. What takes place there ? To be brief, I must refer to the most striking case of all. Take the case of metallic uranium. That gives out something which, like the Röntgen rays, has an influence passing through black paper, and capable of affecting a photographic plate. It is also capable of effecting the discharge of statically-charged electrified bodies. Apparently this goes on indefinitely. You do not need, apparently, to expose the metal to rays of high refrangibility in order that this strange thing should go on. What takes place ? My conjecture is that the molecule of uranium has a structure which may be roughly compared to a flexible chain with a small weight at the end of it. Suppose you have vibrations communicated to such a chain at the top; they travel gradually to the bottom, and near the bottom produce a disturbance which deviates more from a simple harmonic undulation. So, if a vibration is communicated to what I will call the tail of the molecule of uranium, it may give rise to a disturbance in the ether which is not of a regular periodic character. I conceive, then, that you have vibrations produced in the ether, not of such a permanently regular character as would constitute them vibrations of light, and yet not of so simple a character as in the Röntgen rays— something between. And accordingly there is enough irregularity to allow the ethereal disturbance to pass through black paper, and enough regularity on the other hand to make possible a certain amount of refraction*. You can also obtain evidence of the polarisation, and, consequently, of the transverse character of these rays†.

According to the theory of the nature of the Röntgen rays which I have endeavoured very briefly to bring before you, we have here, as I think, a system the various parts of which fit into one another. You start with the Röntgen rays, which consist, as I conceive, of an enormous succession of independent pulses; you pass to the Becquerel rays, which are still irregular, but are beginning to have a certain amount of regularity; and you end with the rays which constitute ordinary light. According to this theory, the absence of diffraction in the Röntgen rays is explained, not by supposing they are rays of light of excessively short wavelength, but by supposing they are due to an irregular repetition of isolated and independent disturbances. So far as I know, the view I have been led to form as to the nature of refraction, and which forms an integral portion of the theory as to the Röntgen rays, is altogether new; so much so that I felt at first rather startled by it; but I found myself fairly driven to it by the ideas I entertain as to the nature of the Röntgen rays, and I am not aware of any serious objection to it.

[* According to the interpretation of Sir George Stokes' views advanced *supra*, p. 272, footnote, this implies that the various molecules give out partially periodic disturbances which are entirely out of connexion with each other, not being correlated by any dependence on temperature.]

[† The Becquerel rays were a recent discovery when this was written; it has since been ascertained that they consist of projected electric particles, together with radiation analogous to that of Röntgen.]

ADDITIONAL NOTE.

The problem of diffraction in the case of a vast system of *independent* very slender pulses deserves to be treated in somewhat greater detail. It is rather simpler than the problem of diffraction in the case of series of undulations such as those which constitute light, because the pulses are to be treated separately and independently, like streams of light from different sources*; and as the whole thickness of a pulse in the case of the Röntgen rays may probably be something comparable with the millionth of an inch, we have no need to inquire what will be the disturbance continually passing across a fixed surface in space; we may treat the shell at any moment as constituting an initial disturbance in the ether, and then examine the efficiency of different parts of the shell in disturbing at a future time the ether at a given point of space in front of the shell.

The thickness of the shell is not necessarily the same at points situated in widely different directions as regards their bearing from the centre, and the same applies to the direction of disturbance. But in any case for a small portion of the shell the thickness may be deemed uniform, and the direction of disturbance sensibly the same as we pass from point to point in a direction tangential to the shell, while it varies with great rapidity, at least as regards its amount, when we pass from point to point in a normal direction, vanishing at the outer and inner boundaries of the shell.

As the disturbance we are concerned with is of the distortional kind only, the disturbance at time t at a point P in front of the shell may be obtained from that at time 0 in the shell in its position which is taken as initial by the last equation in Art. 22 of my paper on diffraction already cited. Let R be a point in the shell of disturbance when in that position which is regarded as initial, r, r' the distances PR, OR; θ, θ' their inclination to OP; ϕ the azimuth round OP of the plane PRO. Then in the formula

[* The radiation from each point-source or single molecule may however be considered as divided up into such spherical shells, each of the breadth of one wave-length, to which these considerations will apply.]

referred to $d\sigma = \sin\theta \, d\theta \, d\phi$.   Also $rd\theta \times \sin(\theta + \theta') = dr'$; and $\sin\theta/\sin(\theta + \theta') = r'/OP = r'/(r + r')$ very nearly.

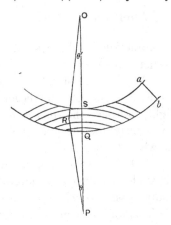

Let OP cut the inner boundary of the shell in S, and let ab or QS, the thickness of the shell, be denoted by λ. In the equation referred to, the term arising from the differentiation with respect to t of the t outside the sign of double integration will be of the order λ/r' as compared with the others, and may, therefore, be neglected. The t outside may be replaced by r/b, and the fraction $r/(r + r')$, being sensibly constant over the range of integration, may be put outside. Our expression then becomes

$$4\pi b\xi = \frac{r'}{r + r'}\iint\left(u_0 - b\frac{d\xi_0}{dr'}\right)_{bt} dr' d\phi *.$$

As the disturbance deemed initial was only a momentary condition of a wave that had been travelling outwards with the velocity b, we must have $u_0 = -b\dfrac{d\xi_0}{dr'}$, and therefore

$$2\pi\xi = -\frac{r'}{r + r'}\iint\left(\frac{d\xi_0}{dr'}\right)_{bt} dr' d\phi.$$

The expression is left in the first instance in this shape in order to show more clearly the manner in which each portion of the disturbance in the state taken as initial contributes towards the future disturbance at P. When there is no obstacle to the transmission we shall have $\int d\phi = 2\pi$, and $\int\left(\dfrac{d\xi_0}{dr'}\right)_{bt} dr' = (\xi_0)_{bt}$ taken

---

* The suffix bt means that the integration is taken over a spherical surface with centre P and radius bt.

between limits. If $bt < PQ$, the sphere round P with radius bt does not cut the disturbed region at all, and the disturbance at P is *nil*. If $bt > PS$, the limits of r' are the distances from O at which the sphere round P cuts the inner and outer limits of the shell, and as the disturbance there vanishes, we have again no disturbance at P. But if bt lies between those limits, and the sphere round P cuts OP in T (which point must lie between Q and S) the limits of r' will be OT to a point in the outer boundary of the shell, where therefore ξ_0 vanishes. Hence the displacement at P is the same as was initially at T, only diminished in the ratio of $r + r'$ to r', as we know it ought to be.

Reverting to the expression for ξ given by the double integral, we see that the only portion of the shell which is efficient in producing a subsequent disturbance at P lies between the sphere round O with radius OQ and the sphere round P with radius PS. If β be the distance from OP of the intersection of these spheres, we have, considering the smallness of the obliquities,

$$\beta^2 = \frac{2rr'\lambda}{r+r'}.$$

If we suppose r and r' to be each 4 inches, and λ the millionth of an inch, we have $\beta = 0.002$ inch, so that at a distance not less than the one-250th of an inch from the projection of the edge of an opaque body intercepting Röntgen rays coming from a point 4 inches off, and received on a screen (fluorescent or photographic) 4 inches on the other side, there would be full effect or no effect according as we take the illuminated or the dark side of the projection. We see then how possible it may be to have an almost complete absence of diffraction of the Röntgen rays if the pulses are as thin as above supposed; and as these rays are started in the first instance in a totally different manner from rays of ordinary light, namely, by the arrival of charged molecules from a cathode at a target instead of by the vibrations of the molecules of ponderable matter, we know of no reason beforehand forbidding us to attribute an excessive thinness to the pulses which the charged molecules excite in the ether.

MATHEMATICAL PROOF OF THE IDENTITY OF THE STREAM LINES
OBTAINED BY MEANS OF A VISCOUS FILM WITH THOSE OF
A PERFECT FLUID MOVING IN TWO DIMENSIONS.

[From the *Report of the British Association*, 1898.]

THE beautiful photographs obtained by Professor Hele-Shaw*
of the stream lines in a liquid flowing between two close parallel
walls are of very great interest, because they afford a complete
graphical solution, experimentally obtained, of a problem which,
from its complexity, baffles the mathematician, except in a few
simple cases.

In the experimental arrangement liquid is forced between
close parallel plane walls past an obstacle of any form, and the
conditions chosen are such that whether from closeness of the
walls, or slowness of the motion, or high viscosity of the liquid,
or from a combination of these circumstances, the flow is regular,
and the effects of inertia disappear, the viscosity dominating
everything. I propose to show that under these conditions the
stream lines are identical with the theoretical stream lines belonging to the steady motion of a *perfect* (*i.e.*, absolutely inviscid) liquid
flowing past an infinitely long rod, a section of which is represented
by the obstacle between the parallel walls which confine the viscous
liquid†.

Take first the case of the steady flow of a viscous liquid between
close parallel walls. Refer the fluid to rectangular axes, the origin
being taken midway between the confining planes, and the axis of
z being perpendicular to the walls. As the effects of inertia are
altogether dominated by the viscosity, the terms in the equations

[* Cf. also *Phil. Trans.* 195 A, 1900, pp. 303—327.]
[† Cf. a similar discussion, *supra*, p. 205.]

of motion which involve products of the components of the velocity and their differential coefficients may be neglected. Gravity, again, need not be introduced, as it is balanced by the variation of hydrostatic pressure due to it. The equations of motion, then, with the usual notation, are simply

$$\frac{dp}{dx} = \mu \left(\frac{d^2u}{dx^2} + \frac{d^2u}{dy^2} + \frac{d^2u}{dz^2} \right),$$

with similar equations for y, v and z, w, μ being the coefficient of viscosity.

In the present case the flow takes place in a direction parallel to the walls, so that $w = 0$, and the third equation of motion gives $\frac{dp}{dz} = 0$, so that p is constant along any line perpendicular to the walls. The velocities u, v vanish at the walls, and along any line perpendicular to the walls are greatest in the middle. As by hypothesis the distance ($2c$) between the walls is insignificant compared with the lateral dimensions of the obstacle, the rates of variation of u and v when x and y vary may be neglected compared with their variation consequent on that of z. Hence the equations of motion become simply

$$\frac{dp}{dx} = \mu \frac{d^2u}{dz^2}, \quad \frac{dp}{dy} = \mu \frac{d^2v}{dz^2} \quad \ldots\ldots\ldots\ldots\ldots(1),$$

which must be combined with

$$\frac{du}{dx} + \frac{dv}{dy} = 0 \ldots\ldots\ldots\ldots\ldots\ldots\ldots(2).$$

Over an area in the plane xy, which is small compared with the obstacle, though large compared with c^2, the whole velocity and each component vary, as we know, as $c^2 - z^2$; so that if u', v' denote the mean components along a line perpendicular to the walls

$$u = \frac{3}{2} u' \left(1 - \frac{z^2}{c^2} \right), \quad v = \frac{3}{2} v' \left(1 - \frac{z^2}{c^2} \right),$$

and (1) and (2) give

$$\frac{dp}{dx} = -\frac{3\mu}{c^2} u', \quad \frac{dp}{dy} = -\frac{3\mu}{c^2} v', \quad \frac{du'}{dx} + \frac{dv'}{dy} = 0 \ \ldots\ldots(3).$$

If ψ be the stream line function, taken, say, with reference to the mean velocities u', v',

$$d\psi = u'dy - v'dx,$$

and the elimination of p from the first two equations (3) gives

$$\frac{d^2\psi}{dx^2} + \frac{d^2\psi}{dy^2} = 0 \quad \dots\dots\dots\dots\dots\dots(4).$$

The general partial differential equation (4), combined with the condition that the boundaries shall be stream lines, serves to determine completely the function ψ. It may be remarked that the lines of equal pressure are the orthogonal trajectories of the stream lines, and can therefore be drawn from the photographs. If we suppose the stream lines equally spaced out in a part of the fluid where the flow is uniform in parallel lines, the velocity at any point will be inversely as the distance between consecutive stream lines *. This statement is subject to a qualification which will be mentioned presently.

Let us turn now to the other problem, that of determining the stream lines for the irrotational motion in two dimensions of a perfect liquid flowing past an infinitely long body, a transverse section of which, by two close parallel planes, would form the obstacle in our thin plate of highly viscous liquid. In this case the stream line function satisfies the same partial differential equation (4) as before, and the conditions at the boundaries are the same, namely, that the boundaries shall be stream lines. Therefore, notwithstanding the wide difference in the physical conditions, the stream lines are just the same in the two cases. In this latter case they cannot be almost realised experimentally by means of an almost perfect fluid on account of the instability of the motion. The orthogonal trajectories of the stream lines are lines of equal velocity-potential, but not in this case lines of equal pressure.

[* When the thin stratum of liquid is curved, and its thickness is not constant, its aggregate steady flow is the analogue of steady electric flow in a current-sheet of the same form, of which the specific superficial conductivity is at each point proportional to the square of the thickness of the stratum, the pressure being the analogue of electric potential—assuming a linear law of viscosity as above.]

It may be objected that the stream lines cannot be the same in the two cases, inasmuch as the perfect liquid glides over the surface of the obstacle, whereas in the case of the viscous liquid the motion vanishes at the surface of the obstacle. This is perfectly true, and forms the qualification above referred to; but it does not affect the truth of the proposition, which applies only to the limiting case of a viscid liquid confined between walls which are infinitely close. Any finite thickness of the stratum of liquid will entail a departure from the identity of the stream lines in the two cases, which, however, will be sensible only to a distance from the obstacle comparable with the distance between the walls, and therefore capable of being indefinitely reduced by taking the walls closer and closer together.

[Extract from Prof. Hele-Shaw's paper to which the above is appended.]

The Effect of Using a Wedge-shaped Section.

The author attempted to solve the problem of obtaining the flow round a solid of revolution by using a wedge-shaped section, the obstruction being also represented by a wedge representing a segment of the body, the thinnest part of the wedge corresponding to the axis of revolution.

Professor Stokes has been good enough to look into this matter, and has found that the partial differential equation which the stream-line function must satisfy in the case of a slender wedge of viscous fluid is

$$\frac{d^2\psi}{dx^2} + \frac{d^2\psi}{dy^2} - \frac{3}{y}\frac{d\psi}{dy} = 0,$$

x being measured parallel and y perpendicular to the edge; whereas, for a perfect fluid flowing axially over a solid of revolution, generated by the revolution round the edge of the wedge of the body interrupting the flow in the wedge of fluid, the equation is

$$\frac{d^2\psi}{dx^2} + \frac{d^2\psi}{dy^2} - \frac{1}{y}\frac{d\psi}{dy} = 0,$$

which is not the same as the other, and therefore the stream lines are not the same in the two cases *.

* Since this paper was read Professor Sir G. G. Stokes has further investigated the matter, and has been able to obtain the equation of the stream lines for the case of a slender wedge of a viscous fluid interrupted by a wedge forming a section of a sphere, which he finds in terms of polar coordinates to be as follows:

$$\left(\frac{a^5}{r} - r^4\right) \sin^4 \theta = \text{constant.}$$

The two following equations, therefore, may, for convenience, be expressed thus:

Case of flow of perfect fluid round a sphere:

$$\left(1 - \frac{a^3}{r^3}\right) r^2 \sin^2 \theta = \text{constant.}$$

Case of slender wedge with spherical sector:

$$\left(1 - \frac{a^5}{r^5}\right) r^4 \sin^4 \theta = \text{constant;}$$

and Professor Stokes remarks that the equation shows, even without plotting, the general character of the difference between the wedge lines and spherical lines.

ON THE DISCONTINUITY OF ARBITRARY CONSTANTS THAT APPEAR AS MULTIPLIERS OF SEMI-CONVERGENT SERIES. (*A Letter to the Editor.*)

[From *Acta Mathematica*, Vol. xxvi, 1902, Abel Centenary Volume, pp. 393—397.]

CAMBRIDGE, 23 *April*, 1902.

DEAR SIR,

I regret that from circumstances which I need not detail the invitation with which you honoured me to write something for the collection of papers which are being put together in commemoration of Abel has remained so long without reply.

At my age you will perhaps hardly expect me to produce something new and original. The subject ought to be one of pure mathematics, for it is in honour of Abel, and most of my work refers to applications of mathematics. There is one thing I thought might perhaps do, but it has, I fear, been too long before the public to make it suitable. However, it is published in the *Proceedings* or *Transactions* of the Cambridge Philosophical Society, which are not, I believe, so widely known as many other serial works. I thought that just a short *résumé* of my results might not be wholly uninteresting.

The subject is the discontinuity of arbitrary constants that appear as multipliers of semi-convergent series, the variable according to powers of which the series proceed being a mixed imaginary. My results are contained in three papers read before the Cambridge Philosophical Society in the years 1857, 1868, and 1889, which are published respectively in the *Cambridge Philosophical Transactions*, Vol. x, p. 105, the *Transactions*, Vol. xi, p. 412, and the *Proceedings*, Vol. vi, p. 362*. In these papers

[* *Ante*, Vol. iv, pp. 77—109 (cf. footnote, p. 80); *ante*, Vol. iv, pp. 283—298 (cf. footnotes, pp. 289, 298) ; *supra*, pp. 221—225.]

I have for the most part confined myself to the complete integral of the differential equation of the second order which is satisfied by Bessel's Functions; an equation which without loss of generality may be put under the form

$$\frac{d^2y}{dx^2} + \frac{1}{x}\frac{dy}{dx} - \frac{n^2}{x^2}y = y \ldots\ldots\ldots\ldots\ldots(1).$$

The variables x and y are taken to be mixed imaginaries, but I have confined myself to the case in which n is real. Although I have, as I said, limited myself to the integral of that particular differential equation, the method is, I believe, of much wider application. I have not taken n integral but (subject to its being real) general. Putting $x = r(\cos\theta + i\sin\theta)$, I suppose the range of r and θ defined by the imparities

$$0 < r < \infty, \quad -\infty < \theta < \infty.$$

It is well known that the integral of (1) may be put under the form

$$y = Ax^n \left(1 + \frac{x^2}{2(2+2n)} + \ldots\right)$$

$$+ Bx^{-n}\left(1 + \frac{x^2}{2(2-2n)} + \ldots\right)\ldots\ldots\ldots(2)$$

$$= AU + BV, \text{ say,}$$

or under the form

$$y = Cx^{-\frac{1}{2}}e^x \left(1 + \frac{1^2 - (2n)^2}{8x} + \ldots\right)$$

$$+ Dx^{-\frac{1}{2}}e^{-x}\left(1 - \frac{1^2 - (2n)^2}{8x} + \ldots\right)\ldots\ldots(3)$$

$$= Cu + Dv, \text{ say.}$$

The series (2) are always convergent, and completely define the function y over the whole range, but are not available for calculation when r is large, as they begin by diverging rapidly. The series (3) are always divergent (save when $2n$ is an odd integer, when they terminate), but begin by converging rapidly when r is large. But it is easy to see that (3) cannot be equivalent to (2) over the whole range of θ unless the constants C, D have different

values in different parts of the range.   I have shown where and
how they change discontinuously.

Of the functions u, v, let that be called the superior which has
the real part of the index of the exponential positive, and that the
inferior which has it negative.   As θ increases in one direction,
the functions u, v become alternately the superior and the inferior.
I have shown that as θ changes, the constant C or D can only
change when the index of the exponential in the function it
multiplies is real and negative; and the change it then suffers is
proportional, other circumstances being the same, to the coefficient
of the other term, which of course is for that value of θ the
superior term.   The way in which the constants change with the
value of θ may be illustrated by a pair of curves of sines drawn
with inks of different colours (or also distinguished by a difference
of marking) in the different parts.   The ordinates may be taken to
represent, for a given value of r, the way in which the real part of

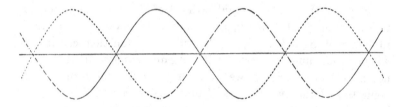

the index changes with θ.   A change of the coefficient is represented by a change in the colour of the ink (or of the marking of
the line) with which the curves are drawn.   The nature of the
change is that in crossing, in the positive direction, one of the
critical points represented by a change of colour or marking,
the coefficient (C or D as the case may be) of the inferior term
is increased by $2i \cos n\pi$ multiplied by the coefficient of the
superior term.

The way in which the paradox of giving a discontinuous expression for a continuous function is explained is this.   A semi-
convergent series (considered numerically, and apart from its
analytical form) defines a function only subject to a certain
amount of vagueness, which is so much the smaller as the
modulus of the variable according to inverse powers of which it
proceeds is larger.   I have shown that, in general (*i.e.* for general

values of θ), the vagueness of the superior function ultimately, as
r is increased, disappears in comparison with the whole value of
the inferior term.   But for the critical values of θ for which the
index of the exponential is real the vagueness of the superior
function becomes sufficient to swallow up the inferior function.
As θ passes through the critical value, the inferior term enters as
it were into a mist, is hidden for a little from view, and comes out
with its coefficient changed.   The range during which the inferior
term remains in a mist decreases indefinitely as the modulus r
increases indefinitely.

We know the value of $Cu + Dv$, that is we know the values of
C and D, for all values of θ provided we know them for a
half period, say for $0 < \theta < \pi$; and it follows from what has been
said that for the determination of C and D in terms of A and B we
may take any value or values of θ within or at the edge of that
range that we find convenient.

Suppose that in the neighbourhood of $\theta = 0$ we have only an
inferior term Dv, C being 0, and that for $\theta = 0$ we can express D
in terms of A and B.   Then Dv will be a function continuous
within the range $-\pi < \theta < \pi$.   Again, suppose that in the
neighbourhood of $\theta = \pi$ we have only an inferior term Cu, D
being 0, and that we express C in terms of A and B for $\theta = \pi$.
Then Cu will express a continuous function within the range
$0 < \theta < 2\pi$.   Now superpose these results, and $Cu + Dv$, in which
C and D have been expressed in terms of A and B, will express a
function continuous within the range $0 < \theta < \pi$.

When there is only an inferior term Dv in the neighbourhood
of $\theta = 0$, then Dv is constant for the range $-2\pi < \theta < 2\pi$, though
it is only within the range $-\pi < \theta < \pi$ that Dv represents the
continuous function y.   Let the known integral of (1) in definite
integrals P, Q be $EP + FQ$; then within the range last mentioned
$Dv = EP + FQ$, and we may determine D in terms of E and F by
putting $\theta = 0$.   Similarly if in the neighbourhood of $\theta = \pi$ there
is only an inferior term Cu, we may determine C in terms of E
and F for the range $0 < \theta < 2\pi$ by putting $\theta = \pi$.   And putting
the two together we have $(0u + Dv) + (Cu + 0v)$, or $Cu + Dv$, equal
to $EP + FQ$ within the range $0 < \theta < \pi$ provided C and D are two
known linear functions of E and F.   And as the linear relations

between E, F and A, B are easily found, we have by eliminating E and F two linear relations between C, D and A, B which will make $Cu + Dv = AU + BV$ within the range $0 < \theta < \pi$, from whence can be found the relations between A, B in the ascending and C, D in the descending series for all values of θ.

But as I showed in my third paper* these relations can be found *directly from the series* (2) and (3), by taking the *superior* terms for $\theta = 0$ and for $\theta = \pi$, the first giving C for the range $-\pi < \theta < \pi$, and the second giving D for the range $0 < \theta < 2\pi$, so that we need not know that it is possible to express the complete integral of (1) by means of definite integrals in the form $EP + FQ$. And this method I believe is of very general application.

I remain, dear sir, with the highest respect,

Yours very faithfully,

G. G. STOKES.

[* *Supra*, pp. 221—225.]

ON THE METHODS OF CHEMICAL FRACTIONATION.
BY W. CROOKES, F.R.S. (*Extract.*)

[From the *Chemical News*, Vol. LIV; read at the British Association, Birmingham Meeting, 1886.]

....

WHEN the balance of affinities, of which I spoke above, seems to be established, and the earths appear in the same proportion in the precipitate and the solution, they are thrown down by ammonia, and the precipitated earths are worked up by some other process so as to alter the ratio between them, when the previous operation can be again employed.

Fractional precipitation by ammonia is the process generally adopted, for although in some cases it is not so powerful as other processes it is more generally applicable. There are three different methods of operating. According to one plan one-third the equivalent quantity of ammonia is added; then, after the precipitate has quite settled, it is filtered, to the filtrate one-third more ammonia is added, and the precipitate is again filtered off. The remainder of the bases present are then fully precipitated. In this way the earths are divided into three parts, which I will designate − 1, 0, and + 1. Each of these lots is then treated as just described, and the thirds are added to the vessel on each side. This plan is rapid in the actual separations, but I think more time is spent in the extra filterings and washings than is saved in the lessened number of operations.

The second plan was one which was proposed by Prof. Stokes, with whom I have frequently discussed the subject of fractionation. I am indebted to him for the following description of his plan:—

Suppose that A and B are the two earths in solution. It seems reasonable to suppose that the tendency at any moment to precipitation is such that the proportion of the number of molecules of A to that of B, that go down, is that of $p \times$ No. of molecules of A in solution to $q \times$ No. of mols. of B in solution, where p and q are constants, the reciprocals of which may be called the affinities of the earths for the acid. The proportion is accordingly supposed to change even during a single pre-cipitation.

The separation is best when the ratio of the number of mole-cules left in solution to the whole number is

$$\frac{qx + py}{x + y} \left(\frac{p^p}{q^q}\right)^{1 \div (q-p)}$$

where x and y are the number of molecules of A and B present in solution before precipitation*.

[* This constitutes an early application of the ideas now associated with chemical dynamics, but the formula does not seem to verify without limitation.

An analogous, but different, problem is treated theoretically and experimentally by Lord Rayleigh, *Phil. Mag.* Nov. 1902, "On the Distillation of Binary Mixtures."

If ξ, η are the numbers of the p, q molecules that have been precipitated at a given instant of the process,

$$\frac{-d\xi}{p(x-\xi)} = \frac{-d\eta}{q(y-\eta)},$$

yielding
$$\left(1 - \frac{\xi}{x}\right)^q = \left(1 - \frac{\eta}{y}\right)^p = \lambda, \text{ say.}$$

The excess is greatest when $\xi - \eta$ is maximum; then $d\xi = d\eta$, giving from the above

$$\frac{x-\xi}{q} = \frac{y-\eta}{p}, \text{ and therefore } \lambda^{\frac{1}{q} - \frac{1}{p}} = \frac{qy}{px}.$$

The ratio of the number of molecules left of both kinds to the whole number is then

$$\frac{p+q}{x+y} \frac{x-\xi}{q},$$

which is
$$(p+q) \left(\frac{p^p}{q^q}\right)^{\frac{1}{q-p}} \left(1 + \frac{y}{x}\right)^{-1} \left(\frac{y}{x}\right)^{\frac{p}{p-q}}.$$

But the *proportionate* separation, which is the absolutely best criterion, is greatest in the solution when $(x-\xi)/(y-\eta)$ is maximum: this is clearly never attained, as matters go on improving or deteriorating to the end. It is

When p and q are very nearly equal—*i.e.*, the earths have nearly the same affinity for the acid—this ratio becomes e^{-1}, where e is the base of the Napierian logarithms, 2·718.... This is much nearer three than two, indicating that nearly two-thirds of the whole should be precipitated. Prof. Stokes therefore proposes the following plan:—

Suppose the filtrate to travel to the right and the precipitate to the left. Taking the contents of any bottle, precipitate ⅔rds of the whole, and advance the filtrate 2 places to the right and the precipitate 1 place to the left, and so on. Theoretically this is the best way to operate, and I have carried on several long series with much success by its means. There is, however, a slight disadvantage, in that the precipitate is twice as much as the earth in solution; and as the largest part of the operation is the feeding of the filters, I was obliged to fall back on a plan which, although not theoretically so good, makes up for diminished speed by requiring less manipulation.

The third plan is to add half the equivalent amount of precipitant to the liquid, and, after full settlement, filter. Starting with, say, 1000 grms. in the zero bottle, transfer 500 to bottle -1 and 500 to bottle $+1$. Then add another 1000 grms. to the 0 bottle, and repeat the operation as in the following table:—

Numbers of the Bottles.

	-6	-5	-4	-3	-2	-1	0	1	2	3	4	5	6
							1000						
A.						500	1000	500					
B.					250	500	500	500	250				
C.				125	250	375	500	375	250	125			
D.			63	125	250	375	375	375	250	125	63		
E.		31	63	156	250	312	375	312	250	156	63	31	
F.	15	31	94	156	234	312	312	312	234	156	94	31	15

best in the precipitate when ξ/η is maximum: this requires $d\xi/\xi = d\eta/\eta$: thus $(x - \xi)/q\xi = (y - \eta)/p\eta$; which leads to intractable equations.

The result stated in the text corresponds to $d\xi/x = d\eta/y$, which makes $\xi/x - \eta/y$, and therefore the difference between the *proportions* of the two bases that are precipitated, or that remain, maximum. This criterion gives a systematic symmetrical method of procedure, the same for both parts into which the mixture is divided, and is probably preferred for that reason.]

After the seventh fractionation the 2000 grms. of earth are spread out amongst 13 bottles in the proportion represented in the bottom line.

The separation of *two* earths by either of these plans is comparatively easy. The precipitation by ammonia depends not directly on the affinities of the earths for the acid, but rather on the excess of affinity of the precipitating ammonia. For if the affinities of the two earths are represented by 100 and 101, and that of ammonia by 150, the affinities on which the precipitation depends would be represented by $150 - 100 = 50$, and $150 - 101 = 49$, the difference of which is 2 per cent. of the larger.

Now if a precipitant of which the affinity for the acid was only 110 were used, the affinities in question would be $110 - 100 = 10$, and $110 - 101 = 9$, and the difference $10 - 9 = 1$, is 10 per cent. of the larger instead of only 2 per cent. Therefore if an alkali, for which the affinity of the acid was only a little greater than that of the earths, were used for precipitation, it is likely that the differences between the two earths would come out more strongly, and the labour of fractional precipitation might be much reduced. Prof. Stokes has suggested that some of the compound ammonias might prove useful as precipitants instead of ammonia. I have not, however, tested this suggestion.

[In reply to an inquiry on this subject Sir W. Crookes has kindly sent the following letter.]

ATHENAEUM CLUB, 21 *July*, 1885.

DEAR MR CROOKES;

On looking again into the matter I find my qualm was needless, and I adhere to what I sent you.

As to the three earths *A, B, C* in that order of sequence, what I meant was that by continuing the process of fractional precipitation you may obtain specimens of *A* and of *C* as pure as you please. But as regards *B* you may obtain a specimen as free as you like from *C*, but then the matter in hand consists of *A* with a little *B*. Or you may obtain a specimen as free as you like from

A, but then the material consists of *C* with a little *B*.   You may reach a stage where the quantity of *B* as compared with *A* and *C* *together* is as great as possible, but there you stick.   If you go to the right you diminish the quantity of say *A* relatively to *B*, but then you increase the relative quantity of *C* more than you diminish that of *A* ; and if you go to the left it is the same thing, merely interchanging *A* and *C*.   You don't improve matters by taking the portion richest in *B* and fractioning again by the same chemical process.   The portion richest in *B* relatively to *A* and *C* together belongs [not] to a particular bottle but to a particular length to which the process is carried on.   The expression for this could be given if it were of any use, but it involves two unknown quantities, viz. the 2 ratios between 3 affinities.   It is true that if you start with a mixture extra rich in *B* the fractioning deteriorates the mixture.   But it does not follow that if you start with a mixture poor in *B*, the mixture (*i.e.* a properly chosen fraction of it) first improves and then deteriorates.   What you do is, you in the one case work down towards, and in the other up towards, a limit.

Yours sincerely,

G. G. STOKES.

If there are *n* substances to separate, the number of different chemical processes which it is necessary to employ is not necessarily as great as *n* – 1.   One process may divide them into groups, another into sub-groups, and so on.   They need not necessarily be separated one at a time.

THE ULTRA-VIOLET SPECTRUM OF RADIUM.
By Sir W. CROOKES, F.R.S. (*Extract.*)

[From the *Proceedings of the Royal Society*, Vol. LXII, pp. 295—304.
August 1, 1903.]

....

MEASUREMENTS are taken of the exact distances apart of the
radium lines and certain adjacent iron lines used as standards.
By using a formula, first communicated to the British Association
by Sir George Stokes*, the wave-lengths of the unknown lines
can be calculated. At the time the formula was given it was
sufficiently accurate for the instrumental means employed for
photographing spectra, but the formula only gave approximate
results, and the accuracy of determination of wave-lengths has
since improved so much that a correction is required to the
original formula. Sir George Stokes, before whom I placed
the difficulty in June, 1895, quickly solved it in a satisfactory
manner. The usual formula requires the positions of two standard
lines of known wave-length, n_1 and n_3, one on each side of the unknown line n_2. To make the small correction Sir George advised
me to take a third line of known wave-length, chosen well-removed
from the selected known lines n_1 and n_3. If chosen in the interval
1—3 it had better not be greatly distant from the middle. There
is, however, very wide latitude for choice in this respect. All these
lines must be photographed and measured in the usual way.
Calculate the approximate wave-length of the unknown line by
the original formula, and then calculate the approximate wave-
length of the third known line by the same formula, *as if it were
unknown*, using the two original standards for this purpose also.
We have now the approximate wave-length of a known line, as
given by the formula, and also its true wave-length. The

[* *Brit. Assoc. Report*, 1849: *ante*, Vol. II, p. 176. Also J. Willard Gibbs,
Silliman's Journal, 1870.]

difference between these two values leads (*infra,* p. 295) to the correction to apply to the approximate value of the unknown line.

It often happens that four or more lines are well placed for use as standards. If any three of these be taken, and from them the positions of the other lines be calculated, it will sometimes be found that there is a small residual error in some of them. In such a case the error can be minimised by adjusting the value of the three primary standards so that the sum of the errors on all the lines is a minimum, and in no individual case is very great. Then, with these three corrected standards, the unknown lines may be calculated with confidence to seven figures.

The following is the method of calculation I now employ:—

$$n_1 \quad \lambda_1$$
$$n_2 \quad \lambda_2$$
$$n_3 \quad \lambda_3$$
$$n_4 \quad \lambda_4$$

n_1 and n_3 are scale positions on the measuring machine of standard lines of known wave-lengths, λ_1 and λ_3.

n_2 is the scale position of the line whose wave-length is required (λ_2).

n_4 is the scale position of an additional standard line whose wave-length is known. It is used for obtaining the correction to apply to the approximate wave-length of λ_2.

E_2 and E_4 are the calculated errors obtained for λ_2 and λ_4, which have to be added to or subtracted from λ_2 or λ_4 to get them accurate.

Rule.

First calculate the approximate value of λ_2 by the following formula:—

$$\frac{\lambda_3^2 \lambda_1^2 (n_3 - n_1)}{\lambda_1^2 (n_2 - n_1) + \lambda_3^2 (n_3 - n_2)} = \lambda_2^2 \text{ (approx.)} \quad \text{.........(1).}$$

Next, in a similar way, find the approximate value of λ_4, using the following formula:—

$$\frac{\lambda_3^2 \lambda_1^2 (n_3 - n_1)}{\lambda_1^2 (n_4 - n_1) + \lambda_3^2 (n_3 - n_4)} = \lambda_4^2 \text{ (approx.)} \quad \text{.........(2).}$$

Then $\qquad\qquad \lambda_4 \text{ (true)} = \lambda_4 \text{ (approx.)} + E_4.$

Now calculate E_2 by the following equation:—

$$\frac{\text{(Approx.) } \lambda_2{}^3 (n_2 - n_1)(n_3 - n_2)}{\text{(Approx.) } \lambda_4{}^3 (n_4 - n_1)(n_3 - n_2)} E_4 = E_2 \ \ldots\ldots\ldots(3).$$

Then   (Approx.) $\lambda_2 \pm E_2 = \lambda_2$.

[In fact, writing y for λ^{-2}, of which the refractive index is a function, and x for n, the graph of the relation between y and x may in this neighbourhood be identified with a parabola. The formulæ (1) and (2) neglect the effect of its curvature, by taking the points 2 and 4 to be on the chord connecting the points 1 and 3. The corrections E_2 and E_4 are thus connected with the distances from the points on this chord to the true points on the curve, by the formula $\delta y = -2E/\lambda^3$; and the formula (3) connecting them is the expression of a well-known geometrical property of a parabola*.]

[* In correspondence with Sir W. Crookes, a graphical method of correction, by use of this parabola, was suggested by Sir George Stokes.

In reply to an inquiry, Mr Newall refers to a formula of Hartmann (Potsdam Obs. xii, abstracted in *Astrophysical Journal*, viii, 1898, p. 218) as giving the most convenient method of interpolation. It appears that for prisms of the usual kinds of crown and flint glass, $\log(\lambda - \lambda_0)$ when plotted against $\log(\mu - \mu_0)$ gives a straight line of gradient $-\frac{5}{8}$, over a very wide range, where λ is wave-length and μ index of refraction. Thus $\lambda = \lambda_0 + C(\mu - \mu_0)^{-\frac{5}{8}}$. Also, if the range of λ is as much as 1000 tenth-metres (Ångström units), and n is merely the direct measurement of the spectrum plate, the simple relation $\lambda - \lambda_0 = A(n - n_0)^{-1}$ holds within $\frac{1}{10}$ of a tenth-metre. The latter formula had been used by Cornu in 1880. Cf. Kayser, *Handbuch der Spectroscopie*, i, § 328. By taking the range small, the various formulæ should give the same results.]

APPENDIX.

EXAMINATION PROBLEMS.

[Sir George Stokes acted as an Examiner for the Mathematical Tripos at Cambridge in the years 1846 to 1849 inclusive. As a matter of historical interest, some of the papers of problems which he set as Moderator, either alone or in conjunction with the other Moderator, are here reprinted.

After he became Lucasian Professor of Mathematics in the year 1849 it was his duty in most years to set one of the papers in the examination for the Smith's Prizes, in which the best mathematicians of the year again competed shortly after the Tripos Examination. These papers were continued until 1882, when the mode of award of the Smith's Prizes was altered. They are reprinted here on account of the valuable theorems and points of view which they contain, and also of the light which they throw on the course of their author's ideas. They are of additional interest in relation to the history of mathematical studies at Cambridge.]

MATHEMATICAL TRIPOS.

THURSDAY, *January* 8, 1846.   1—4.   (PROBLEMS.)

1. Prove that
$$(A^2 + B^2 + C^2 + \ldots)(a^2 + b^2 + c^2 + \ldots) > (Aa + Bb + Cc + \ldots)^2,$$
unless A, B, C... are proportional to a, b, c...

2. There are n planes of which no two are parallel to each other, no three parallel to the same right line, and no four pass through the same point; prove that the number of lines of intersection of the planes is $\frac{1}{2}n(n-1)$, and that the number of points of intersection of those lines is $\frac{1}{6}n(n-1)(n-2)$.

3. A person engages in play against a bank, playing under a slight disadvantage, but retaining the option, which is not allowed to

the bank, of ceasing to play when he pleases, and he determines accordingly to cease after a certain run of luck; show by general reasoning that no possible arrangement on his part can throw the advantage on his side, or even diminish the odds against him.

4. ABC is a great circle of the sphere, AA', BB', CC' are arcs of great circles drawn perpendicular to ABC, and reckoned positive when they lie on the same side of it; prove that the condition of A', B', C' lying in a great circle is

$$\tan AA' \sin BC + \tan BB' \sin CA + \tan CC' \sin AB = 0.$$

5. A given number of persons take wine together, each bowing to each of his friends in succession; if all the persons bow simultaneously each time, find the chance that a given person shall bow to a given number of his friends just as they are bowing to him.

6. Prove that the equation $(x-1)\, \epsilon^x = (x+1)\, \epsilon^{-x}$ has two, and only two real roots, which are equal in magnitude and of opposite signs.

7. Find the volume of the solid the equation to the surface of which is $\dfrac{x^2}{a^2} + \dfrac{y^2}{b^2} + \dfrac{z^4}{c^4} = 1$.

8. Two equal weights P, P' are connected by a string which passes over a rough fixed horizontal cylinder; compare the forces required to raise P, according as P is pushed up or P' pulled down.

9. A body describes a circular orbit about a centre of force situated in the centre of the circle; prove that the form of the orbit will be stable or unstable according as the value of $\dfrac{d \log P}{d \log u}$ for $u = a$ is less or not less than 3, P being the central force, u the reciprocal of the radius vector, and $1/a$ the radius of the circle.

10. AB are the centres of two equal circles, and AP, BQ are two radii which are always perpendicular to each other; find the curve which is always touched by the right line PQ, and explain the result when $AB^2 = 2AP^2$.

11. The curve whose polar equation is $r = a$ versin θ rolls on its axis, find the equation to the curve traced out by the pole; show geometrically how the latter curve comes to be a closed curve and find its area.

12. BAB', CAC' are two circles, of radii a, b respectively, intersecting in A at an angle α; Ax, Ay are right lines bisecting the angle between the tangents at A and its supplement respectively, and the arcs AB, AC, AC' are taken equal to each other: if P, Q are the points of ultimate intersection of the lines BC, BC' with Ax, Ay respectively, as AB is indefinitely diminished, prove that the triangle PAQ is equal to $\dfrac{4 \sin \alpha \cdot a^2 b^2}{a^2 \sim b^2}$.

13. A very small capillary tube which is slightly conical is inserted vertically into a liquid which moistens its surface; prove that there are in general two positions of equilibrium of the liquid, and show in what case the equilibrium is stable.

14. When objects are viewed by means of pencils of light which are refracted through an isosceles prism suffering an internal reflection at the base, the objects are seen quite free from chromatic dispersion, provided the prism be not very large; prove that this must happen in the general case, that is, when the refraction is not supposed to take place in a plane perpendicular to the edges of the prism.

15. Two equal particles are connected by two given strings without weight, which are placed like a necklace on a smooth cone with its axis vertical and vertex upwards; find the tensions of the strings.

16. If P be any point in a surface, a, b the principal radii of curvature at P, r the radius of curvature of the normal section made by a plane inclined at an angle θ to the principal section to which a refers, $PQ = s$ an indefinitely small arc in this section, prove that if D be the minimum distance of the normals at P and Q, c the distance from P of their point of nearest approach,

$$D^2 = \frac{\sin^2\theta \cos^2\theta\,(a-b)^2 s^2}{b^2\cos^2\theta + a^2\sin^2\theta}, \quad c = \frac{ab\,(b\cos^2\theta + a\sin^2\theta)}{b^2\cos^2\theta + a^2\sin^2\theta}.$$

Express also $(D/s)^2$ and c in terms of a, b and r.

17. The equations to a system of lines in space, straight or curved, contain two arbitrary parameters; show how to find whether the lines can be cut at right angles by a system of surfaces, and when they can, show how to find the equation to that system.

Examples. (1) Let the lines be a system of right lines each of which intersects two given right lines which are perpendicular to each other but do not intersect. (2) Let the equations to the system of lines be $Ax^\alpha = By^\beta = Cz^\gamma$, where A, B, and C are arbitrary.

18. A rigid body is attached to a fine wire, and made to perform small oscillations by the torsion of the wire; regarding for simplicity the length of the wire as infinite compared with the dimensions of the body, form the equation one of the roots of which gives the time of oscillation.

Show that the time of oscillation which would be obtained by considering only the moment of the force of torsion, and the moment of inertia of the body about a vertical axis passing through its centre of gravity, would in general be erroneous unless that axis be a principal axis. Find also the time of oscillation to a second approximation, supposing the force of torsion very small.

19. If $A_1 \sin x + A_2 \sin 2x \ldots + A_n \sin nx + \ldots$ is the expansion of an arbitrary function of x, $f(x)$, between the limits $x = 0$ and $x = \pi$, and if $f(x)$ is continuous, and $f'(x)$ not infinite between those limits, prove that $\qquad f(0) = \tfrac{1}{4}\pi\,(L + L'), \quad f(\pi) = \tfrac{1}{4}\pi\,(L - L'),$ L or L', according as n is odd or even, being the limit of nA_n as n increases beyond all limit.

20. Show that according to the common hypotheses made in the corpuscular theory of light the refractive index of a given substance, for light coming from a given star, ought to be increased by the motion

of the Earth by the quantity $a/\mu\,(\mu^2 - 1)\cos\lambda\sin(l - s)$ nearly, where a is the coefficient of aberration, λ the latitude of the star, s its longitude, and l the longitude of the Earth.

21. $x'Ax$ is a stretched string, of which the portions Ax, Ax' are of different thicknesses; prove that any small transversal vibrations travelling from x to A will, on arriving at A, be partly reflected and partly transmitted to Ax', and that the displacements due to the incident, reflected and transmitted vibrations are to each other as $1 + \mu : 1 - \mu : 2$, where μ is the ratio of the velocity of propagation in Ax to that in Ax'.

22. A biaxal crystal presents naturally two plane faces perpendicular to two of the axes of elasticity; the crystal being immersed in fluid, a pencil of homogeneous light is transmitted through it across those faces, in a plane perpendicular to them both, and the minimum deviation (D) of the extraordinary emergent pencil is observed. Show that the observation furnishes towards the determination of the principal indices of refraction the equation

$$(\mu_1{}^2 - \mu^2)(\mu_2{}^2 - \mu^2) = \mu^4 \sin^2 D,$$

where $\mu_1\,(= v/a)$, $\mu_2(= v/b)$ are two of the principal indices required, and μ is the refractive index of the fluid.

FRIDAY, *January* 8, 1847. 1—4. (PROBLEMS.)

1. A boat's crew row $3\frac{1}{2}$ miles down a river and back again in 1 h. 40 m.; supposing the river to have a current of 2 miles per hour, find the rate at which the crew would row in still water.

2. If a and x both lie between 0 and 1, prove that $\dfrac{1 - a^x}{1 - a} > x.$

3. In the figure of Euclid, Book I. Prop. 35 (*Parallelograms on the same base and between the same parallels are equal*), if two diagonals be drawn to the two parallelograms respectively, one from each extremity of the base, and the intersection of the diagonals be joined with the intersection of the sides (or sides produced) in the figure, prove that the joining line will bisect the base.

4. A piece of paper a mile long is rolled into a solid cylinder; find approximately the diameter of the cylinder, supposing 240 leaves of such paper to have a thickness of one inch.

5. PT, QT are two equal tangents to a parabola, P and Q being the points of contact; if PT, QT be both cut by a third tangent, prove that their alternate segments will be equal.

6. Two concentric ellipses which have their axes in the same directions intersect, and four common tangents are drawn so as to form a rhombus, and the points of intersection of the ellipses are joined so as to form a rectangle; prove that the product of the areas of the rhombus and rectangle is equal to half the continued product of the four axes.

7. If $\phi f(x) = \phi F(x)$ for all values of x from a to b, and if c be a quantity not less than a nor greater than b such that $f(c) = F(c)$ and $f'(c)$, $F'(c)$ have opposite signs, prove that $\phi(y)$ is necessarily a maximum or minimum when $y = f(c)$.

8. An imperfectly elastic ball is projected with a given velocity from a given point in a smooth inclined plane; find the direction of projection in order that the ball may cease to hop just as it returns to the point of projection.

9. A small plane touches a self-luminous paraboloid of revolution at its vertex, and is then moved parallel to itself along the axis produced; prove that the illumination of the plane varies inversely as its distance from the focus.

10. Three plane mirrors, A, B, C are parallel to the same straight line; find the position of a luminous point, which is so distant that parallax may be neglected, in order that the rays reflected from A may be parallel to those which are reflected from B and C in succession.

11. Two heavy particles are connected by a light string, and immersed in a fluid whose specific gravity is intermediate between those of the particles, and which revolves uniformly about a fixed axis, and is not acted on by gravity; find the position of equilibrium of the particles relatively to the fluid, and show that when the equilibrium is possible it is stable.

12. A ray of light passes from air into a transparent prism of either single or double refraction, and emerges after any number of internal reflections; apply the principle of quickest propagation to prove that the emergent ray or rays are inclined to the edges of the prism at the same angle as the incident ray.

13. A system of ellipses is described such that each ellipse touches two rectangular axes, to which its axes are parallel, and that the rectangle under the axes of the ellipses is constant; prove that each ellipse is touched by two rectangular hyperbolas, the rectangle under the transverse axes of which is equal to the rectangle under the axes of any one of the ellipses.

14. There are two systems of curves which cut each other everywhere at right angles; OM, NP are two curves of one system, ON, MP are two of the other, and the arc $OM = x$, $ON = y$, $NP = \xi$, $MP = \eta$, and ρ, ρ' are the radii of curvature of PN, PM at the point P; prove that

$$\rho \frac{d^2\xi}{dx\,dy} = \rho' \frac{d^2\eta}{dx\,dy} = \frac{d\xi}{dx} \frac{d\eta}{dy},$$

the differential coefficients being taken on the supposition that O remains fixed while P alters.

15. The radius vector of any point in the surface of a solid differs from a constant by a small quantity of the order a, and the angle which it makes with the normal is of the same order; prove that if small quantities of the order a^2 are neglected the sphere which has the same volume as the given solid is also that which has the same surface.

16. A uniform flexible and inextensible heavy string AB is laid in the form of a circular arc on a smooth horizontal plane; a given impulsive force being applied at A in the direction of the tangent, find the impulsive tension at any point, and show that the direction of the initial motion of A makes with the direction of the force an angle whose tangent $= \dfrac{\epsilon^a - \epsilon^{-a}}{\epsilon^a + \epsilon^{-a}}$, a being the angle which AB subtends at the centre.

17. Tangent planes to the surface whose equation, referred to rectangular co-ordinates, is $\dfrac{x^3}{a^3} + \dfrac{y^3}{b^3} + \dfrac{z^3}{c^3} = 1$ pass through a point P; prove that a sphere can be described through the curve of contact provided P lie in a certain right line passing through the origin.

18. The equations to a system of right lines in space contain two arbitrary parameters; prove that when the roots of a certain quadratic are real and unequal, there are two planes passing through a given line of the system which contain consecutive lines.

19. Find the value of $\displaystyle\int_0^\infty \{\epsilon^{-(a+a\sqrt{-1})x} - \epsilon^{-(b+\beta\sqrt{-1})x}\} \dfrac{dx}{x}$, where a and b are positive, but a and β positive or negative; and show that it is wholly real when $a/a = \beta/b$.

20. Supposing the luminiferous ether to gravitate to the Sun, and to be incompressible and at rest, and supposing the velocity of light to vary by a small quantity proportional to the pressure of the ether, show that the observed times of the eclipses of Jupiter's satellites will be affected by an inequality expressed by $c \log \tan \frac{1}{2}(E + S) \cot \frac{1}{2}E$, where c is constant, and E, S are two angles of the triangle ESJ.

21. A heavy particle is attached to a light string which passes over a horizontal cylinder; supposing the particle to perform small finite oscillations in a plane perpendicular to the axis of the cylinder, find approximately the correction to the time of oscillation due to the finite arc of oscillation, and show that it vanishes when the radius of the cylinder is to the length of string hanging down as $\sqrt{3}$ to 2.

22. The refractive index (μ) of a transparent medium varies continuously from point to point; prove by means of the principle of quickest propagation, or otherwise, that the differential equations to a ray of light within the medium are

$$\frac{ds}{dx}\left(\frac{d\mu}{dx} - \mu\frac{d^2x}{ds^2}\right) = \frac{ds}{dy}\left(\frac{d\mu}{dy} - \mu\frac{d^2y}{ds^2}\right) = \frac{ds}{dz}\left(\frac{d\mu}{dz} - \mu\frac{d^2z}{ds^2}\right),$$

where the differential coefficients of μ are partial.

23.   An infinite mass of homogeneous incompressible fluid acted on by no forces is at rest, and a spherical portion of the fluid is suddenly annihilated; find the instantaneous alteration of pressure at any point of the mass, and prove that the cavity will be filled up in the time $\left(\dfrac{6\rho}{\varpi}\right)^{\frac{1}{2}} a \displaystyle\int_0^1 \dfrac{\lambda^4 d\lambda}{\sqrt{1-\lambda^6}}$, where a is the initial radius of the sphere, and ϖ the pressure at an infinite distance, which is supposed to remain constant.

24.   A narrow hoop is rolled along a rough horizontal plane in such a manner as to move nearly in a vertical plane, but make small oscillations on each side of it; find the time of oscillation, and show that oscillations of this sort are not possible unless the velocity of the centre of the hoop be greater than that due to one-third of the radius.

SATURDAY, *January* 9, 1847.   9—11½.   (PROBLEMS.)

(G. G. STOKES and J. C. ADAMS.)

1. AB is a common chord of the segments ACB, $ADEB$ of two circles and through C, any point in ACB, are drawn the straight lines ACE, BCD; prove that the arc DE is invariable.

2.   If CP, CD be any conjugate diameters of an ellipse $APBDA'$, and BP, BD be joined, and also AD, $A'P$, these latter intersecting in O, show that $BDOP$ is a parallelogram, and that its greatest area is $ab\,(\sqrt{2}-1)$.

3.   Find the limiting value of $x\epsilon^{-x^2}\displaystyle\int_0^x \epsilon^{x^2} dx$ when $x=\infty$.

4.   A regular octahedron is inscribed in a cube, so that the corners of the octahedron are in the centres of the faces of the cube; prove that the volume of the cube is six times that of the octahedron.

5.   A bucket partly filled with water is attached to a weight by a string which passes over a fixed pulley; supposing the water to revolve with a given angular velocity as the bucket is ascending or descending, find the form of the free surface.

6.   A body acted on by a central force P is moving in a medium whose resistance $= c$ (velocity); prove that

$$\frac{d^2r}{dt^2} + P - \frac{h^2}{r^3}\,\epsilon^{-2ct} + c\,\frac{dr}{dt} = 0,$$

where h is an arbitrary constant.

7.   A plane drawn through a given point is illuminated by two self-luminous spheres; find the position of the plane when the illumination at the given point is a maximum.

8. OAA_1 is a spherical triangle, right-angled at A_1; the arc A_1A_2 of a great circle is drawn perpendicular to OA, A_2A_3 is drawn perpendicular to OA_1, and so on; prove that A_nA_{n+1} vanishes when n becomes infinite, and find the value of $\cos AA_1 . \cos A_1A_2 . \cos A_2A_3...$ *to infinity.*

9. Having given the times by a sidereal clock at which three known stars are observed to have the same zenith distance, the absolute zenith distance being unknown, find the latitude of the place of observation and the error of the clock.

10. A circular disc is suspended by a fine wire attached to the centre, and immersed horizontally in a fluid; the wire being suddenly turned through a given angle, determine the motion of the disc, supposing each element of the surface acted on by a friction varying as the velocity, and show that the successive arcs described from rest to rest are in geometric progression. Show also that if the friction exceed a certain quantity the disc will not come to rest at all.

11. A sphere touches each of two right lines which are inclined to each other at a right angle, but do not intersect; prove that the locus of its centre is a hyperbolic paraboloid.

12. A slender rod suspended horizontally by two equal parallel strings attached to two points equidistant from its ends oscillates round a vertical line; find the time of a small oscillation.

If in the position of equilibrium the strings are inclined at equal angles to the vertical, show that the time of oscillation is the same as it would be if the strings were parallel, of a length equal to the projection of either of them on a vertical line, and at a distance equal to a mean proportional between their distances at the points of suspension and attachment respectively.

13. A given quantity of incompressible fluid is contained in an elastic spherical envelope, and just fills the space inclosed without stretching the envelope; if the particles of the fluid be acted on by a repulsive force varying inversely as the square of the distance from the centre, find the absolute force when the space originally occupied by the fluid is left a vacuum.

14. A circle always touches the axis of z at the origin, and passes through a fixed straight line in the plane of xy; find the equation to the surface generated. Show that the origin is a singular point, and that in its immediate neighbourhood the surface may be conceived to be generated by a circle having its plane parallel to that of xy, and its radius proportional to z^2.

15. P is a point in the base of a tetrahedron $VABC$, of which V is the vertex, so taken that the volume of the parallelepiped constructed on VP as diagonal, and having three of its edges coincident with VA, VB, VC, is a maximum; supposing the base of the tetrahedron to turn in all directions round a given point, while the edges adjacent to V remain fixed in position, find the locus of P.

16. A sphere touches an elliptic paraboloid at the vertex, and has its diameter a mean proportional between the parameters of the

principal sections of the paraboloid; supposing the curve of intersection of the sphere and paraboloid to be projected on the tangent plane at the vertex, find the area of the curve of projection.

17. A tube of small bore, in the form of a logarithmic spiral, revolves with a uniform angular velocity about an axis passing through its pole and perpendicular to its plane, which is horizontal, and contains a particle which moves freely in it; supposing the initial velocity of the particle relatively to the tube to be equal to the velocity of the point of the tube in contact with the particle, show that the path of the particle is another logarithmic spiral.

18. Show that there cannot be any curve such that if tangents PT, QT be drawn at any points P and Q to meet in T, the angle which PT subtends at a fixed point S may always bear a constant ratio to that subtended by QT, except when that ratio is one of equality, and prove that in this case the curve is a conic section having S for its focus.

TUESDAY, *January* 18, 1848.   9—12.   (PROBLEMS.)

(G. G. STOKES and T. GASKIN.)

1. AB, CD are any two chords of a circle passing through a fixed point O, EF any chord parallel to AB; join CE, DF meeting AB in the points G and H, and DE, CF meeting AB in the points K and L; show that the rectangle $OG \cdot OH = OK \cdot OL$.

2. In a given circle inscribe a rectangle equal to a given rectilineal figure.

3. A cannon-ball is moving in a direction making an acute angle θ with a line drawn from the ball to an observer; if V be the velocity of sound, and nV that of the ball, prove that the whizzing of the ball at the different points of its course will be heard in the order in which it is produced, or in the reverse order, according as $n < > \sec \theta$.

4. Compare the brightness of the Earth as seen from Venus with the brightness of Venus as seen from the Earth, supposing the sizes and reflecting powers of the two bodies equal.

5. With two conjugate diameters of an ellipse as asymptotes a pair of conjugate hyperbolas is constructed; prove that if one hyperbola touch the ellipse the other will do so likewise; prove also that the diameters drawn through the points of contact are conjugate to each other.

6. Show that the curve which trisects the arcs of all segments of a circle described upon a given base is an hyperbola whose eccentricity = 2.

7. Let D be a point in the axis minor of an ellipse whose eccentricity is e, S the focus, O the centre of curvature at the extremity of the axis minor; with centre D and radius $= DS/e$ describe a circle; shew that this circle will touch the ellipse or fall entirely without it according as D is nearer to or farther from the centre than the point O.

8. PSp is any focal chord of an ellipse, A the extremity of the axis major; AP, Ap meet the directrix in two points Q, q; shew that $\angle QSq$ is a right angle.

9. In a right-angled spherical triangle, shew that
$$\sin a \tan \tfrac{1}{2} A - \sin b \tan \tfrac{1}{2} B = \sin (a - b).$$
Shew also that if E be the spherical excess
$$\sin \frac{E}{2} = \frac{\sin \tfrac{1}{2} a \sin \tfrac{1}{2} b}{\cos \tfrac{1}{2} c}; \quad \cos \frac{E}{2} = \frac{\cos \tfrac{1}{2} a \cos \tfrac{1}{2} b}{\cos \tfrac{1}{2} c}.$$

10. Prove that the remainder after n terms of the infinite series $\dfrac{1}{1^a} + \dfrac{1}{2^a} + \dfrac{1}{3^a} \ldots$, where $a > 1$, lies between $\dfrac{1}{(a-1)(n+\tfrac{1}{2})^{a-1}} \left(= \displaystyle\int_{n+\frac{1}{2}}^{\infty} \frac{dx}{x^a} \right)$ and $\dfrac{1}{(a-1)(n+1)^{a-1}}$, approaching much nearer to the former limit when n is large.

11. If a and na be the respective distances of a satellite and of the Sun from a planet, p and mp the periodic times of the satellite and planet, which are supposed to describe circles round the planet and the Sun respectively, shew that the orbit of the satellite will always be concave towards the Sun provided n be greater than m^2.

12. A uniform slender rod passes over the fixed point A and under the fixed point B, and is kept at rest by the friction at the points A and B; determine the limiting positions of equilibrium.

13. Four uniform slender rods AB, BC, CD, DA, rigidly connected, form the sides of a quadrilateral figure such that the angle A is a right angle, and the points B, C, D are equidistant from each other; when the whole is suspended at the angle A, determine the position of equilibrium.

14. A given inelastic mass is let fall from a given height on one scale of a balance, and two inelastic masses are let fall from different heights on the other scale, so that the three impacts take place simultaneously; find the relations between the masses and heights in order that the balance may remain permanently at rest.

15. A body of given elasticity is projected with a given velocity, and rebounds n times at a horizontal plane passing through the point of projection; determine the direction of projection so that the angle between the direction of projection and the direction of the ball immediately after the last impact may be the greatest possible.

16. If a body be projected with a given velocity about a centre of force which $\propto 1/(\text{dist.})^{-2}$, shew that the axis minor of the orbit described will vary as the perpendicular from the centre of force upon the

direction of projection; and determine the locus of the centre of the orbit described.

17.   If there had been no stars how might the absolute periodic times of the Earth and planets have been determined, even if the equator had coincided with the ecliptic?

18.   Find the geometrical focus of a pencil of rays refracted through a hollow glass sphere whose external and internal radii are r, r' respectively.

19.   A star map is laid down on the gnomonic projection, the plane of projection being parallel to the equator; give a graphical solution of the problem, to determine the time at a known place by observing when two stars laid down in the map are in the same vertical plane.

20.   A cylindrical vessel with its axis vertical is filled with fluid, which issues from a great number of small orifices pierced in the side; find the surface which touches all the streams of spouting fluid.

21.   If three chords be drawn mutually at right angles through a fixed point within a surface of the second order whose equation is $u = 0$, shew that $\Sigma \, 1/Rr$ will be constant, where R and r are the two portions into which any one of the chords drawn through the fixed point is divided by that point.

Prove also that the same will be true if instead of the fixed point there be substituted any point in the surface whose equation is $u = c$.

22.   Two bodies, P, P' describe round a central body S circular orbits lying in one plane, the orbit of P being within that of P'; prove that the disturbing force of P' on P, when wholly central, and additious, will be equal to the disturbing force when P, P' are on opposite sides of S provided SP' be a mean proportional between SP and $SP + SP'$.

WEDNESDAY, *January* 19, 1848.   9—12.

1.   In the given right lines AP, AQ are taken variable points p, q such that $Ap : pP :: Qq : qA$; prove that the locus of the point of intersection of Pq, Qp is an ellipse which touches the given right lines in the points P, Q.

2.   A parallelepiped is cut by three systems of parallel planes, given in number, parallel to the three pair of opposite faces respectively; find the total number of parallelepipeds formed in every way.

3.   The corner of a sheet of paper is turned down so that the sum of the edges turned down is constant; find the equation to the curve traced out by the vertex of the angle; find also the area of the curve.

4. A string of given length is attached to the extremities of the arms of a straight lever without weight, and passes round a small pulley which supports a weight; find the position of equilibrium in which the lever is inclined to the vertical, and prove that the equilibrium is unstable.

5. Two equal strings of length l are attached to the fixed points A, B and C, D respectively, which if joined would form a horizontal rectangle; a sphere whose diameter $= AB$ is laid symmetrically upon the strings; find the position of equilibrium and the tension of either string, supposing $l > AC + \frac{1}{2}\pi AB$. Shew also how the problem is to be solved when this condition is not fulfilled.

6. A cannon ball is fired at a mark at a place whose north latitude is l; shew that in consequence of the Earth's rotation the vertical plane containing the axis of the cannon must be inclined at an angle of $15t \sin l$ seconds to the left of the vertical plane passing through the mark, t being the time of flight expressed in seconds.

7. An imperfectly elastic homogeneous rough sphere is projected obliquely, without rotation, against a fixed plane; if i, i' be the angles of incidence and reflection, λ the coefficient of elasticity for direct impact, and ρ the ratio of the tangential forces of restitution and compression, prove that

$$2\rho = 5 - 7\lambda \tan i' \cot i.$$

8. Tangents to a system of similar and concentric ellipses are drawn at a given perpendicular distance from the centre; find the locus of the point of contact, and shew that the area of the curve is equal to that of an ellipse which has the same greatest and least diameters.

9. If $f(x)$ be positive and finite from $x = a$ to $x = a + h$, shew how to find the limit of

$$\left\{ f(a) \cdot f\left(a + \frac{1}{n}h\right) \dots f\left(a + \frac{n-1}{n}h\right) \right\}^{\frac{1}{n}},$$

or $n = \infty$, and prove that the limit in question is less than $\dfrac{1}{h} \displaystyle\int_a^{a+h} f(x)\,dx$, assuming that the geometric mean of a finite number of positive quantities which are not all equal is less than the arithmetic.

Hence prove that $\epsilon^{\int_0^1 u\,dx} < \displaystyle\int_0^1 \epsilon^u \, dx$, unless u be constant from $x = 0$ to $x = 1$.

10. Find the locus of the foot of the perpendicular let fall from the origin on the tangent plane to the surface $xyz = a^3$; point out the general form of the required surface, and find the whole included volume.

11. A closed vessel is filled with water containing in it a piece of cork which is free to move; if the vessel be suddenly moved forwards by a blow, shew that the cork will shoot forwards relatively to the water.

12. A closed vessel is filled with water which is at rest, and the vessel is then moved in any manner; apply the principle of the conservation of areas to prove that if the vessel have any motion of rotation no finite portion of the water can remain at rest relatively to the vessel.

13. A parallelogram is constructed by drawing tangents at the extremities of two conjugate diameters of an ellipse; prove that the diagonals of the parallelogram form a second system of conjugate diameters, and that the relation between the two systems is reciprocal.

14. A walks to Trumpington and back by Grantchester in $1\frac{1}{4}$ hr., starting between 2 o'clock and $2\frac{3}{4}$, B walks the same distance in the same direction in $1\frac{1}{2}$ hr., starting between 2 and $2\frac{1}{4}$; find the chance that A overtakes B before he gets home.

15. Find the equation to that involute of a cycloid which passes through the cusp, and shew that in the immediate neighbourhood of the cusp it becomes the curve $2a\,(4y)^3 = (3x)^4$, a being the radius of the generating circle.

16. Two given masses are connected by a slightly elastic string, and projected so as to whirl round; find the time of a small oscillation in the length of the string.

Give a numerical result supposing the masses to weigh 1 lb., 2 lbs. respectively, and the natural length of the string to be 1 yard, and supposing that it stretches $\frac{1}{10}$th inch for a tension of 1 lb.

17. Light proceeding from a given point P suffers any number of reflections and refractions; if consecutive rays of a given colour come out parallel, in the direction determined by the angular co-ordinates θ, ϕ, shew that $\dfrac{d\theta}{d\mu}$, $\dfrac{d\phi}{d\mu}$ may be obtained by differentiating as if the differently coloured rays which severally come out parallel to their consecutives started from P in the same direction.

Application. In the case of the rainbow of the pth order, given

$$D = p\pi + 2\phi - 2\,(p+1)\,\phi', \quad \sin\phi = \mu\sin\phi',$$

find the order of the colours.

18. A spherical wave of light is incident directly on a lens; find approximately the retardations of the several portions of the wave, and prove in this way the common equation $\dfrac{1}{v} - \dfrac{1}{u} = \dfrac{1}{f}$.

19. A sphere is composed of an immense number of small free particles, equally distributed, which gravitate to each other without interfering; supposing the particles to have no initial velocity, prove that the mean density about a given particle will vary inversely as the cube of its distance from the centre.

20. A plane moves so as always to enclose between itself and a given surface S a constant volume; prove that the envelope of the system of such planes is the same as the locus of the centres of gravity of the portions of the planes comprised within S.

21.   A rough sphere rolls within a hollow cylinder with its axis vertical, so as to be in contact with the curved surface and the flat bottom; find the reactions and frictions in terms of the angular velocity with which the sphere goes round, and explain the indeterminateness of the problem.

22.   Prove geometrically, or otherwise, that if g be the attraction which a particle m exerts on a point in a closed surface S, θ the angle between the direction of g and the normal, $d\omega$ an element of S,

$$\iint g \cos\theta d\omega = 4\pi m, \quad \text{or} = 0,$$

according as m is within or without S, the attraction of m at the distance r being m/r^2.

Extend this result to the case of a finite mass cut by S, and thence prove, by taking for S an elementary parallelepiped, that if V be the potential of any mass for an internal particle,

$$\frac{d^2V}{dx^2} + \frac{d^2V}{dy^2} + \frac{d^2V}{dz^2} = -4\pi\rho.$$

SMITH'S PRIZE EXAMINATION PAPERS.

January, 1850.

1.   If a parallelogram be inscribed in the inner of two similar, concentric, and similarly situated ellipses, and its sides be produced to meet the outer, and the adjacent points of intersection belonging to each pair of parallel lines be joined, shew that the quadrilateral figure formed by producing these joining lines will be a parallelogram, having its corners situated in a third ellipse, similar to the two former, and independent of the original parallelogram.

2.   If p be the probability *a priori* that a theory is true, q the probability that an experiment would turn out as indicated by the theory even if the theory were false, shew that after the experiment has been performed, supposing it to have turned out as was expected, the probability of the truth of the theory becomes $\dfrac{p}{p + q - pq}$.

3.   Shew that the limit of the continued product

$$\left(1 - \frac{\theta}{m}\right)\left(1 - \frac{\theta}{m-1}\right)\ldots\ldots\left(1 - \frac{\theta}{1}\right)\theta\left(1 + \frac{\theta}{1}\right)\left(1 + \frac{\theta}{2}\right)\ldots\ldots\left(1 + \frac{\theta}{n}\right),$$

when m and n become finite together, is $\pi^{-1}\lambda^\theta \sin\pi\theta$, λ being the limit of n/m.

4.   Describe Horner's method of approximating to the real roots of equations, explaining any abbreviations of which the method is susceptible in practice, and illustrating your explanation by a numerical example.

5. Shew that a definite meaning may be attached to the expression $\int_{-a}^{a} \frac{dx}{x}$ by supposing x to pass from $-a$ to $+a$ through a series of imaginary values, and that the result is $(2n+1)\pi\sqrt{-1}$, where n is an integer. Also give a geometrical illustration of the multiplicity of values.

6. Prove that the definite integral $\int_{0}^{\infty} \sin ax \cdot x^{n-1} dx$ is convergent if n lies between the limits -1 and $+1$, and otherwise divergent. Also express the value of the integral in finite terms by means of the tabulated function $\Gamma(n)$.

7. A surface is referred to the polar co-ordinates r, θ, ϕ: required to express the direction-cosines of the normal at any point by means of the partial differential coefficients of r, the three lines to which the direction of the normal is referred being the radius vector, and two lines perpendicular to the radius vector, and lying respectively in and perpendicular to the plane in which θ is measured.

8. If u be a function of the rectangular co-ordinates x, y, z, and $\frac{d^2u}{dx^2} + \frac{d^2u}{dy^2} + \frac{d^2u}{dz^2}$ be denoted by ∇u, and if T be a small space about the point (x, y, z), which in the end is supposed to vanish, U its volume, dS an element of its surface, dn an element of a normal drawn outwards, prove that

$$\nabla u = \text{limit of } \frac{\iint \frac{du}{dn} dS}{U}.$$

Hence obtain the expression for ∇u in polar co-ordinates, by taking for T the elementary volume $dr \cdot rd\theta \cdot r \sin\theta d\phi$.

9. A cylindrical wire is formed into a spring, which in its natural state has the form of a helix, the tangents to which are nearly perpendicular to its axis: given the radius of the cylinder in which the axis of the wire lies, and the modulus of torsion of the wire, find how much the spring will stretch when pulled by a small given force.

10. A body revolves in an ellipse of small eccentricity about a centre of force in one of the foci; examine the changes in the eccentricity and direction of the line of apsides produced (1) by a small central disturbing force acting when the body is in the neighbourhood of a given point of the orbit, (2) by a sudden small change in the absolute force, which takes place when the body is at that point.

Shew that the precise points in an orbit of any eccentricity at which the latter kind of disturbance would produce no change (1) in the distance between the foci, (2) in the ratio of that distance to the major axis, are (1) the extremities of the latus rectum passing through the second focus, (2) the extremities of the minor axis.

11. A sphere is held in contact with the interior surface of a rough hollow cylinder with its axis vertical, and is then projected horizontally in a direction parallel to the tangent plane at the point of contact; determine the motion.

12. If V be a function of the rectangular co-ordinates x, y, z which vanishes at an infinite distance from a given finite closed surface S, prove that

$$\iiint V \cdot \nabla V \cdot dx\,dy\,dz + \iint V \frac{dV}{dn}\,dS$$
$$+ \iiint \left\{ \left(\frac{dV}{dx}\right)^2 + \left(\frac{dV}{dy}\right)^2 + \left(\frac{dV}{dz}\right)^2 \right\} dx\,dy\,dz = 0,$$

where the triple integrals extend to all infinite space outside S, and the double integral extends to the whole surface S, the meaning of ∇ and n being the same as in question 8.

13. If the normal component of the attraction of any mass be given throughout its surface, or else if the value of the potential throughout the surface be given, prove that the attraction on a particle external to the surface will be determinate, the law of attraction being that of the inverse square of the distance.

14. Shew that the equality of pressure in all directions in a fluid, whether at rest or in motion, is a necessary consequence of the hypothesis that the mutual pressure of two adjacent portions of a fluid is normal to the surface of separation.

Mention any phenomena which you conceive either to confirm the accuracy or to prove the inaccuracy of the above hypothesis, considering separately the cases of rest and motion.

15. Find, according to the usual suppositions, the rate at which air is admitted into a partially exhausted receiver, through a small orifice; examine the result obtained by supposing the exhaustion perfect, and explain the paradox.

16. Prove that the velocity of propagation of a long wave of small height in a uniform canal of any shape is equal to that acquired by a heavy body in falling through a space equal to the area of a transverse section of the fluid divided by twice the breadth at the surface.

17. Shew that the differential equation to a ray of light propagated through a medium of variable density, which is symmetrical with respect to the plane of xy, which is that of the ray, is

$$V - U\frac{dy}{dx} = \frac{\dfrac{d^2y}{dx^2}}{1 + \left(\dfrac{dy}{dx}\right)^2},$$

where U, V denote the partial differential coefficients $\dfrac{d\log\mu}{dx}$, $\dfrac{d\log\mu}{dy}$, μ being the refractive index.

Apply this equation to shew that the astronomical refraction (r) can be expressed in a series of the form

$$A \tan z - B \tan z \sec^2 z \ldots\ldots,$$

where z is the apparent zenith distance.

18. What phenomena shew that the velocity of propagation of light in vacuum is the same for all colours?

19.   If it be assumed that the true wave surface in a biaxal crystal is symmetrical with respect to three planes at right angles to one another, that it is expressed by an equation of the fourth degree, and that each of its principal sections consists partly of a circle, the radii of the three circles being different, shew that it is the wave surface of Fresnel.

By what experiments has the third assumption been confirmed with respect to light of each particular degree of refrangibility?

20.   A prism is cut from a doubly refracting substance, and a pencil of light is transmitted through it, the plane of incidence being perpendicular to the edge: the angle of incidence, the angle of the prism, and the deviation of one of the emergent pencils being observed, obtain formulæ for determining the direction and velocity of the corresponding wave within the crystal, and put the formulæ in a convenient shape for numerical calculation.

January, 1851.

1.   Find the number of polygons of n sides, in an extended sense of the word *polygon*, that are formed by n indefinite right lines in a plane, supposing no two of the lines to be parallel, and no three to meet in a point.

2.   Three circles are described, each touching one side of a triangle and two sides produced, and a new triangle is formed by joining their centres.   The new triangle is then treated like the first, and this process is carried on indefinitely.   Prove that the triangles so formed tend indefinitely to become equilateral.

3.   A rigid body is acted on at each element of a closed curve by a force in the direction of a tangent, and proportional to the length of the element.   Prove that the system of forces is equivalent to a couple, acting in a plane for which the area enclosed by the projection of the curve is a maximum, and having a moment proportional to this maximum area.

4.   Prove, in the manner of Newton (*Principia*, Lib. I. Prop. I.), that if a body move in any manner under the action of forces directed to any number of points in a fixed straight line, the volume described by the triangle which has a given portion of the line for base and the body for vertex varies as the time.

5.   If the probability of the occurrence of an error lying between x and $x + dx$ in the observation of a certain quantity be proportional to $\epsilon^{-hx^2} dx$, shew that in the long run the mean of the squares of the errors of observation will be to the square of the mean error as π to 2.

6. If Ox, Oy, Oz, and Ox', Oy', Oz', be two systems of rectangular axes, and l, m, n, be the cosines of the angles $x'Ox$, $x'Oy$, $x'Oz$, l', m', n', the same for y', and l'', m'', n'', the same for z', prove that

$$l'' = mn' - m'n, \quad m'' = nl' - n'l, \quad n'' = lm' - l'm,$$

the positive directions of the axes being so chosen that the direction of revolution $x'y'z'x'$ ... is the same as $xyzx$ ....

7. Prove the formulæ

$$\int_s^\infty \cos\frac{\pi}{2}s^2\,ds = N\cos\frac{\pi}{2}s^2 - M\sin\frac{\pi}{2}s^2, \quad \int_s^\infty \sin\frac{\pi}{2}s^2\,ds = M\cos\frac{\pi}{2}s^2 + N\sin\frac{\pi}{2}s^2,$$

where $M = \dfrac{1}{\pi s} - \dfrac{1\,.\,3}{\pi^3 s^5} + \dfrac{1\,.\,3\,.\,5\,.\,7}{\pi^5 s^9} - \ldots,\quad N = \dfrac{1}{\pi^2 s^3} - \dfrac{1\,.\,3\,.\,5}{\pi^4 s^7} + \ldots$

8. If u be a function of the oblique coordinates x, y, z, and

$$Q = A\,\frac{d^2u}{dx^2} + B\,\frac{d^2u}{dy^2} + C\,\frac{d^2u}{dz^2} + 2D\,\frac{d^2u}{dydz} + 2F_{,}\,\frac{d^2u}{dzdx} + 2F\,\frac{d^2u}{dxdy},$$

shew that there exists a system of rectangular coordinates, x', y', z', and an infinite number of systems of oblique coordinates, for which Q takes the form

$$A'\,\frac{d^2u}{dx'^2} + B'\,\frac{d^2u}{dy'^2} + C'\,\frac{d^2u}{dz'^2}.$$

9. Eliminate by differentiation the transcendental function from the equation

$$y = 1 - \frac{x^2}{1\,.\,2} + \frac{x^4}{1\,.\,2^2\,.\,3} - \frac{x^6}{1\,.\,2^2\,.\,3^2\,.\,4} + \ldots$$

and thence prove that the large roots of the equation $y = 0$ form ultimately an arithmetic series having a common difference $\dfrac{\pi}{2}$.

10. Integrate the simultaneous equations

$$\frac{dx}{x + cy - bz} = \frac{dy}{y + az - cx} = \frac{dz}{z + bx - ay}.$$

11. Find the most general values of u, v, w which satisfy at the same time the five following partial differential equations:

$$\frac{du}{dx} = \frac{dv}{dy} = \frac{dw}{dz},$$

$$\frac{dv}{dz} + \frac{dw}{dy} = \frac{dw}{dx} + \frac{du}{dz} = \frac{du}{dy} + \frac{dv}{dx} = 0.$$

12. Explain why the achromatism of an object-glass consisting of two lenses cannot be rendered perfect. Illustrate by a figure the distribution of the foci of the various colours along the axis of a pencil when a compensation is effected as far as possible.

Supposing that you had it in your power to measure the refractive index of each kind of glass for any one or more of the principal fixed lines of the spectrum, what measures would you take, and what numerical values would you substitute for $\delta\mu$ and $\delta\mu'$ in the ordinary formulæ, so as to produce the best effect?

13.   In a Kater's pendulum, supposing the times of vibration about the two axes to be slightly different, investigate an expression for the time of vibration which must be employed in the calculation, in order that the deduced length of the seconds' pendulum may be correct, the distance of the centre of gravity from either axis being supposed approximately known.

14.   A rigid body at rest is struck in the direction of a line not passing through the centre of gravity: find the conditions under which the initial motion will be simply one of rotation.

15.   A rigid body is suspended symmetrically by two fine parallel wires. A vertical line drawn through the centre of gravity is equidistant from the suspending wires, and is a principal axis of the body. The body being turned round this axis through a small angle, and then left to itself, it is required to find the time of a small oscillation, taking into account the force arising from the torsion of the wires.

16.   A finite portion of an infinite mass of heterogeneous elastic fluid, acted on by no external forces, is in motion in any manner: apply the ordinary equations of motion of a fluid to prove that the increase of vis viva during the time dt is equal to $2Udt$, where

$$U = \int_{-\infty}^{\infty} \int_{-\infty}^{\infty} \int_{-\infty}^{\infty} p \left(\frac{du}{dx} + \frac{dv}{dy} + \frac{dw}{dz} \right) dx\,dy\,dz.$$

Prove also directly that the expression under the integral signs, multiplied by dt, is the work done during the elementary time dt by the expansion of the fluid occupying the elementary volume $dx\,dy\,dz$.

17.   In the calculation of the phenomena of diffraction produced when a screen containing one or more apertures is placed before a lens through which a luminous point is viewed in focus, prove that the error arising from the neglect of the spherical aberration of a direct pencil, or the astigmatism of an oblique pencil, is a small quantity of the fourth order, the obliquities of the several rays being regarded as small quantities of the first order.

18.   Explain the mode in which the wave-lengths corresponding to the principal fixed lines of the spectrum have been accurately measured, investigating the requisite formula.

19.   A pure plane-polarized spectrum is analyzed, and the analyzer is turned till the light is extinguished. A plate of selenite, of such a thickness as to give a difference of retardation amounting to several waves' lengths, is then interposed, behind the analyzer, with its principal planes inclined at angles of 45° to the plane of primitive polarization. Deduce from theory the appearance presented as the analyzer is turned round through 90°.

20.   A uniform flexible string of infinite length, subject to a given tension, is continually acted on at a given point by a given variable force in a transverse direction: determine the motion. Also examine specially the case in which the force is expressed by $c \sin nt$.

February, 1852.

[N.B. *Only one question is to be answered out of each pair.*]

1 A. *ABC* is a triangle inscribed in a circle, and *AB, AC* are produced to meet in *D, E*, the tangent at the extremity of the diameter passing through *A*; prove that a circle may be described about the quadrilateral figure *BDEC*.

1 B. *O* is a fixed, and *P* a variable point, and in *OP*, produced if necessary, a point *I* is taken such that $OP . OI = a^2$: if *I* be called the *image* of *P*, shew that the angle of intersection of any two curves in space which intersect will be equal to the angle of intersection of their images.

2 A. In any telescope in which the refraction at the object-glass is centrical, prove that the magnifying power is equal to the ratio of the clear aperture of the object-glass to the diameter of the bright image in front of the eye-piece.

Would this image be formed if the telescope were directed to a single luminous point?

2 B. Shew how to find whether the equilibrium of an irregular floating body is stable or unstable.

3 A. A plane mirror is moved by clock-work about an axis parallel to its own plane and to the axis of the earth, at such a rate as to make one turn in 48 hours; shew that it will reflect the Sun's light in a fixed direction, changes of declination &c. being neglected.

What choice of directions does a heliostat of this construction afford? If the direction of the reflected light is to be horizontal, find for a given place and a given day its inclination to the meridian.

3 B. Prove that the attraction of a uniform spherical shell on an external particle varies inversely as the square of the particle's distance from the centre of the shell, the law of attraction being that of the inverse square of the distance.

From Newton's demonstration of this theorem, deduce a geometrical solution of the following problem: To divide the shell into two parts, by a plane perpendicular to the line joining the particle with the centre, such that the attractions of the segments may be in a given ratio.

4 A. Three straight lines in a plane *A, B, C* meet in a point, and three others *A', B', C'* also meet in a point; prove that in general (AA', BB'), (AC', CB'), and (BC', CA') will meet in a point, the symbol (AA', BB') denoting the straight line joining the point of intersection of *A* and *A'* with that of *B* and *B'*.

How many such systems exist when the first six lines are given?

4 B. From a fixed point *O* a perpendicular *ON* is let fall on the tangent plane at any point *P* of a curved surface, and from *ON*, produced if necessary, a length *OP'* is cut off, such that $ON . OP' = k^2$, *k* being a given constant; prove that the locus of *P'* is a surface from which the first surface may be got back by the same construction, the points *P, P'* being corresponding points on the two surfaces.

5 A.   When the sum of the series

$$A_0 + A_1 x + A_2 x^2 + A_3 x^3 + \ldots$$

is known, shew how the sum of the series

$$a_0 A_0 + a_1 A_1 x + a_2 A_2 x^2 + a_3 A_3 x^3 + \ldots$$

may be found, where a_0, a_1, $a_2 \ldots$ are given multipliers which recur according to a certain cycle.

Example.   Find the sum of the infinite series

$$\frac{x}{1} - \frac{x^4}{1 . 2 . 3 . 4} + \frac{x^7}{1 . 2 . 3 . 4 . 5 . 6 . 7} - \ldots$$

5 B.   Given the equations of a system of curves in space containing two arbitrary parameters, shew how to find whether the curves admit of being cut orthogonally by a system of curved surfaces; and how in that case to find the equation of the system.

Examples.   Let the equations of the system of curves be

(1) $x = A e^{-nz}$, $y = B e^{nz}$;

(2) $x^2 + y^2 = A$, $y = x \tan n (z - B)$;

where A, B are the arbitrary parameters.

6 A.   Explain, in a manner as elementary as is consistent with rigour, the manner in which the rotation of the Earth may be exhibited by means of a pendulum.

6 B.   A rigid body is suspended by a string, regarded as infinitely long and light, which is attached to a point in one of the principal planes through the centre of gravity: explain the nature of the motion which takes place when the body is slightly disturbed in the most general manner.   Find also the time of oscillation, and the direction of the axis of rotation, in that kind of oscillation in which the point of attachment moves in a direction perpendicular to the principal plane.

Form the quadratic which determines the time of oscillation when the point of attachment has any arbitrary position.   Find also in this case the direction of the axis about which there must be no initial angular velocity in order that there may be no continual revolution of the body.

7 A.   Water is flowing gently and regularly through a cylindrical pipe inclined at an angle a to the horizon: supposing the fluid to be retarded by an internal friction producing a tangential pressure proportional to the rate of sliding, and just counteracting the effect of gravity, determine the motion, supposing the film of fluid immediately in contact with the pipe to be at rest.   Shew also that the mean velocity in the pipe is to the mean velocity in a broad stream flowing over a perfectly even bed having a uniform slope β as $3a^2 \sin a$ to $8b^2 \sin \beta$, where a is the radius of the pipe, and b the depth of the stream, measured in a direction perpendicular to its bed.

Is the motion of running water in practical cases of the kind here supposed?   If not, what do you conceive to be the cause of the difference, and what the nature of the actual motion?

7 B.   An infinite uniform stretched string, not acted on by gravity, is loaded at a certain point (taken for origin) with a given mass.   An indefinite series of small transverse disturbances, expressed by

$$c \sin k \,(at + x),$$

is continually propagated from an infinite distance, and is incident on the mass: determine the motion.

Determine also the simultaneous motions of the string and the mass when the latter is acted on by a given small disturbing force f, and the former is subject to no disturbances except those which travel from the mass outwards.   Examine in particular the case in which $f = c \sin kat$ from $t = -\infty$ to $t = 0$, and $f = 0$ from $t = 0$ to $t = \infty$.

8 A.   Give a general explanation of the fringes seen about the shadow of an opaque body bounded by a straight edge; of the continual and rapid decrease of illumination on receding from the geometrical shadow inwards; of the fluctuations of illumination outside, and of the increasing rapidity and decreasing amount of those fluctuations on receding from the geometrical shadow.

Shew that the fringes, regarded as existing in space, and considered only in a plane drawn through the luminous point perpendicular to the diffracting edge, form a system of hyperbolas starting from the edge; and find the law according to which the breadth of a given fringe depends upon the distances from the luminous point to the diffracting edge, and from the latter to the screen on which the fringe may be supposed to be received.

How did Fresnel measure the distances of the fringes from the geometrical shadow?

8 B.   Shew that a tube filled with sirup of sugar, followed by a Fresnel's Rhomb, when interposed between a polarizing plate and double-image prism will present the same general phenomena as a plate of selenite without the rhomb, so far as regards the changes of colour produced by turning the prism round, but that the actual tints seen in the two cases will not precisely correspond.

How may the accuracy of the rhomb be tested by this arrangement?

February, 1853.

1.   Through any point P in the diagonal AC of a parallelogram $ABCD$ are drawn any two straight lines meeting the sides AB, AD respectively, in E, F, and the opposite sides in G, H; prove that EF is parallel to GH.

2.   Prove that the six planes bisecting the dihedral angles of a triangular pyramid meet in a point.

3.   Examine the effect of the disturbing force of S on the nodes of P's orbit.   (NEWTON, Princip. Lib. I. Prop. 66.)

4.   When a body describes an ellipse round a centre of force in one of the foci, if a line be drawn from a fixed point always parallel to the direction of motion, and proportional to the velocity of the body, the extremity of the line will trace out a circle.

5.   Given the specific heat of a gas when (1) the pressure p (2) the density ρ is constant, find the specific heat when p and ρ vary together in such a manner that $\dfrac{dp}{d\rho}$ has a given value; and interpret the result when $\dfrac{dp}{d\rho} = \dfrac{p}{\rho}$.

6.   If an indefinitely small spherical portion of a homogeneous incompressible fluid in motion be suddenly solidified, shew that in addition to its motion of translation it will revolve round its centre with an angular velocity of which the components are

$$\tfrac{1}{2}\left(\frac{dw}{dy} - \frac{dv}{dz}\right), \quad \tfrac{1}{2}\left(\frac{du}{dz} - \frac{dw}{dx}\right), \quad \tfrac{1}{2}\left(\frac{dv}{dx} - \frac{du}{dy}\right).$$

7.   Assuming the laws of reflexion and refraction, shew that in the case of a ray reflected or refracted any number of times $\Sigma\mu s$ is ultimately constant, when the points of incidence on the several surfaces are made to vary arbitrarily by indefinitely small quantities, s denoting the length of that portion of the path which lies within the medium whose refractive index is μ.

Apply this principle to find the general equations of a refracted ray, when a pencil of rays emanating from a given point is incident on a surface bounding a medium of different refractive power.

8.   Give a full explanation of the formation of the primary and secondary rainbows on the principles of geometrical optics, and state any circumstances of the phenomenon which require for their explanation a more refined theory.

9.   Two surfaces touch each other at the point P; if the principal curvatures of the first surface at P be denoted by $a \pm b$, those of the second by $a' \pm b'$, and if ϖ be the angle between the principal planes to which $a + b$, $a' + b'$ refer, δ the angle between the two branches at P of the curve of intersection of the surfaces, shew that

$$\cos^2\delta = \frac{a^2 + a'^2 - 2aa'}{b^2 + b'^2 - 2bb'\cos 2\varpi}.$$

10.   A function is tabulated for a series of equidifferent values of the variable; investigate a formula for deducing the value of the differential coefficient of the function for a value of the variable intermediate between those found in the tables.

Ex.   Given $u = 1\cdot8733395$, $1\cdot8748744$, $1\cdot8764069$, $1\cdot8779372$, for $x = 6\cdot51$, $6\cdot52$, $6\cdot53$, $6\cdot54$, find $\dfrac{du}{dx}$ for $x = 6\cdot514$.

11.   Integrate the differential equation

$$axdy^2 + bydx^2 = (xdy - ydx)^2,$$

and find its singular solutions.

12. Explain the method of integrating the partial differential equation

$$L\frac{du}{dx} + M\frac{du}{dy} + N\frac{du}{dz} + \ldots = V,$$

where $L, M, N \ldots V$ are functions of $x, y, z \ldots u$. Illustrate your explanation, in the case of two independent variables, by reference to geometry.

If

$$x\frac{du}{dx} + y\frac{du}{dy} + z\frac{du}{dz} + \ldots = nu,$$

shew that u is a homogeneous function of $x, y, z \ldots$ of n dimensions.

13. Shew that

$$\int_0^\infty \sin ax \left\{ 1 - \frac{x^2}{2^2} + \frac{x^4}{2^2 4^2} - \ldots \right\} = 0 \text{ or } = \frac{1}{\sqrt{a^2 - 1}}$$

according as a, which is supposed to be positive, is less or greater than 1.

14. Compare the quantities of heat received from the Sun during one day at different places on the Earth's surface when the Sun has a given north declination, supposing the whole or a given fraction of the incident rays to be absorbed.

Shew that this quantity has two maxima, one at a latitude north of that at which the Sun is vertical at noon, the other at the north pole. Shew also that when the Sun's declination exceeds 17° 39′ more heat is received at the north pole than at the equator.

Given

$\cot^{-1}\pi = 17°\,39'$; $\cos 2h = -\tan^2 25°\,0'$, when $\tan 2h = 2h$ and $0 < 2h < 270°$.

15. Given the direction of an extraordinary ray within a crystal of Iceland spar, determine by a geometrical construction the direction of the bounding plane by which it will emerge without deviation. Find also, for a given bounding plane, the direction of the incident ray when the course of the extraordinary ray is a prolongation of that of the incident ray.

16. A heavy vertical circle is mounted so as to admit of sliding in a vertical plane down a smooth inclined plane, but is supported. A uniform smooth heavy string, whose length is less than half the circumference, is then laid on the circle so as to rest in equilibrium. The support being now suddenly removed, find at the commencement of the motion the tension of the string at any point and the pressure on the curve.

17. A heavy elastic string, uniform in its natural state, is attached at one extremity to a fixed point, and at the other sustains a heavy particle, and the system performs periodic oscillations in a vertical direction; shew that the time of oscillation is determined by a transcendental equation of the form $x \tan x = c$.

Discuss the roots of this equation, and shew how any required root may readily be calculated with the assistance of trigonometrical tables.

February, 1854.

1.   Straight lines AP, BP pass through the fixed points A, B, and are always equally inclined to a fixed line; shew that the locus of P is a hyperbola, and find its asymptotes.

2.   A number of equal vessels communicate successively with each other by small pipes, the last vessel opening into the air.   The vessels being at first filled with air, a gas is gently forced at a uniform rate into the first; find the quantity of air remaining in the nth vessel at the end of a given time, supposing the gas and air in each vessel at a given instant to be uniformly mixed.

3.   Separate the roots of the equation
$$2x^3 - 9x^2 + 12x - 4.4 = 0,$$
and find the middle root to four places of decimals by Horner's method, or by some other.

4.   Investigate a formula in Finite Differences for transforming a series the terms of which (at least after a certain number) are alternately positive and negative, and decrease slowly, into one which is generally much more rapidly convergent.
   Example.   Find the sum of the series
$$1\cdot4142 - \cdot7071 + \cdot5303 - \cdot4419 + \cdot3867 - \cdot3480 + \cdot3190 - \cdot2962 + \dots$$

5.   Given the centre and two points of an ellipse, and the length of the major axis, find its direction by a geometrical construction.

6.   Integrate the differential equation
$$(a^2 - x^2)\,dy^2 + 2xy\,dy\,dx + (a^2 - y^2)\,dx^2 = 0.$$
Has it a singular solution?

7.   In a double system of curves of double curvature, a tangent is always drawn at the variable point P; shew that, as P moves away from an arbitrary fixed point Q, it must begin to move along a generating line of an elliptic cone having Q for vertex in order that consecutive tangents may ultimately intersect, but that the conditions of the problem may be impossible.

8.   If X, Y, Z be functions of the rectangular co-ordinates x, y, z, dS an element of any limited surface, l, m, n the cosines of the inclinations of the normal at dS to the axes, ds an element of the bounding line, shew that
$$\iint \left\{ l\left(\frac{dZ}{dy} - \frac{dY}{dz}\right) + m\left(\frac{dX}{dz} - \frac{dZ}{dx}\right) + n\left(\frac{dY}{dx} - \frac{dX}{dy}\right) \right\} dS$$
$$= \int \left(X\frac{dx}{ds} + Y\frac{dy}{ds} + Z\frac{dz}{ds} \right) ds,$$
the differential coefficients of X, Y, Z being partial, and the single integral being taken all round the perimeter of the surface*.

[* This fundamental result, traced by Maxwell (*Electricity*, I, § 24) to the present source, has of late years been known universally as Stokes' Theorem.   The same

9. Explain the geometrical relation between the curves, referred to the rectangular co-ordinates x, y, z, whose differential equations are

$$\frac{dx}{P} = \frac{dy}{Q} = \frac{dz}{R},$$

and the family of surfaces represented by the partial differential equation

$$P\frac{dz}{dx} + Q\frac{dz}{dy} = R.$$

10. Write a short dissertation on the theoretical measure of mass. By what experiments did Newton prove that masses may be measured by their weights? Independently of such experiments, how may it be inferred from the observed motions of the heavenly bodies that the mutual gravitation of two bodies depends only on their masses, and not on their nature? In what two different senses is the term *weight* used?

11. What are the conditions to be satisfied in order that two moving systems may be dynamically as well as geometrically similar?

If it be desired to investigate the resistance to a canal boat moving 6, 8, 10 miles an hour by experiments made with a small model of the boat and canal, if the boat be 36 feet long and its model only 2 ft. 3 in., what velocities must be given to the latter?

12. A rod is suspended at two given points by unequal light elastic strings, in such a manner that the rod is horizontal and the strings are vertical in the position of equilibrium; the rod being slightly disturbed in a vertical plane, in such a manner that no displacement or velocity is communicated to the centre of gravity in a horizontal direction, it is required to determine the motion.

13. Shew how to determine the time of rotation of the Sun about his own axis, and the position of his equator.

14. Rays coming from a luminous point situated in the axis of a large convex lens, and beyond the principal focus, are received after transmission through the lens on a screen held perpendicular to the axis, which is moved from a little beyond the extremity of the caustic surface to a little beyond the geometrical focus: compare, according to the principles of geometrical optics, the illumination at different points of the screen, and at different distances of the screen from the lens, the lengths of any lines in the figure being regarded as known.

kind of analysis had been developed previously in particular cases in Ampère's memoirs on the electrodynamics of linear electric currents. And in a letter from Lord Kelvin, of date July 2, 1850, relating to such transformations, which has been found among Stokes' correspondence, the theorem in the text is in fact explicitly stated as a postscript. The vector which occurs in the surface integral had been employed by MacCullagh, who recognized its invariance, about 1837, in optical dynamics. It reappeared in Stokes' hands in 1845 (*ante*, Vol. I, p. 81) as twice the differential rotation in the theory of fluid motion and formed the basis of the mathematical theory of viscosity of fluids, as also at a later time of Helmholtz's theory of vortex motion: its application to vibrations of solid elastic media was developed by Stokes in 1849 (*ante*, Vol. II, p. 253). Thus the theorem, though first stated by Lord Kelvin, relates to a quantity which, as regards physical applications, may be claimed to be Stokes' vector.]

Give a sketch of the method of finding the illumination in the neighbourhood of a caustic according to the theory of undulations. What is the general character of the result, and in what natural phenomenon is it exhibited?

15. A glass plate, the surface of which is wetted, is placed vertically in water; shew that the elevation of the fluid varies as the sine of half the inclination of the surface to the horizon, and compare its greatest value with the elevation in a capillary tube of given diameter. Find also the equation of the surface.

16. Explain the different modes of determining the Mass of the Moon.

17. Plane polarized light is transmitted, in a direction parallel to the axis of the crystal, across a thick plate of quartz cut perpendicular to the axis, and the emergent light, limited by a screen with a slit, is analyzed by a Nicol's prism combined with an ordinary prism; describe the appearance presented as the Nicol's prism is turned round, and from the phenomena deduce the nature of the action of quartz on polarized light propagated in the direction of the axis.

January 30, 1855.

1. A straight line drawn through the middle point of one side of a triangle divides the two other sides, the one internally and the other externally, in the same ratio.

2. If P, Q' be the points in one of two concentric and confocal ellipsoids which correspond to the points P', Q respectively in the other, prove that $PQ' = P'Q$.

3. Shew how to find whether $f(x, y, z) = 0$ is or is not a particular case of $u = \phi(v)$, where u and v are given functions of x, y, z.

4. A long cylinder, loaded on one side, rests with its axis not horizontal on a rough inclined plane; find the condition that equilibrium may be just possible.

5. Find the volume of the solid generated by the revolution of the closed part of the curve $x^3 - 3axy + y^3 = 0$ about the line $x + y = 0$.

6. If A, B be two terrestrial stations at no great distance, find approximately the difference between the north azimuth of B as seen from A, and the south azimuth of A as seen from B.

7. Two smooth perfectly elastic balls are suspended by threads from two points in the same horizontal plane, in such a manner that when the balls are at rest their centres lie in a horizontal plane, and the surfaces are just in contact. If the balls be withdrawn in different planes from their position of rest, and let go simultaneously, shew that the motion will be periodic, the effect of the rotatory inertia of the balls and of the finite arc of oscillation on the time of vibration being neglected.

8. Find the condition of achromatism of an eye-piece composed of a single block of glass worked at the ends into spherical surfaces, and used with an object-glass of great focal length.

9. Given the force of gravity at the top and bottom of a mine of known depth, and the density of the intervening mass, find the mean density of the Earth.

10. If from a variable point in a conic section perpendiculars be let fall on the sides of any fixed, inscribed polygon with an even number of sides, the product of the perpendiculars let fall on one set of alternate sides will be to the product of the perpendiculars let fall on the other set in a constant ratio.

11. A wheel, of which an axle projecting on each side forms a part, is supported in a vertical plane by having the axle on each side resting on a pair of friction wheels, each of which is just like the first wheel, and is similarly supported, and so on indefinitely; compare the inertia of the whole system, in relation to a rotation of the first wheel, with that of the first wheel alone.

12. Find in any manner the approximate effect of the Sun's disturbing force on the mean motion of the Moon in a month, and thence deduce the Annual Equation.

13. If a body move round the Sun in a circular orbit, the radius of which is slowly diminishing in consequence of a resisting medium, find the amount of work spent on the medium as the body moves from one given distance to another; and compare the amount spent in passing from an infinite distance to close to the surface with the work spent when the body impinges on the Sun, and is thereby reduced to rest.

14. Shew that

$$\int_{-\infty}^{\infty} e^{-\left(x^2\cos 2\theta + \frac{a^2}{2x^2}\sin 2\theta\right)} \frac{\cos}{\sin}\left\{x^2\sin 2\theta + \frac{a^2}{2x^2}\cos 2\theta\right\} dx = \pi^{\frac{1}{2}} e^{-a} \frac{\cos}{\sin}(\theta + a),$$

θ being comprised between the limits $\pm\frac{\pi}{4}$.

15. Investigate the series of notes which may be produced by the vibration of the air within a tube closed at one end and open at the other. Why is the supposition usually made as to the condition at the open end inexact, and in what manner do the results of theory, on this supposition, differ from those of experiment?

16. Find numerically what would be the ellipticity of the Earth if it consisted of a sphere in which the density was a function of the distance from the centre, covered by a comparatively shallow superficial crust having a density equal to $\frac{5}{11}$ths of the mean density.

17. Under what experimental circumstances is the reflected system of Newton's rings white-centred, either with common or polarized light, and what point of theory in each case is confirmed by the phenomenon?

18. A solid of revolution is made to revolve with great rapidity round its axis, which is so mounted that the solid is free to turn in all directions round .its centre of gravity. If a force be applied perpendicularly to the axis, a considerable resistance is experienced, and the instantaneous axis moves perpendicularly to the direction of the force; but if the axis be mounted so as to be moveable in one plane only, and a force be applied in that plane, the solid goes round just as if it had no motion of rotation originally. Explain this.

19. In a system of curves in space whose equations contain two arbitrary parameters, shew that the pencil of tangents drawn at the points where the curves are cut by a small plane perpendicular to one of them consists ultimately of a pencil of parallel lines altered (1) by being made to converge to two focal lines in rectangular planes, (2) by being twisted; and that the analytical condition of the possibility of cutting the lines orthogonally by a system of surfaces expresses that there is no twisting. Shew also geometrically how the twisting would render it impossible to cut the lines in the manner described.

20. Write a short dissertation on the evidence, or want of evidence, of the truth of the laws of double refraction in biaxal crystals which result from the theory of Fresnel. Do you conceive these laws to be rigorously or only approximately true; to be applicable to each colour in particular, or only to white light as a whole?

January 30, 1856.

1. If the distances of the points A, B, C, D in a straight line from a point O in the same line be in harmonic progression, prove that the rectangle under the extreme segments AB, CD of AD is equal to the rectangle under the middle segment and the mean of the three.

2. Required a point which is at the same time the middle point of each of two chords of two given circles respectively, these chords both passing through the same given point. Is the solution of the problem always geometrically possible?

3. Are there any objections to the representation of imaginary branches of a curve by branches lying in a perpendicular plane?

Shew that the asymptotes of a given circle are independent of the arbitrary choice of rectangular co-ordinate axes.

4. Trace the curve
$$\left(\frac{x-a}{x}\right)^2 + \left(\frac{y-b}{y}\right)^2 = 1.$$

5. Shew that the family of curves
$$\left(\frac{x^2}{c^2-a^2} + \frac{y^2}{c^2-b^2}\right)^2 = \frac{4}{c^2}\left(\frac{a^2x^2}{c^2-a^2} + \frac{b^2y^2}{c^2-b^2}\right)$$
has the same envelope whether a or b be the variable parameter.

How do consecutive curves of the system lie when the points of contact with the envelope are imaginary?

6. The equation of a surface is given implicitly by explicit expressions for x, y, z in terms of two parameters; required the expressions for the direction-cosines of the normal.

7. Find the lines of magnetic force in the case of a pair of poles of equal strength, regarded as points, (1) when the poles are of the same name, (2) when they are of opposite names. Shew how the lines may be graphically constructed.

8. If a triangle be circumscribed about a conic, and the points of contact be joined with the opposite vertices, and tangents drawn at the points of intersection with the conic so as to form a second triangle, prove that the relation between the two triangles will be reciprocal.

9. Shew how to find the lines on a surface at which one of the principal curvatures vanishes. How does the form of the surface in general alter in passing across such a line? Is the line in question a line of curvature?

10. If $f(x) = x^n + p_1 x^{n-1} \ldots + p_{n-1} x + p_n$, where $p_1 p_2 \ldots p_n$ are imaginary, and if $f(x + \sqrt{-1} y) = P + \sqrt{-1} Q$, where P and Q are real, examine the forms of the intersections of the surfaces $z = P$, $z = Q$ with the plane $z = 0$, and thence shew that the equation $f(x) = 0$ has n roots of the form $x + \sqrt{-1} b$.

11. What is meant by the *complete primitive*, the *general primitive*, and the *singular primitive equation* of a partial differential equation of the first order between three variables? Illustrate the relation between these three forms by reference to geometry.

Explain the mode of integrating the equation $f(x, y, z, p, q) = 0$.

12. What is meant by the irrationality of dispersion, and how are its effects manifested in an object-glass? Shew that they may be got rid of (to the lowest order of approximation) by combining a crown-glass lens with a flint-glass lens of proper strength, placed a proper distance down the tube of the telescope. What would be the objection to such a construction?

13. Two rods equally inclined to the vertical meet in an angle, which opens upwards; required the smallest angle of a double cone which if laid symmetrically upon the rods will roll away from the angle.

14. Find the latitude at sea from two altitudes of the Sun and the time between, correcting for the ship's change of place.

15. Explain why a solid iron plate should be stronger than a compound plate made up of thinner plates rivetted together, and having the same aggregate thickness.

16. Two equal streams of light interfere, but the difference of path is too great to allow of the exhibition of colours. The mixed streams being limited by a screen with a slit parallel to the direction which the fringes would have if they appeared, and being then analyzed by a prism, explain the appearance presented; and shew

how to deduce from observation the absolute retardation, supposing the wave-lengths for two given points of the spectrum to be known, and the difference of path to have occurred in air.

17.   What is the relation between the invariable plane of a dynamical system and the set of impulsive forces by which the system may have been originally set in motion?   Would there exist an invariable plane in the case of a system composed of the fragments of a body which exploded having been previously at rest?

A homogeneous ellipsoid at rest, fixed at its centre, being struck in the direction of a given line, find the angle between the initial instantaneous axis and the normal to the invariable plane.

18.   Obtain the equation of steady motion in hydrodynamics, shewing clearly in what sense and under what restrictions it is true.

19.   Shew that the effect of the term $\frac{15}{4} me \sin\{(2-2m-c)pt-2\beta+a\}$ in the expression for the Moon's longitude is equivalent to that of a periodic fluctuation in the excentricity and longitude of the perigee. Account for these fluctuations by general reasoning.

20.   A ray is incident in a given direction on a doubly refracting medium; explain clearly the steps of the physical reasoning which leads to a geometrical construction for determining the directions of the refracted rays when the velocity of plane waves within the medium is known as a function of the direction.

February 3, 1857.

1.   If $a > b > c$, then $\dfrac{a}{c} + \dfrac{b}{a} + \dfrac{c}{b} > \dfrac{a}{b} + \dfrac{b}{c} + \dfrac{c}{a}$.

2.   Shew that every imaginary plane contains one and but one real straight line.

3.   Prove that the mean of n positive quantities which are not all equal is greater than the nth root of their product.

4.   When the sum of a series according to ascending powers of x is known, shew how to deduce the sum of the series obtained by taking terms of the former at regular intervals.

Example.   From the series

$$\tan^{-1} x = x - \frac{x^3}{3} + \frac{x^5}{5} - \dots$$

deduce the value of $1 - \frac{1}{5} + \frac{1}{9} - \frac{1}{13} + \dots$

5.   Obtain a series for the ready calculation of $\displaystyle\int_x^\infty e^{-x^2} dx$ for large values of x.

6. Given that

$$y \frac{dz}{dx} = x \frac{dz}{dy}, \text{ and } cz = (x - y)^2 \text{ when } x + y = a,$$

required the relation between x, y, z.

7. Trace the curve

$$(x + y)^2 (x^2 + y^2) - a^3 x - b^3 y = 0,$$

and shew how it passes into what it becomes when $b = a$.

8. If the equation of a curve be given implicitly by two equations of the form $x = \phi(t)$, $y = \psi(t)$, where $\phi(t)$, $\psi(t)$ are free from radicals, shew how to find (1) the double points; (2) the cusps, regarded as singular double points; (3) the cusps, by a method not introducing the double points.

Apply your method to the example

$$a^2 x = (t + a)^2 (t - b), \qquad b^2 y = (t + b)^2 (t - a).$$

9. Explain the generation of a developable surface (1) by lines, (2) by planes of which it is the envelope; and point out the mutual relations of the generating lines, the enveloping planes, and the cuspidal edge. Explain also the form of the surface about such an edge.

10. Explain the effect of the sun's disturbing force in rendering the moon's orbit oval, supposing the undisturbed orbit circular; and find the ratio of the axes of the oval orbit.

11. A uniform flexible string is suspended by one end, and performs small movements in a vertical plane; form the differential equation of the motion, and thence obtain, in a series, the transcendental equation whose roots determine the periodic times of the various possible symmetric oscillations.

12. Obtain the equations of motion of a rigid body moveable about a fixed point, in terms of the angular velocities about the principal axes through that point.

Prove that the rotation of a free body revolving round a principal axis through its centre of gravity is stable or unstable according as that axis is one of greatest or least, or else one of mean, moment of inertia.

13. Shew that a small lunar atmosphere would affect the duration of an occultation, but not sensibly affect the apparent diameter of the moon.

14. How do you account for the remarkable effect of wind on the intensity of sound?

15. Find the condition of achromatism of an eye-piece composed of a solid block of glass with a thin lens of different glass cemented to it on the end next the eye, the surfaces being worked spherical.

16.   Find the attraction of a prolate spheroid on an internal particle.

A mass of homogeneous fluid is subject to the mutual gravitation of its particles, and to a repulsive force tending from a plane through its centre of gravity and varying as the perpendicular distance from that plane ; shew that the conditions of equilibrium will be satisfied if the surface be a prolate spheroid of a certain ellipticity, provided the repulsive force be not too great.

17.   Account for the spectra formed by a fine grating ; and supposing the grating placed obliquely to the incident light, find an expression giving the length of a wave of light in terms of quantities which may be observed.

18.   If a continuous medium be continuously displaced, shew that the most general displacement of an element of the medium consists of a displacement of translation, the same as that of a point P taken in the element, a rotation round some axis through P, and three elongations along three rectangular axes passing through P.

19.   Expressing the equations of motion of a fluid in a form in which the particle is supposed to remain the same in differentiations with respect to the time, and supposing the density either constant or a function of the pressure, and the forces such that $Xdx + Ydy + Zdz$ is a perfect differential, obtain first integrals of the three equations resulting from the elimination of the pressure.   Are these equations altogether independent of each other?   What important theorem may be proved by means of these integrals?

20.   Describe fully some one experiment by which it may be shewn that two streams of light from the same source, polarized in rectangular planes, and afterwards brought to the same plane of polarization, do or do not interfere according as the light from the primitive source is or is not polarized.

N.B.   It is not to be assumed that the colours of crystalline plates in polarized light are due to interference.

February 2, 1858.

1.   If three circles pass each through one corner of a triangle and the points of bisection of the adjacent sides, they will meet in a point.

2.   If XYZ be a spherical triangle having each side a quadrant, XPN, YPM great circles cutting the sides in N, M, and if the sides ZM, &c. of the quadrilateral $ZMPN$ be denoted by x, η, ξ, y, find the relations between x, y and ξ, η.   If the sides of the quadrilateral be small, up to what order of small quantities will the opposite sides be equal?

3. Shew generally how to form, by elimination, the equation whose roots are the sums of every two roots of a given equation, and apply the general method to obtain the actual result in the case of the equation

$$x^3 + ax^2 + bx + c = 0.$$

4. In the single moveable pulley with weights not in equilibrium, find by direct application of D'Alembert's Principle the accelerating force and the tension of the string, neglecting the rotatory inertia of the fixed and of the moveable pulley.

5. A mirror of given aperture and focal length, and of small curvature, has the form of a prolate spheroid ; shew that the aberration for parallel rays varies inversely as the major axis.

6. A sailor provided with a chronometer, sextant, and artificial horizon lands on a small island for a short time on a moonless night, the sky being cloudy towards the north ; what observations would he take to determine the geographical position of the island? Reduce the deduction of the required from the observed quantities to the solution of spherical triangles, without writing down the formulæ.

7. Find the condition which must be satisfied in order that

$$Udx + Vdy + Wdz$$

may be the exact differential of a function of two independent variables, the equation connecting x, y, z being

$$pdx + qdy + rdz = 0,$$

which is supposed to be integrable as an equation between three independent variables.

8. Find the cusps of the curve defined by the equations

$$b^2(a^2 - b^2)x^2 = (2b^2 - \theta)^2(a^2 - \theta) ; \quad a^2(a^2 - b^2)y^2 = (2a^2 - \theta)^2(\theta - b^2),$$

where θ is the variable parameter.

9. If a curve be treated as defined by explicit expressions for the co-ordinates x, y involving a variable parameter, double points do not appear as singularities, but if by an equation $f(x, y) = 0$ they do. Account for this.

10. An infinite cylinder revolves uniformly round its axis in an infinite mass of fluid, which is thus made to revolve in cylindrical shells in a permanent manner ; find the angular velocity of any shell, assuming that the tangential pressure arising from friction varies as the rate of sliding, and that the shell in contact with the cylinder revolves with it.

11. A chain fixed at two points to a vertical axis revolves uniformly about it ; find the differential equation of the curve which it forms by the condition that the function which expresses the total work of the forces shall be a maximum, and shew how the arbitrary constants are to be determined.

12. If the particles of a continuous medium be slightly and continuously displaced, shew that the relative displacement about a given point is symmetrical with respect to a system of rectangular axes.

13. Shew that the equilibrium of a free body acted on by forces varying according to the inverse square of the distance is essentially unstable.

14. A luminous point is viewed in focus through a telescope, and the object-glass is then covered by a screen with a small rectangular aperture; describe and account for the appearance presented.

15. A solid of revolution revolves about its axis of which one point is fixed, and the axis is constrained to move in a given manner by a force passing through it; find in terms of given quantities the moments of the constraining force.

16. A chain having initially the form of a closed plane curve very nearly a circle whirls in its own plane round its centre of gravity; determine the motion so far as to form the linear partial differential equation on which it depends.

17. Describe fully an experiment by which it is shewn that in total internal reflexion the phase of vibration of light polarized perpendicularly to the plane of incidence is accelerated, but in metallic reflexion retarded, relatively to that of light polarized in the plane of incidence.

18. Determine the motion of the bullet in an air-gun, supposing the mass of the compressed air small but not infinitely small, and neglecting friction.

February 1, 1859.

1. Shew how to deduce from any formula in Spherical Trigonometry the corresponding formula in Plane Trigonometry; and from the formula giving the relation between four consecutive parts of a spherical triangle deduce the corresponding formula for a plane triangle, proving also the latter independently.

2. Explain fully some one method of approximating to the real roots of an equation.

3. If one logarithmic spiral roll on another, what will be the curve traced out by its pole?

4. Trace the curve

$$x^n - ax - by + y^n = 0,$$

n being an integer; and shew what the curve approaches towards indefinitely when n, being either odd or even, increases indefinitely.

5. In the first Definition of Newton's *Principia* occurs the passage, "Innotescit ea [massa] per corporis cujusque pondus : nam ponderi proportionalem esse reperi per experimenta pendulorum accuratissime instituta, ut posthac docebitur." Describe these experiments, and discuss their bearing.

6. Examine the form of the surface

$$x^3 + y^3 + z^3 = a^3,$$

and find the nature of its section by the tangent plane at the point in which it is cut by the axis of z.

7. If an ordinary refracting telescope (not Galileo's) with a sufficiently high magnifying power be directed to a dark chimney, the vertical edge of which is seen against the sky in the centre of the field of view, and the right or left half of the object-glass be covered by a screen, the edge of the chimney will be seen fringed with green or purple, and will be sharper in the former case than in the latter: explain this.

8. If U be the variable *vis viva* of a limited portion, consisting of the same set of particles, of a homogeneous incompressible fluid acted on by forces X, Y, Z such that $Xdx + Ydx + Zdx = dV$, from the direct expression for U deduce, by means of the ordinary equations of fluid motion, an expression for dU/dt involving double integrals only, and not involving the arbitrary direction of the co-ordinate axes.

9. Shew that the equation

$$\frac{d^2y}{dt^2} = f\left(\frac{dy}{dx}\right)\frac{d^2y}{dx^2}$$

has a particular first integral of the form

$$\frac{dy}{dt} = \phi\left(\frac{dy}{dx}\right),$$

from whence may be obtained a second integral (or what is equivalent to such) involving one arbitrary function.

10. Examine the effect of the oblateness of a planet on the motion of its satellite.

11. A surface being defined by the equations

$$x = f_1(\lambda, \mu), \qquad y = f_2(\lambda, \mu), \qquad z = f_3(\lambda, \mu),$$

shew how to find whether it has a cuspidal edge, and in such case to obtain the equation of the edge under the form

$$\phi(\lambda, \mu) = 0.$$

Example. Apply your method to the surface

$$bcx = (\lambda - a)^2(\mu - a), \qquad cay = (\lambda - b)^2(\mu - b), \qquad abz = (\lambda - c)^2(\mu - c),$$

and find the equations of the projections of the cuspidal edge on the co-ordinate planes.

12. There are n parallel reflecting surfaces at each of which the same given fraction of the light incident upon it is reflected, and the rest transmitted; if light of intensity unity be incident upon the set, find the intensities of the reflected and the transmitted lights.

13. If the light reflected from a moderately thin plate of mica be limited by a slit and analyzed by a prism, the spectrum will be seen traversed by dark bands at regular intervals; account for these,

and determine the number seen under given conditions between two given points of the spectrum.

14.   An infinitely thin coin or ring moves on a rough horizontal plane in such a manner as to preserve a constant inclination to the horizon; determine the motion, and shew that the following relation subsists between the angular velocity (ω) of the body about its axis, the angular velocity (ϖ) of the horizontal tangent, the radius (a), the radius of gyration about the axis (k), and the inclination (i):

$$(a^2 + k^2)\,\omega\varpi + \tfrac{1}{2}k^2 \cos i \,.\, \varpi^2 = ag \cot i.$$

15.   A free uniform elastic rod, symmetrical with respect to a longitudinal plane, performs small transversal vibrations parallel to the plane of symmetry.   Assuming that flexure calls into play a couple, tending to straighten the rod, which varies as the curvature, form the partial differential equation of the motion: and thence shew that there are two series of simple isochronous vibrations, to one of which correspond 2, 4, 6 ... and to the other 3, 5, 7 ... nodal points.

Write short dissertations on the following subjects:

(1)   On a geometrical interpretation of $\sqrt{-1}$, generalizing thereby some simple proposition relative to one dimension so as to obtain a proposition relative to two dimensions, and pointing out the insufficiency of pure geometry for the complete interpretation, in passing to imaginary quantities, of results relative to two dimensions.

(2)   On the different methods of determining the mass of the Moon.

(3)   On Fresnel's theory of double refraction, as far as to the determination of the propagation of plane waves; and on the extent to which it is a rigorous dynamical theory.

January 30, 1860.

1.   If two circles intersect at right angles in two given points, prove that the sum of the squares of the reciprocals of the radii of the circles will be constant.

2.   State and prove the relation connecting the numbers of solid angles, edges, and plane faces of any polyhedron.

3.   Shew that the six planes passing each through one edge of a triangular pyramid and bisecting the opposite edge meet in a point.

4.   Each element of a transparent sphere which is denser than the surrounding medium is uniformly self-luminous; find what fraction of the whole light produced emerges from the sphere.

5. Trace the curve

$$(a^2y^2 + b^2x^2 - a^2b^2)(b^2y^2 + a^2x^2 - a^2b^2) = na^4b^4,$$

examining particularly the cases (1) when $n = 1$, (2) when n is very small, positive or negative, (3) when $n = -\dfrac{(a^2 - b^2)^2}{4a^2b^2}$.

6. An elastic string, uniform in its natural state, whirls uniformly round its middle point; find the stretched length of any portion, and the greatest angular velocity which can be given to the string without breaking it.

7. A portion of matter, at first indefinitely diffused, becomes arranged, by the mutual gravitation of its particles, in a sphere whose density at the distance r from the centre is a given function $f(r)$ of r; find the total work of the attraction in passing from the first state to the second.

8. Assuming the formula for the refraction of a direct centrical pencil through a thin lens, shew that in general there are either two positions or none of the conjugate foci for which the lens will be aplanatic. Can an aplanatic lens if reversed remain aplanatic for the same positions of the conjugate foci?

9. Find the deflection from a vertical line of a body let fall from a tower of given height in a given latitude, neglecting the resistance of the air.

10. Shew how to determine the Moon's parallax by observations made at two fixed observatories which differ widely in latitude, stating precisely the nature of the actual observations, and the corrections which must be applied to them.

11. If a uniform ring of attracting matter existed at a small distance round the Sun, how would its effect on the motions of the planets, supposed to have their orbits in the plane of the ring, differ from that of a corresponding increase in the Sun's mass?

12. A fixed vertical cylinder, closed at the top except as to a small orifice, is provided with a smooth piston which is held at rest in a given position. The piston being released determine its motion, so far as to form the differential equation on which it depends, neglecting the inertia of the air except in so far as it regulates the flow through the orifice.

13. X, Y, Z, the components of a force R, are given functions of the rectangular co-ordinates x, y, z, and a particle moves under the action of a force whose direction is perpendicular to that of R, and to that of the particle's motion, and whose magnitude is proportional to the product of R by the component of the velocity in a direction perpendicular to R; form the differential equations of motion. What will be the path of the particle in the particular case in which X, Y, Z are constant, and to what will it approach indefinitely, whether in this or in the general case, when R is indefinitely increased?

14. Solve the functional equations

$$\phi(x+y) = \phi(x) + \frac{\phi(y)\{\psi(x)\}^2}{1 - \phi(x)\phi(y)},$$

$$\psi(x+y) = \frac{\psi(x)\psi(y)}{1 - \phi(x)\phi(y)}.$$

15. Shew that the system of pressures in different directions in the immediate neighbourhood of a given point within a strained elastic solid is symmetrically arranged with respect to three rectangular axes, on a plane perpendicular to any one of which there is no tangential pressure. Shew also that if the three principal pressures have not all the same sign, the pressure is wholly tangential for any tangent plane of a certain elliptic cone.

16. A plate of glass has one face replaced by three planes cut so as to form a solid angle, contained by three equal dihedral angles of not much less than 180°; light from a luminous point is transmitted through the plate placed at some distance, and is viewed at some distance on the other side of the plate through a lens or eye-piece; investigate the appearance produced.

Write short dissertations on the following subjects:

(1) On the transformation of multiple integrals.

(2) On the methods which have been employed for determining the mean density of the Earth.

(3) On the evidence in favour of the optical theory of transversal vibrations.

January 29, 1861.

1. Prove that $(Mx + Ny)^{-1}$ cannot be an integrating factor of $M + N\frac{dy}{dx} = 0$, unless the equation be homogeneous.

2. Trace the curve

$$(x^2 + y^2)(a^2c^2x^2 + b^4y^2) - 2a^2b^2c^2x^2 - (a^2 + c^2)b^4y^2 + a^2b^4c^2 = 0,$$

and examine its change of form as b passes through the value a or c. Find also the area, supposing b to lie outside the limits a and c.

3. In what manner would you proceed if you wished to obtain numerically the sums of the following slowly convergent infinite series?

(1) $1 - 2^{-\frac{1}{2}} + 3^{-\frac{1}{2}} - 4^{-\frac{1}{2}} + \ldots\ldots,$

(2) $1 + 2^{-\frac{3}{2}} + 3^{-\frac{3}{2}} + 4^{-\frac{3}{2}} + \ldots\ldots$

4. If $F(x) = \sqrt{\dfrac{2}{\pi}} \displaystyle\int_0^\infty \phi(a) \cos axda, \quad f(x) = \sqrt{\dfrac{2}{\pi}} \int_0^\infty \psi(a) \sin axda,$
shew that the functions F and ϕ, and also f and ψ, are reciprocal; and thence from the known values of $\displaystyle\int_0^\infty e^{-g^2} \cos cxdx$ and $\displaystyle\int_0^\infty e^{-g^2} \sin cxdx$ deduce those of two other definite integrals.

5. In a rainbow of any order, shew that a small pencil of rays in the primary plane in the middle of its course within the drop has its focus at the middle point of incidence, or else at an infinite distance, according as the number of reflections is odd or even.

6. The altitudes of two known stars are taken with a theodolite which is clamped in azimuth, find the latitude of the place, and point out the circumstances which are favourable to accuracy in the determination.

7. In a heliostat of such a construction as to involve but one reflection, shew that a line drawn through any fixed point perpendicular to the plane of the mirror will describe in the 24^h an elliptic cone, the section of which by the plane of the equator is a circle which is described uniformly. Examine the case in which the rays are reflected towards a point the south polar distance of which is equal to the north polar distance of the Sun.

8. Express the position of the invariable plane of a system of bodies not acted on by forces external to the system in terms of quantities depending on the motions of and the motions about the centres of gravity of the several bodies.

9. A planet circulates in a slightly resisting medium; find the effect of the resistance on the angle described in a long time, the undisturbed orbit being supposed circular.

10. A uniform flexible and inextensible string at rest in space is pulled by an impulsive force applied at one end in the direction of the tangent; shew that the impulsive tension (T) is determined by the equation $d^2T/ds^2 = T/R^2$, where R is the radius of absolute curvature; and find the components of the initial velocity at any point in the directions of the tangent, radius of absolute curvature, and normal to the osculating plane.

11. A slender fluid ring revolves uniformly round a centre of force situated at its centre, the force varying inversely as the square of the distance; find approximately the form of a section of the ring.

12. If a mass of steel be permanently magnetized in any manner, shew that the effect of the Earth's magnetism upon it will be the same as if it were replaced by a slender bar, fixed relatively to the mass, and uniformly and longitudinally magnetized. What is meant by the *magnetic moment* of a magnet?

13. When a small flame is viewed by reflection in a slightly tarnished looking-glass some way off, at a moderate angle of incidence, a series of coloured bands is seen accompanying the image; explain the formation of these bands, and calculate their forms.

14. A small solid sphere is contained in an infinite mass of air not acted on by external forces; assuming that when the sphere moves through the air at rest, along a straight line, with a small velocity v, the resistance is expressed by $a\dfrac{dv}{dt} + bv$, determine the motion of the sphere produced by an infinite succession of plane waves of sound, in which the disturbance is expressed by a sine or cosine.

15. The motion of an indefinitely extended and slightly disturbed homogeneous elastic medium, not acted on by external forces, nor subject in the position of equilibrium to internal tensions, being determined according to the method of Lagrange by the equation

$$\iiint \rho \left(\frac{d^2u}{dt^2}\,\delta u + \frac{d^2v}{dt^2}\,\delta v + \frac{d^2w}{dt^2}\,\delta w \right) dx\,dy\,dz = \iiint \delta V dx\,dy\,dz,$$

where u, v, w are the displacements parallel to the rectangular axes of x, y, z, and V is a function of the nine differential coefficients of u, v, w with respect to x, y, z, it is required to express by means of the function V the normal and tangential tensions at any point on planes parallel to the co-ordinate planes.

———————

Write short dissertations on the following subjects :

(1) On the signification and integration of the equation
$$Pdx + Qdy + Rdz = 0.$$

(2) On Clairaut's Theorem.

(3) On the colours of thin crystalline plates in polarized light.

———————

January 28, 1862.

1. Shew that the elimination of x, y, z from the equations
$$(y - a)(z - a) = ea^2 + fa + g,$$
$$(z - b)(x - b) = eb^2 + fb + g,$$
$$(x - c)(y - c) = ec^2 + fc + g,$$
$$(x - b)(y - c)(z - a) = (x - c)(y - a)(z - b),$$
leads to
$$(b - c)(c - a)(a - b)(e - 1) = 0.$$

2. Examine the general form of the surface
$$xyz - a^2x - b^2y - c^2z + 2abc = 0,$$
and shew that it has a conical point. Shew also that each of the planes passing through the conical point and a pair of the intersections with the axes touches the surface along a straight line.

3. Shew generally how to find whether a set of given functions of at least as many independent variables are independent of each other or connected by a functional relation.

Apply the general method to determine whether $uy - vx$ is a function of $x^2 + y^2$, $u^2 + v^2$, and $ux + vy$.

4. Compare the most rapid variation in the Moon's apparent semi-diameter due to the Earth's rotation with that due to her motion in her orbit, giving a rough numerical result.

5. In the system of curved lines defined by the differential equations

$$\frac{dx}{P} = \frac{dy}{Q} = \frac{dz}{R},$$

if those lines be selected which pass through a small closed curve, so as to generate an elementary tube, and a point move round the tube always in a direction perpendicular to its generating lines, shew that it will trace out a spiral in which the distance of consecutive threads, taken with its proper sign, is

$$\frac{P\left(\dfrac{dQ}{dz} - \dfrac{dR}{dy}\right) + Q\left(\dfrac{dR}{dx} - \dfrac{dP}{dz}\right) + R\left(\dfrac{dP}{dy} - \dfrac{dQ}{dx}\right)}{P^2 + Q^2 + R^2} S,$$

S being the area of a section of the tube.

6. When gas of a given kind is discharged from a reservoir into a vacuum, through a long uniform capillary tube, it is found that the rate of discharge varies directly as the square of the pressure in the reservoir, and inversely as the length of the tube. Assuming that the pressure just within the tube at either end may be deemed the same as outside the mouth, and that the effects of inertia generally may be neglected, and supposing that the resistance, referred to a unit of length of the tube, is some function of the velocity and density, find the form of that function.

7. A comet moving in a parabolic orbit makes a near approach to a small planet; point out from general considerations the circumstances under which the orbit is rendered elliptic or hyperbolic.

8. A heavy particle is suspended from a fixed point by an elastic string, and performs small oscillations in a vertical direction; supposing the string uniform in its natural state, and of small finite mass, shew that the time of oscillation will be approximately the same as if the string were without weight, and the mass of the particle were increased by one-third of that of the string.

9. Shew that the bright circle in front of the eye-piece of any telescope but Galileo's, seen when the telescope is adapted for distinct vision and turned towards the sky, may be regarded indifferently as the image of the object-glass, or the common section of the various pencils proceeding from distant points; and shew that the magnifying power is equal to the ratio of the diameter of the object-glass to that of the bright circle.

10.   Give a full explanation of the phenomenon of the Rainbow. How are white rainbows accounted for?

11.   Deduce the equation of steady motion in Hydrodynamics from the general equations in their ordinary form.   Under what circumstances may the difference between the squares of the velocities at *any* two points in the fluid, at which the pressures are known, be inferred from the equation?

12.   Assuming the equation

$$\frac{d^2 r\phi}{dt^2} = a^2 \frac{d^2 r\phi}{dr^2}$$

applicable to the propagation of sound to or from a centre, determine the motion of a mass of air initially at rest, but slightly and uniformly condensed throughout a sphere having its centre at the origin; selecting for special examination the case in which t exceeds a certain value.

13.   Give a general explanation of the variations of illumination in the neighbourhood of the boundary of the geometrical shadow of an opaque screen terminated by a straight edge, and exposed to a beam of light emanating from a luminous point; and shew how the distance of any fringe from the edge of the geometrical shadow is measured experimentally.

According to theory, the fringe seen at any distance on the bright side of the edge arises from the interception of a portion of a wave equal to that which produces the illumination at an equal distance on the dark side; and yet the fringes on the bright side are visible to a distance far greater than that at which the illumination on the dark side ceases to be sensible: account for this.

Write short dissertations on the following subjects:

(1)   On integration with respect to an imaginary variable, and the evaluation thereby of definite integrals.

(2)   On the principle of dynamical similarity. (*Principia*, Lib. II. Prop. 32.)

(3)   On the evidence, theoretical and experimental, of the correctness of Fresnel's interpretation of his formulæ for the intensity of reflected light, when applied to the case of total internal reflection.

February 3, 1863.

1.   Define a plane by means of the right line; and in accordance with your definition reduce to its most elementary form the axiom that a straight line which passes through two points in a plane lies wholly in the plane.

2. If the focus of a conic section be joined with any point, and likewise with the intersection of the polar of that point and the directrix, the joining lines will be perpendicular to each other.

3. If the chance of hitting a given small circle at the distance r from the point of a target aimed at vary as e^{-hr^2}, find the inner radius of the annulus (1) of given small breadth, (2) of given area, which is most likely to be hit by a person aiming at its centre.

4. If two normals to a curve in space move along the curve in such a manner as to generate developable surfaces, the angle of intersection of those surfaces at the curve will be constant.

5. Find the attraction of an infinite circular cylinder (1) on an external, (2) on an internal particle, the attraction varying inversely as the square of the distance.

6. Find the conditions which must be satisfied in order that

$$Wdw + Xdx + Ydy + Zdz = 0$$

may be integrable as a total differential equation with three independent variables.

What will be the number of such independent conditions in the case of a total differential equation with n variables?

7. Two thin spherico-cylindrical lenses are combined, with the axes of the cylinders inclined at any angle; shew that for direct centrical pencils they are equivalent, aberration being neglected, to a single lens of the same kind; and find the power and angular position of this lens.

8. Explain how a secondary spectrum may be obtained by two prisms of the same kind of glass; and find the conditions that two prisms of different kinds of glass may so achromatise each other as not to leave even a secondary spectrum.

9. If three systems of surfaces cut each other orthogonally, every surface of each system will be cut by the surfaces of the two other systems along its lines of curvature.

10. Describe the experiment with a partially exhausted receiver by which it has been attempted to determine directly the ratio of the specific heat of air at a constant pressure and at a constant volume, giving the necessary formulæ.

If the exhaustion be not small, shew that it will make a difference in the observed result, independently of any communication of heat to the sides of the vessel, whether the air be admitted suddenly, or only gradually though rapidly.

11. Explain the nature of the lunar disturbances which result from the oblateness of the Earth. Does their magnitude depend on the law of density within the Earth, or only on the Earth's ellipticity?

12. A soap-bubble of uniform thickness is filled with a gas of such density that the weight of the whole is equal to that of the air displaced; find the form of the bubble, which is supposed to differ but little from a sphere.

13. An inextensible but slightly flexible elastic wire is formed into a circular ring, and made to rest in a vertical position on one point; find the form which it will assume.

14. Find according to the usual theory the series of notes which can be given by a tube (1) closed at one end and open at the other, (2) open at both ends; and point out in what respect chiefly the suppositions made in the theory deviate from the actual state of things.

Describe and explain the phenomenon of resonance.

15. An infinitely long stretched string, not acted on by gravity, is loaded with a uniform closely fitting tube, free to move about its centre of gravity, which is fixed; a given indefinite series of transversal vibrations in one plane being propagated towards the tube on one side, determine the motion.

16. A doubly refracting crystal is in optical contact with an ordinary medium of greater refractive power, and the angle of incidence within the ordinary medium, in a given plane, is increased until one of the limits of total internal reflection is reached; point out the limiting positions of the refracted ray and wave; and if the limiting angle of incidence be denoted by $\sin^{-1}\dfrac{\mu}{\mu'}$, where μ' is the refractive index of the ordinary medium, indicate the signification of μ.

Give the actual formulæ for finding the limiting angle in the case of the extraordinary ray of a uniaxal crystal.

February 4, 1864.

1. Develope $\left(\dfrac{1+\sqrt{1-x^2}}{2}\right)^n$, in ascending powers of x; and shew that the first term $(=1)$ of the expansion is included in the expression for the general term.

2. A symmetrical function $S\,a^\alpha b^\beta c^\gamma \ldots$ of the roots of an equation, where α is not less than any other index, is expressible as a function of the degree α of the coefficients.

3. Explain and prove the theorem "Given the value of a function of the roots of an equation, the value of any cotypical function is expressible rationally in terms of the coefficients." Shew that the theorem fails in particular instances.

4. Transform the binary cubic function $ax^3 + 3bx^2y + 3cxy^2 + dy^3$ into its canonical form, $= (\lambda x + \mu y)^3 + (\lambda' x + \mu' y)^3$.

5. The equations of any four lines in the same plane, such that no three of them meet in a point, may be taken to be $x = 0$, $y = 0$, $z = 0$, $w = 0$, where $x + y + z + w = 0$.

6. State, and prove analytically, the general theorem obtainable by projection from the theorem "the middle points of the three diagonals of a complete quadrilateral lie in a line."

7. Explain the meaning of the expressions, (1) polar of a point in relation to a conic; (2) rth polar of a point in relation to a curve of the mth order.

Deduce from your definition a geometrical construction of the second polar of a point in relation to a system of three lines.

8. Shew that the curve $\sqrt[4]{x} + \sqrt[4]{y} + \sqrt[4]{z} = 0$ has three double points; and trace the curve.

9. Shew that a cubic curve which passes through eight of the nine intersections of two given cubic curves, passes through the ninth intersection.

Mention any special cases of the theorem.

10. A quadric surface is cut by two parallel planes; find an analytical expression for the ratio of corresponding lines in the two sections respectively.

11. Shew à *posteriori* that the tangent plane of a quadric surface cuts the surface in a pair of lines.

12. What is the nature of the locus represented by the equation

$$F(\gamma y - \beta z, \ az - \gamma x, \ \beta x - \alpha y) = 0,$$

where x, y, z are trilinear coordinates *in plano*?

State the corresponding theorem or theorems *in solido*.

13. Find the equation of the cone having a given vertex and passing through the section of a given quadric surface by a given plane.

Shew that the cone in question has double contact with the cone of the same vertex, circumscribed about the quadric surface.

If the vertex is on the quadric surface, what is the theorem which replaces the last-mentioned theorem?

Deduce any theorems relating to Stereographic Projection.

14. Explain what is meant by the 'order' of a system of algebraical equations; and write down a system of equations representing a cubic curve in space. Give also a geometrical construction for the curve.

15. Given any three lines, and a fourth line touching the hyperboloid through the three lines; then will each one of the four lines touch the hyperboloid through the other three lines.

16. Find an expression for the element of area included between the four curves

$$u = f(x, y), \quad u + \delta u = f(x, y), \quad v = \phi(x, y), \quad v + \delta v = \phi(x, y),$$

where x, y are ordinary rectangular coordinates.

17. If $c = U$ be an integral of the system of differential equations

$$\frac{dx}{X} = \frac{dy}{Y} = \frac{dz}{Z},$$

shew that the function U satisfies a certain partial differential equation; and, by means of the result, explain the theory of the integration of a partial differential equation $pX + qY - Z = 0$.

Write short dissertations on the following subjects:

(1) Elimination.

(2) The Partition of Numbers.

(3) The Cartesian Ovals.

(4) The Γ Function.

January 31, 1865.

1. If a fraction in its lowest terms with a prime denominator p which is neither 2 nor 5 be reduced to a decimal, the number of digits in the circulating period will be either $p - 1$ or a submultiple of $p - 1$. Prove this, and the proposition in the theory of numbers from which it immediately follows.

2. Determine n numbers such that the consecutive integers 1, 2, 3, 4..., as many as possible, may be formed from them by addition, subtraction, or omission.

3. A substance A, in small quantity, when acted on by a mixture of two solvents X, Y which separate after agitation dissolves in the proportion, for equal volumes, of x to y, and a substance B similarly treated in the proportion of x' to y'; supposing that the substances dissolve independently, and that a mixture of the quantities a of A and b of B is dissolved in the volume U of X, and the solution washed n times with fresh volumes V of Y, find the quantities remaining in solution in X; and supposing V and n disposable, find to what extent the substance which is relatively less soluble in X may be got rid of without entailing more than a given loss of the other substance.

4. If $V = C$, C being an arbitrary parameter, be the equation of a system of surfaces such that the distance between any two consecutive surfaces is the same at all points, prove that

$$\left(\frac{dV}{dx}\right)^2 + \left(\frac{dV}{dy}\right)^2 + \left(\frac{dV}{dz}\right)^2 = \phi(V),$$

ϕ denoting an arbitrary function.

5. Shew that the system of straight lines meeting two given straight lines cannot be cut orthogonally by a system of surfaces; and find the equations of the system of curves lying in planes parallel to both the given lines by which the system of straight lines is so cut.

6. A mass of liquid not acted on by gravity revolves uniformly round a fixed axis, and contains, revolving with it, two small solids connected by a string, one of the solids being denser and the other rarer than the liquid; find the condition of equilibrium, and discuss the cases which may arise.

7. Shew that the least surface which can be drawn on a given closed curve of double curvature has its principal curvatures at any point equal in magnitude and opposite in direction.

8. A double convex lens having radii r, s, and a small finite thickness t, is placed with its posterior surface in contact with a fixed ideal plane; shew that when the lens is reversed the position of the focus for parallel rays will be altered by the quantity $\dfrac{r-s}{r+s}\dfrac{t}{\mu}$.

9. White light is refracted through a compound prism formed of n prisms placed in contact; if $i_1, i_2, \ldots i_n$ are the angles, $\mu_1, \mu_2, \ldots \mu_n$ the refractive indices of the prisms, $\phi_1, \phi_2, \ldots \phi_{n+1}$ the successive angles of incidence, $\rho_1, \rho_2, \ldots \rho_{n+1}$ the corresponding angles of refraction, shew that the angular length of the spectrum, regarded as small, is given by the expression

$$\frac{1}{\cos \rho_{n+1} \cos \rho_n} \times$$
$$\left\{\sin i_n\, \Delta\mu_n + \frac{\cos \phi_{n+1}}{\cos \rho_{n-1}}\left[\sin i_{n-1}\,\Delta\mu_{n-1} + \frac{\cos \phi_n}{\cos \rho_{n-2}}\left(\sin i_{n-2}\,\Delta\mu_{n-2} + \&c.\right)\right]\right\}.$$

10. A spherical shell of homogeneous gravitating liquid initially at rest is left to itself; find the pressure at any point during the collapse.

11. In a compound pendulum consisting of two heavy particles attached to light strings, and performing small vibrations in one plane, if the upper mass be much the greater of the two it is evident that for one of the two simple motions which coexist the periodic time must be nearly the same as if the under mass were removed, and for the other nearly the same as if the upper mass were fixed. The latter periodic time changes continuously from greater to less than the former as the under string changes from longer to shorter than the upper; and therefore for some intermediate length the periodic times must be equal.

Expose the fallacy of this reasoning; and prove that whatever be the lengths and masses the two periodic times are necessarily unequal*.

12. Prove that the expansion of $f(x)$ by Maclaurin's theorem is convergent or divergent according as the modulus of x is less or greater than that of the least value of x which renders $f(x)$ or its derivative infinite or discontinuous.

Apply this theorem to the examples $\tan^{-1}x$, $\log_e(1+x)$, e^x, e^{-1/x^2}.

[* Cf. *ante*, Vol. IV, p. 334.]

13.   Assuming that the capillary forces at the surface of a liquid are equivalent to a superficial tension, find the work of those forces when the surface is altered in form; and from the principle that when a system is in stable equilibrium the work of all the forces is a maximum, deduce the differential equation of the curve which by its revolution generates the surface of the liquid within a vertical cylindrical tube.

14.   Describe the appearance of the boundary of the shadow of an opaque screen with a straight edge exposed to a stream of light coming from a luminous point; and indicate the whole of the measurements, and (when not obvious) the mode of taking them, which are required to compare theory and experiment.

15.   A pencil passing through a feebly doubly refracting plate is defined by two small holes through which it has to pass, the holes being situated in a line perpendicular to the plate and on opposite sides of it; shew that, whatever be the law of double refraction, when the thickness of the plate and the distances of the holes vary, the angle in air between the two pencils which can pass varies as $h/\{h + \mu\,(k + k')\}$ where h is the thickness of the plate, and k, k' the distances of the holes from the surfaces respectively next them.

16.   A spherical shell vibrates as a bell, communicating vibrations to the surrounding air; supposing that the excursion of any point of the surface is represented by $c\,(\cos^2\theta - \tfrac{1}{3})\sin nt$, determine the vibration at any distance in the air; and compare the intensity at a great distance with what it would have been if the air had been divided into an infinite number of infinitely slender filaments, by conical surfaces having the centre for vertex.

Apply the result to illustrate the use of sounding-boards.

Write short dissertations on the following subjects:

(1)   On a geometrical interpretation of imaginary quantities, and its completeness or incompleteness.

(2)   On the nature and use of series which are at first convergent, though ultimately divergent.

(3)   On the foundations of Hydrodynamics.

(4)   On the consequences of the mean motions of two planets being nearly in the ratio of two low integers.

January 30, 1866.

1.   Find the volume of a tetrahedron in terms of (1) the coordinates of the summits, (2) the lengths, inclination, and shortest distance, of two opposite edges; and deduce an expression for the shortest distance between two given lines.

2. Find the envelope of a sphere passing through a given point and having its centre on a given circle. Shew that the sections of the surface by planes parallel to the plane of the circle are Cartesians; and investigate the form of the surface in the neighbourhood of the given point.

3. A particle moves in a plane under the action of forces which depend only on the position of the particle, and the coordinates (x, y) are given functions of a variable parameter θ (so that the orbit is a given curve): find the relation which exists between the forces.

4. If a particle move in a circle under the action of a central force, the centre of force being a given point, find the law of force: determine also the motion of the particle, and take notice of any peculiarity of the motion in the case where the centre of force is outside the circle.

5. Shew that infinitesimal rotations impressed upon a solid body may be compounded together according to the rules for the composition of forces.

6. If $A = \int x^2 dm$, $H = \int xy\, dm$, $B = \int y^2 dm$, where dm is an element of mass and the integrations are extended over any solid body whatever; shew that $AB > H^2$.

7. If y is a function of x given by the equation $y = x + ay^n$, investigate a series for $F\left[x \dfrac{d}{dx} \right] x^a y$; and thence shew that

$$\left[x \frac{d}{dx} \right]^{n-1} y = na \left[\frac{n}{n-1} x \frac{d}{dx} - \frac{2n-1}{n-1} \right]^{n-1} x^{n-1} y.$$

8. Consider a ray of light refracted at a curve, and let Q be a point on the incident ray, G the point of incidence, N a point on the normal, and q a point on the refracted ray: and let QG, QGN denote the length QG and the area of the triangle QGN respectively, and similarly for qG, qGN: shew that the equation

$$(qG)^2 . (QGN)^2 - \mu^2 (QG)^2 . (qGN)^2 = 0,$$

which is rational of the second order in regard to the coordinates (x, y) of the point q, contains as a factor the equation of the refracted ray; and explain the signification of the other factor.

Shew that the determination of the caustic for rays proceeding from a point and refracted at a circle, depends on the calculation of the discriminant of a sextic function.

9. Indicate a process for finding the relation between the lengths of the sides and diagonals of a quadrilateral; and (taking the relation in question to be known) find the condition of equilibrium for two forces acting along the diagonals of the quadrilateral formed by four strings of given lengths.

10. Explain the principle or principles on which the solutions of the following questions depend:

(1) The balls drawn at m successive trials from an urn containing $m + n$ balls are all of them white; what is the chance that the ball drawn at the next trial will be white?

(2)   What is the chance that a penny which has turned up heads m times in succession, will turn up heads at the next trial?

11.   If $T = \frac{1}{2}(a\xi^2 + 2h\xi\eta + b\eta^2)$, where a, h, b are given functions of (x, y), so that T is a given function of (x, y, ξ, η); then writing

$$(u = a\xi + h\eta, \quad v = h\xi + b\eta),$$

and denoting by H the value of T expressed as a function of (x, y, u, v), shew that

$$\frac{dT}{dx} = -\frac{dH}{dx}, \quad \frac{dT}{dy} = -\frac{dH}{dy}.$$

12.   Find the locus of a point such that its coordinates (x, y, z) and (x_1, y_1, z_1) in regard to two sets of rectangular axes through the same origin are proportional to each other.

13.   State the relations which exist between two circles, and between three circles, in regard to their radical axes, orthotomic circles, and centres of similitude; and give a construction, or constructions, for the circles which touch three given circles.

14.   Given the four cones

$$
\begin{aligned}
-cy^2 + bz^2 - fw^2 &= 0, \\
cx^2 \qquad\quad - az^2 - gw^2 &= 0, \\
-bx^2 + ay^2 \qquad\quad - hw^2 &= 0, \\
fx^2 + gy^2 + hz^2 \qquad\quad &= 0,
\end{aligned}
$$

and the four conics which are the sections of these by the planes $x = 0$, $y = 0$, $z = 0$, $w = 0$ respectively; then any line touching three of the four cones touches the fourth cone; and any line meeting three of the four conics meets the fourth conic.

15.   There exist on any quadric surface curves of the order $m + n$ which meet each generating line of the one kind in m points, and each generating line of the other kind in n points.

16.   The Moon is considered to move round the Earth in a circular orbit in the plane of the equator, and it is assumed that, when regard is had to the tide, the Moon has upon the Earth the same action as it would have upon a homogeneous sphere loaded at the extremities of an equatorial diameter with two equal masses, small in comparison with the mass of the sphere—the diameter in question being such that the radius directed to the mass which is nearest the Moon lies always at an angle of 45° to the East of the radius vector of the Moon. Shew that the effect will be a diminution of the Earth's angle of rotation, varying as the square of the time; and estimate roughly the amount of this diminution, taking each of the small masses to be to the mass of the Earth as 1 to 4000 millions.

January 29, 1867.

1. Find the equations of two planes passing through the two lines

$$\frac{x-a}{l} = \frac{y-b}{m} = \frac{z-c}{n}, \quad \frac{x-a'}{l'} = \frac{y-b'}{m'} = \frac{z-c'}{n'},$$

respectively, and through their shortest distance.

2. What difficulty presents itself in the application of the ordinary formulæ of spherical trigonometry to the small triangles which occur in the survey of a country? Explain the methods which have been adopted to obviate this difficulty, without investigating the requisite formulæ.

3. The motion of a plane which always coincides with a fixed plane is determined by the condition that two points in the former move along two straight lines in the latter; shew that the motion of the moveable plane may be defined in one way by supposing a circle fixed relatively to the moving plane to roll within a circle of twice the radius fixed in the fixed plane, and in an infinite number of ways by supposing two points in the moving plane to move along two rectangular axes in the fixed plane.

4. If a, b, c; a', b', c'; a'', b'', c'' be the rectangular coordinates of the extremities of three edges of a parallelepiped which meet in the origin, shew that the volume is equal to

$$ab'c'' - ab''c' + a'b''c - a'bc'' + a''bc' - a''b'c.$$

5. Prove that

$$\int_0^\infty \left(x^{n-1} - \frac{x^{n+1}}{2^2} + \frac{x^{n+3}}{2^2 \cdot 4^2} - \frac{x^{n+5}}{2^2 \cdot 4^2 \cdot 6^2} + \ldots \right) dx$$

$$= \frac{1}{\sqrt{\pi}} \cos\frac{n\pi}{2} \frac{\Gamma(n)\,\Gamma\left(\dfrac{1-n}{2}\right)}{\Gamma\left(1 - \dfrac{n}{2}\right)},$$

n being supposed to lie between the limits 0 and 1.

6. Shew generally how to find the curves on a given surface which are traced from point to point in the direction of the intersection of the tangent plane at the point with the surface; and find them in the case of a helicoid.

7. Prove that a plane cutting two of the principal planes of Fresnel's wave-surface along real or imaginary ray axes cuts the third plane along a ray axis, and that the four imaginary planes thus formed each touch the surface along a plane curve.

8. A small instrument is provided with sights or a telescope mounted equatorially, except that the whole is placed on a stand moveable in azimuth; supposing the declination of the sun known, but the plane of the meridian unknown, shew how the true time may

be found by a single observation of the sun. At what time of day would the method fail?

9. A rigid body is suspended by two equal and parallel threads attached to it at two points symmetrically situated with respect to a principal axis through the centre of gravity, which is vertical, and being turned round that axis through a small angle is left to perform small finite oscillations; investigate the reduction to indefinitely small oscillations, and examine how the correction is affected by connecting the threads at equidistant points by cross bars without weight, placed like the steps of a ladder.

10. A string of length $(n+1)l$ and insensible mass, stretched between two fixed points with a force T, is loaded at intervals l with n equal masses m, not under the influence of gravity, and is slightly disturbed; if $\dfrac{T}{lm} = c^2$, prove that the periodic times τ of the n simple transversal vibrations which in general coexist are given by the formula

$$\tau = \frac{\pi}{c} \operatorname{cosec} \left(\frac{i}{n+1} \frac{\pi}{2} \right),$$

on putting in succession $i = 1, = 2, = 3 \dots, = n$.

Deduce the series of notes which a uniform unloaded string can give out.

11. A prism is mounted on a graduated instrument in the usual way, on a support which is supposed not to move with the telescope, and the deviation of light of a definite kind by refraction, in an arbitrary position of the prism, is measured, as well as the deviation by reflexion from the first surface; find the refractive index in a form adapted to logarithmic computation.

In case the prism be cut from a doubly refracting medium, shew precisely what is determined by the observation.

12. A soap-bubble film at first rectangular is bounded by the sides of a cylindrical wet tube, and by two diametral wires; if one of the wires be turned round the axis of the tube, and the film be left to settle itself, shew that the conditions of its equilibrium will be satisfied by the helicoidal form, the weight of the film being neglected.

13. An electrical point is placed within a spherical cavity in a conducting solid; find the distribution of electricity on the surface of the cavity, and prove that the attraction of the induced electricity at any point in the interior is that due to a certain quantity of electricity condensed in a certain exterior point.

14. If V be a function of x, y, z which varies continuously outside a closed surface S, and vanishes at an infinite distance, prove that

$$\iiint V \left(\frac{d^2V}{dx^2} + \frac{d^2V}{dy^2} + \frac{d^2V}{dz^2} \right) dx\,dy\,dz$$
$$+ \iiint \left\{ \left(\frac{dV}{dx} \right)^2 + \left(\frac{dV}{dy} \right)^2 + \left(\frac{dV}{dz} \right)^2 \right\} dx\,dy\,dz + \iint V \frac{dV}{dn} dS = 0,$$

the triple integrals extending to all space outside S, and the double integral to the whole surface S, and n being the normal measured outwards.

Apply the result to prove that if either the potential of a gravitating mass, or the normal component of its attraction, be given throughout a closed surface containing the whole of the mass, the attraction at any external point will be determinate.

15. Elliptically polarized and partially plane-polarized light present the same appearance when analyzed by a Nicol's prism which is turned round, nor can they be distinguished when a thick plate of selenite is interposed, but when the interposed plate is thin the appearance in the two cases is different; account for this, and deduce from theory the nature of the difference.

16. If a continuous system of points move in planes parallel to that of xy, so that, u, v being the components of the velocity,

$$u\,dx + v\,dy = d\phi,$$

where ϕ expressed in polar coordinates $= or^3 \sin 3\theta$, examine the nature of the instantaneous motion of those points which at a given moment lie in a plane perpendicular to the axis of x.

Apply the result to the solution of the following problem:—A closed vessel of the form of a right equilateral triangular prism, filled with homogeneous liquid, is struck so as to generate in the vessel an angular velocity ω about its axis; find the moment of the impulsive couple.

January 28, 1868.

1. If the intersections of a straight line with the sides of a triangle be joined with the opposite corners, and the corners of the triangle formed by the joining lines be joined with the corresponding corners of the original, the three joining lines will meet in a point or be parallel.

2. Prove that the series

$$\frac{x}{1} + \frac{x^2}{2} + \frac{x^3}{3} + \frac{x^4}{4} + \dots,$$

where $x = \cos\theta + \sqrt{-1}\sin\theta$, is convergent, unless θ is of the form $2n\pi$.

3. Integrate the differential equation

$$\left(y - x\frac{dy}{dx}\right)\frac{d^2y}{dx^2} = 4\left(\frac{dy}{dx}\right)^2.$$

4. Examine the reasoning employed in the process of integrating the partial differential equation $rt - s^2 = 0$, illustrating the steps of the process by reference to geometry.

5. A system of straight lines is defined by intersecting each of two given straight lines inclined to each other at a right angle; find the nature of the system of curves lying on a symmetrically situated sphere which cut orthogonally the lines of the system.

6. Write a short dissertation on the primary idea of *mass*, and on its measure, theoretical and practical.

7. A rigid body moveable about a fixed point, and at rest, being struck by a given blow, prove that the *vis viva* is greater than it would have been if the body had been constrained to turn round any axis not coinciding with the instantaneous axis.

8. Two magnetic needles, the dimensions of which are very small compared with the distance between them, are placed in given relative positions; find the resultant force and couple arising from their mutual action; and deduce the position of equilibrium of one of the needles supposing it free to turn round its centre of gravity, and not to be influenced by the earth's magnetism.

9. A homogeneous liquid, not acted on by external forces, is in motion in two dimensions; if $\psi = \int(vdx - udy)$, find the differential equation which ψ must satisfy if the motion be steady; and apply the result to determine whether a given family of curves can possibly be a system of lines of motion.

Ex.   Take the cases (1) of the ellipses
$$a^2y^2 + b^2x^2 = C,$$
(2) of the parabolas
$$y^2 = Cx,$$
C being the variable parameter.

10. If a function of the angular coordinates θ, ϕ, which, as well as its differential coefficients of the first order with respect to θ and ϕ, varies continuously with change of angular position, be expanded in a series of Laplace's Functions, the function of the order i will be numerically less than C/i, C being a constant.

11. Calculate what would be the ellipticity of the earth if it consisted internally of *spherical* strata of equal density, and the oblateness were due merely to degradation at the surface in the lapse of ages, the density of the superficial parts being taken $= 0 \cdot 4 \times$ the mean density.

12. A uniform stretched string of length l is slightly displaced in one plane so that the ordinate $(y) = 2f(x)$, $f(x)$ being a function given between the limits $x = 0$, and $x = l$, and vanishing at those limits; determine the subsequent motion, (1) by determining the arbitrary functions in the integral of $\dfrac{d^2y}{dt^2} = c^2\dfrac{d^2y}{dx^2}$, (2) by employing the principle that the general disturbance is composed of a series of "simple" periodic disturbances; and prove directly the equivalence of the two results.

13. Light from a luminous point passes through a small circular aperture, and is received on a screen at some distance; compare the illumination at the centre of the image with what it would be if the screen containing the aperture were removed; and trace the variation of appearance of the central spot as the receiving screen is moved from a great distance towards the aperture.

14. Assuming that the equations of motion of a homogeneous isotropic elastic solid, not acted on by external forces, and slightly disturbed, are

$$\rho \frac{d^2u}{dt^2} = A \frac{d}{dx}\left(\frac{du}{dx} + \frac{dv}{dy} + \frac{dw}{dz}\right) + B\left(\frac{d^2u}{dx^2} + \frac{d^2u}{dy^2} + \frac{d^2u}{dz^2}\right),$$

$$\rho \frac{d^2v}{dt^2} = \&c., \qquad \rho \frac{d^2w}{dt^2} = \&c.,$$

where u, v, w are the components of the displacement, deduce the laws of propagation of plane waves in an indefinitely extended medium of the kind; and infer, without exact calculation, the nature of the motion which would ensue if the medium were slightly but arbitrarily disturbed throughout a limited space, and then left to itself.

15. From Huyghens's construction for determining, by means of the wave surface, the course of refracted waves and rays, deduce the corresponding construction in which the surface of wave-slowness is employed.

Obtain also, in terms of the ordinates of the latter surface, an expression for the difference of retardation of the two transmitted streams, when light is incident in a given direction on a crystalline plate bounded by parallel surfaces.

WEDNESDAY, *February* 2, 1870.

1. Two ellipses E, E' have a common major axis; S, H are the foci of E; S', H', those of E'; S, S' lying on the same side of the centre; through S, S', and towards the same side, are drawn the parallels SP, $S'P'$, cutting E, E' in P, P'; and through P, P', perpendicular to the major axis, are drawn PQ', $P'Q$, cutting the adjacent halves of E', E in Q', Q; prove that HQ, $H'Q'$ will be parallel.

The foci S, S' might have been taken on the same side or on opposite sides of the centre, the parallels SP, $S'P'$ might have been drawn towards the same side or towards opposite sides, and each of the lines PQ', $P'Q$ might have been drawn to cut either half of the ellipse E' or E, making 16 cases; shew in which of these the theorem will hold good.

2. If $u = m + e \sin u$, and $u - m$ be developed in a series of sines of m and its multiples, express the coefficient of $\sin im$ by a definite integral.

3. In a system of surfaces with one variable parameter, shew generally how to find the locus of a point at which consecutive surfaces are parallel.

Apply your general method to the family of paraboloids $xy + pz = a^2$, where p is the parameter.

4. When n is a positive integer we have evidently

$$1 . 2 . 3 \ldots 2n = 2^{2n} . 1 . 2 \ldots n . \tfrac{1}{2} . \tfrac{3}{2} \ldots (n - \tfrac{1}{2});$$

prove that this equation, when expressed by means of the function Γ, is true for any positive value of n.

5. A large convex lens is to be tested as to the goodness of the glass of which it is made by putting a small luminous object in the axis at any distance greater than the focal length, and placing the eye so that the whole lens is seen illuminated; but for the sake of seeing the lens more distinctly and somewhat magnified the observer uses an opera glass; shew where this must be placed, and how adjusted, in order that the proper position for the eye may be at a given convenient distance in front of the eye-lens, and the large lens may be seen distinctly.

6. $m + n$ perfectly similar voltaic elements are connected in series of m and n, forming two batteries, the positive poles of which are connected, by long wires, of lengths a, b, with one end of a third long wire, of length c, which leads to a short wire connecting the negative poles; find the currents which pass along the wires a, b, c, respectively, the internal resistance of one element being equal to that of a length r of wire of the same kind, and the resistances of the short connecting wires being neglected.

If c be so large as to render the expression for the current in either battery negative, what really takes place?

7. A soap-bubble film is stretched between two circular wires the centres of which are in a line perpendicular to their planes; shew that a section of the film by a plane through the axis of revolution will be the common catenary.

Considering the case in which the circles are equal, shew that according to the distance of their planes there will be two positions of equilibrium or none, and that in the former case the equilibrium in one position will be stable and in the other unstable. What will take place if the wires be slowly drawn asunder till the equilibrium becomes impossible?

8. A uniform sphere in free space is divided into two by a concentric spherical surface so that the parts are capable of rotating independently, but their relative motion is checked by a friction proportional, at any point, to the relative velocity of the surfaces there in contact. Given in magnitude and direction the initial angular velocities of the two parts respectively, determine the motion, and examine the ultimate state of the system.

9. Investigate the motion of a long wave of small finite height travelling in one direction in a rectangular canal, and shew that the form of such a wave changes as it proceeds.

10. Demonstrate the possibility of expanding a finite but otherwise arbitrary function of the angular coordinates θ, ϕ, which recurs when θ, ϕ are so changed as to bring us back to the direction from which we started, in a converging series of Laplace's Functions.

11. A spectrum is viewed through a telescope in the usual way, and the object-glass is covered by a screen containing a small rectangular aperture half covered by a plate giving a retardation of a good many wave-lengths, the edge of the plate and two sides of the aperture being parallel to the edge of the prism; investigate the appearance produced, and explain how it is that it may make a material difference which half of the aperture is covered.

12. Polarized light is allowed to pass perpendicularly through a crystalline plate with its neutral axes inclined at 45° to the plane of polarization, and after being limited by a slit is analysed by an ordinary prism followed by a Nicol's prism; investigate the appearance presented as the latter is turned round, neglecting the modification of the polarization produced by refraction through the ordinary prism.

How would the appearance be changed if for the ordinary crystalline plate were substituted a thick plate of quartz cut perpendicularly to the axis?

WEDNESDAY, *February* 1, 1871.

1. Express $\sin n\theta$ and $\cos n\theta$ in series according to ascending powers of $\sin\theta$, n being any real quantity, and θ lying between

$$-\frac{\pi}{2} \text{ and } +\frac{\pi}{2}.$$

Deduce the expansions of $\sin x$ and $\sin^{-1} x$.

2. Explain the distinction between the extent to which a function is determined (1) by an equation of finite differences, (2) by a differential equation, (3) by a functional equation, illustrating your explanation by reference to the function a^x.

3. From Fourier's theorem deduce the reciprocity of the functions f, ϕ in the equations

$$f(x) = \sqrt{\frac{2}{\pi}} \int_0^\infty \phi(a)\cos axda, \quad f(x) = \sqrt{\frac{2}{\pi}} \int_0^\infty \phi(a)\sin axda,$$

in which x is supposed to be positive; and give examples of the application of the equations to the evaluation of definite integrals.

4. Form the differential equation of the curves on a surface which touch at each point the intersection of the surface by the tangent plane at that point.

In the case of the surface generated by the revolution of a rectangular hyperbola about one asymptote, shew that the projection of these curves on the plane of the other asymptote will be a double series of logarithmic spirals.

5. The altitudes of the sun and moon are observed simultaneously at sea; find the ship's place, knowing the error of the chronometer.

6. In a compound pendulum consisting of masses m, m' attached to strings of lengths l, l', in which of course the most general small motion in one plane consists of two harmonic vibrations superposed, if the upper mass m be very large compared with the under mass m', it is clear that one of the two periodic times (that corresponding to the mode of vibration in which m is nearly at rest) must be very nearly the same as in a simple pendulum of length l', and the other very nearly the same as in a simple pendulum of length l. By a continuous variation of l', the former may be made to pass continuously from less to greater than the latter, and therefore for some value of l' nearly equal to l the two must be equal. But when a system is in stable equilibrium (as is clearly the case here) the equation the roots of which give the times of vibration cannot have equal roots, for that would imply the transitional condition between stable and unstable.

Point out precisely the fallacy which leads to the above contradiction *.

7. In a spectroscope with any number of prisms of the same kind of glass, having the same or different angles, each prism being placed in its position of minimum deviation, and being just large enough to take in the pencil of one colour required to fill the telescope, shew that the total dispersion varies as the sum total of the lengths of the bases of the prisms †.

8. An eye-piece composed of any number of thin lenses separated by any intervals is used with an object-glass of comparatively great focal length; find the law of formation of the expression for the focal length of the equivalent simple lens in terms of the focal lengths and intervals; and thence (or in any other way) prove that the magnifying power will not be changed by inverting the eye-piece.

9. Shew that the attraction of the Earth, assumed to be an ellipsoid of revolution of small ellipticity, on an external particle is the same as if its mass were condensed into a ring in the plane of the equator; and find the diameter of that ring.

10. Connect on general dynamical principles the change in the length of the day due to luni-tidal friction with the corresponding change in the moon's mean motion.

11. A sphere immersed in an infinite mass of homogeneous liquid initially at rest is moved with a given velocity; determine the instantaneous motion of the fluid; and shew that it may be regarded as the limit of the motion compounded of two equal motions, each alike in all directions, tending one towards and the other from a centre, when the distance between the centres decreases indefinitely and the motion towards or from either increases indefinitely.

[* See *ante*, Vol. IV. p. 334.]

[† See Lord Rayleigh, "On the optical power of spectroscopes," *Phil. Mag.* 1879–80, *Scientific Papers*, I. p. 425, where this law is established from first principles, independently of minimum deviation for a spectroscope with a collimator, and its importance is emphasized.]

12. The equations of small motion of a viscous fluid being

$$\frac{1}{\rho}\frac{dp}{dx} = -\frac{du}{dt} + \mu \left(\frac{d^2u}{dx^2} + \frac{d^2u}{dy^2} + \frac{d^2u}{dz^2} \right),$$

$$\frac{1}{\rho}\frac{dp}{dy} = -\frac{dv}{dt} + \mu \left(\frac{d^2v}{dx^2} + \frac{d^2v}{dy^2} + \frac{d^2v}{dz^2} \right),$$

$$\frac{1}{\rho}\frac{dp}{dz} = -\frac{dw}{dt} + \mu \left(\frac{d^2w}{dx^2} + \frac{d^2w}{dy^2} + \frac{d^2w}{dz^2} \right),$$

it is required to determine the small oscillations of a spherical vessel filled with such fluid, and oscillating by the tension of a suspending wire. How may the result of observation be applied to the determination of μ?

13. Give Dr Young's explanation of the fringes seen about the boundary of the shadow of an opaque body exposed to light coming from a luminous point; shew what features of the phenomenon it explains, and what it fails to explain; indicate the difference between the results of this theory, and of the complete explanation given by Fresnel; and shew how that difference may be tested by experiment.

14. A compound quartz plate is composed of two wedges of equal angles, cut one from a right-handed and the other from a left-handed crystal so that one cut face in each is perpendicular to the axis, cemented together along the oblique faces so that the perpendicular faces are parallel; and being viewed perpendicularly, at the distance of distinct vision, is used in conjunction with a Nicol's prism for the scrutiny of plane-polarized light; examine the appearance presented; and point out the relative advantages of such a plate and of one composed of slender quartz wedges cut in a direction parallel, or nearly so, to the axis, united with their axes crossed.

WEDNESDAY, *January* 31, 1872.

1. Prove that the limit of $\Gamma(x+1) e^x x^{-x-\frac{1}{2}}$ for $x = \infty$ is $(2\pi)^{\frac{1}{2}}$.

2. Find $\displaystyle\int_0^\infty \frac{\sin ax\,dx}{e^x - e^{-x}}, \quad \int_0^\infty \frac{\cos ax\,dx}{e^x + e^{-x}}.$

3. Shew generally how to find the system of conical surfaces which cut orthogonally those of a given system the equation of which contains one arbitrary parameter, the generating lines of both systems passing through the origin.

Example. Take the system of elliptic cones having the same principal planes which is cut by a plane perpendicular to one of the principal axes in a system of similar ellipses.

4. An infinitely thin ring rests on a smooth horizontal cylinder which it passes round, and has then communicated to it an angular velocity round a vertical axis passing through the point of contact; determine the motion.

5. Explain how it is that a *resisting* medium, even though acting for a short time only, would *accelerate* the mean motion of a planet.

6. Give a general explanation of the superiority of an eye-piece of two lenses over a simple lens when a good field is required. By what construction could you throw the alteration of direction of the axes of the pencils and the alteration of convergency on the two lenses respectively? What practical objections would there be to this construction?

7. Apply the equation

$$-\frac{\epsilon}{a}\int_0^a \rho a^2 da + \frac{1}{5a^3}\int_0^a \rho\,\frac{d\,.\,a^5\epsilon}{da}\,da + \frac{a^2}{5}\int_a^{a_1} \rho\,\frac{d\epsilon}{da}\,da + \frac{\omega^2 a^2}{8\pi} = 0$$

obtained in the theory of the figure of the earth, to determine the ellipticities of the outer surface and of the common surface, supposing the whole mass made up of two parts having different given densities. Indicate the mode of proceeding when there is any finite number of such parts.

8. If $f(x)$ be a function of x which is continuous between the limits $x=0$ and $x=\pi$, and which vanishes at the second but not at the first of those limits, and if $f(x)$ be expanded (as it may) between those limits in a series of the form $\Sigma A_n \sin nx$, obtain the expansion of $f''(x)$ in a similar form in terms of that of $f(x)$, supposing $f(0)$ known.

9. Apply the above result to the solution of the following problem. A uniform string is stretched with a given tension between two points, of which one remains fixed, and the other is slightly disturbed in a transverse direction, so that the displacement at the time t is expressed by the given function $F(t)$, where $F(t)=0$ for $t<0$ or $>T$; required to determine at any time t later than T the various harmonic vibrations of the string.

10. Find the distribution of electricity on an infinite conducting plane to which an electrical point is presented.
A large metallic disk, especially if uninsulated, interposed between an electroscope and a body capable of being charged prevents the electroscope from indicating the charge; shew that this does not require us to suppose that electrical attraction or repulsion differs from gravitation in not acting through certain ponderable substances, such as metals.

11. From the equations of motion of a homogeneous perfect fluid acted on by forces which satisfy the conditions

$$\frac{dY}{dz}=\frac{dZ}{dy}, \quad \frac{dZ}{dx}=\frac{dX}{dz}, \quad \frac{dX}{dy}=\frac{dY}{dx},$$

and initially in motion in any arbitrary manner, prove that the particles lying initially in the same vortex line lie in the same vortex line throughout the motion, a vortex line meaning a line drawn at a given instant from point to point in the direction of the axis of instantaneous molecular rotation.

12. A series of long waves is propagated along a uniform canal, arising from a small disturbance at the mouth which varies as a given function of the time; assuming that the effect of friction may be represented in a general way by a retarding force varying as the velocity, determine the motion.

13. If a glass grating be used without a collimator for viewing pure diffraction spectra, the light being incident perpendicularly on the middle of the grating, if the slit be at a finite distance shew that the observing telescope must be focused for a point a little nearer than the slit; and find the distance of the virtual images of the fixed lines in the side spectra.

If the grating be inclined in the plane of diffraction, shew that on one side the approximation of the virtual foci will be much greater than before*.

14. If polarized white light be extinguished by an analyzer, and a thick crystalline plate cut not nearly perpendicular to an optic axis be interposed perpendicularly to the light which it transmits, and be turned round in its own plane, there is in general a restoration of white light. Explain this; find the maximum quantity which can be restored; shew that a thin plate is capable of restoring a larger quantity of light, though it is no longer white; and point out in a general way the thickness which gives the greatest restoration.

WEDNESDAY, *February* 4, 1874.

1. Supposing $\phi(x, y) = 0$ to be the result of the elimination of $p\left(= \dfrac{dy}{dx}\right)$ between $f(x, y, p) = 0$, and $\dfrac{df}{dp} = 0$, explain the geometrical relation between the curve which is the locus of the equation $\phi(x, y) = 0$ and the family of curves represented by the complete integral of

$$f(x, y, p) = 0,$$

both (1) when $\phi = 0$ does, and (2) when it does not satisfy the differential equation.

2. Integrate the partial differential equation

$$\frac{d^2z}{dx^2} - 2\frac{d^2z}{dx\,dy} + \frac{d^2z}{dy^2} = 0,$$

and determine the arbitrary functions by the conditions

$$bz = y^2 \text{ when } x = 0, \text{ and } az = x^2 \text{ when } y = 0.$$

3. Supposing u_n to decrease slowly and non-periodically as n increases, obtain a formula for the transformation of

$$u_0 - u_2 + u_3 - u_5 + u_6 - u_8 + \dots$$

into a rapidly converging series.

[* Cf. Rowland's concave gratings.]

4. On what grounds do we conclude that weight is a measure of mass?

Supposing nothing to be known as to the relation between weight and mass, shew how to compare the masses of two small bodies (1) by a method as simple as may be in conception, (2) by one as accurate as may be in execution.

5. A thin wire is interpolated in a galvanic circuit of known resistance; find the resistance of the interpolated wire in order that the quantity of heat generated in it in a given time may be a maximum.

6. Supposing the faces of a small prism to be spherical instead of plane, find the primary and secondary foci of a refracted pencil, the axis of the pencil lying in the plane passing through the centres of curvature of the faces.

Deduce the distances of the foci in the cases (1) of a prism with plane faces, (2) of a thin lens traversed by an oblique centrical pencil.

7. A detached metallic pendulum is swung in front of the compensated pendulum of a clock, to which its vibrations are referred by the usual method of coincidences, and the temperature of the air is given by a thermometer hung in the neighbourhood. Supposing the temperature of the apartment to undergo periodic fluctuations, find an expression for the difference between the temperature of the experimental pendulum and that indicated by the thermometer. If the fluctuations be slow, shew that the former will be sensibly equal to that indicated by the thermometer at a time earlier by the constant interval T. [N.B. In this part of the question the motion of the pendulum is not taken into account.]

Shew how to determine T experimentally by warming the pendulum, swinging it, and observing the coincidences as it cools.

8. In the motion of a fluid, if $L = 0$ be the equation of a bounding surface, prove that

$$\frac{dL}{dt} + u\,\frac{dL}{dx} + v\,\frac{dL}{dy} + w\,\frac{dL}{dz} = 0, \text{ when } L = 0.$$

How does it appear that this condition ensures that the particles lying at one instant in the surface $L = 0$ lie in that surface throughout the motion, although the condition itself merely relates to the direction of relative instantaneous motion, and would be satisfied for instance, in the case of a fixed boundary, were the particles supposed to move in paths touching the boundary?

In the case of motion in two dimensions for which

$$u\,dx + v\,dy = \tfrac{1}{2}d\left(y^2 - x^2\right),$$

apply the above differential equation to obtain the general equation of lines made up of the same particles; and thence shew that the particles which once lie in a curve of the nth order continue to lie in a curve of the nth order.

9. When a liquid moves in two dimensions, without molecular rotation, round a corner where two branches of the boundary meet at an angle a (measured on the side of the liquid) which is $<180°$, shew that the particle at the corner is in a position of permanent or only instantaneous rest according as $a \gtrless 90°$ or $>90°$.

What takes place when $a > 180°$?

10. A luminous point is viewed in focus through a telescope of which the object-glass is covered by a screen with mn similar and equal apertures, having their corresponding points situated in the intersections of a system of m parallel and equidistant lines by another system of n parallel and equidistant lines; shew that the intensity will be represented by the product of three factors, corresponding respectively (1) to the intensity for a single aperture of the given kind, (2) to that for a row of m, and (3) to that for a row of n equal and very small apertures, the two rows being situated at the intersections of all the lines of each system by a single line of the other.

11. A crystal of any kind with a plane face is cemented by a dense cement to the under side of a rectangular block of flint-glass more refractive than the crystal, and of known index; the limits of total internal reflection in a known plane are observed, and for each the index of the crystal is deduced as if it were an ordinary medium; shew precisely what the velocity which is the reciprocal of this index represents in the actual case, in relation to the wave surface.

12. A uniform flexible suspended string, regarded as extending indefinitely upwards, is slightly disturbed in one plane, and left to itself; determine the motion; and prove directly the possibility of expressing an arbitrary function in the form to which we are led in the solution of the problem.

WEDNESDAY, *February* 3, 1875.

1. Expand $(\sin^{-1} x)^2$ in a series according to ascending powers of x.

2. Sum to n terms the series

$$1 + \frac{a}{b} + \frac{a(a+1)}{b(b+1)} + \frac{a(a+1)(a+2)}{b(b+1)(b+2)} + \ldots\ldots$$

3. Find the function f from the equation $f(ax) = bf(x)$, assuming $f(x)$ not to depend on periodic functions. What is the nature of the solution when there is no such restriction?

4. The expression for the radius of absolute curvature

$$\frac{1}{\rho^2} = \left(\frac{d^2x}{ds^2}\right)^2 + \left(\frac{d^2y}{ds^2}\right)^2 + \left(\frac{d^2z}{ds^2}\right)^2$$

is sometimes very shortly obtained geometrically, by the consideration of an ultimate rhombus; explain carefully the legitimacy of this process.

5.   Evaluate the modulus of $\Gamma\left(\frac{1}{2} + \sqrt{-1}a\right)$.

6.   Prove that the expansion of $f(x)$ in a series according to ascending powers of the real or imaginary variable x is convergent, so long as the modulus of x is less than the least of those of the values of x for which $f(x)$ or $f'(x)$ is infinite or discontinuous.

7.   What is the origin of the diurnal tide, and at what times does it vanish or become greatest? Supposing that in tidal observations at a given port only the times and heights of high and low water were recorded, how would you proceed to determine for any day the coefficient of the diurnal tide?

8.   A moist rectangular lamina is suspended with its edges vertical and horizontal by a string passing over a pulley and attached to a weight by which the lamina is exactly counterpoised, and it is placed so as all but to touch a sheet of water. If contact be established the lamina will be drawn down; find the position of equilibrium, and point out the origin of the work done in overcoming the buoyancy of the water, and also in raising the water near the plate. [N.B. The problem is to be treated as one of two dimensions.]

9.   Find the differential equations of the path of a ray in a medium of variable density; and prove that at a given point in the medium the centre of absolute curvature of any ray lies in a given plane.

10.   In the irrotational motion of a liquid in two dimensions under the action of conservative forces, the motion derived from the former by turning at each instant the direction of motion of each particle in one direction through 90° without changing its velocity will also be a possible irrotational motion, the conditions at the boundaries being altered so as to suit the new motion.
If the former motion be that produced from rest in an infinite mass of liquid by the motion within it of a body bounded by a cylindrical surface, could the latter be similarly produced from rest, and if so, by what contrivance?

11.   Two insulated conducting spheres are charged with electricity of the same kind; supposing for simplicity the charge on one to be feeble, find the distance between the spheres beyond which there is repulsion and within which there is attraction.
Can you suggest an explanation of the fall of rain which is often found to succeed a flash of lightning?

12.   A pure spectrum is viewed through a narrow aperture parallel to the lines of equal refrangibility, which is divided lengthways into two equal parts by a retarding plate covering one-half. Supposing the retardation to amount to several wave-lengths and neglecting the loss by reflection, calculate the appearance presented.
Is it a matter of indifference which half be covered?

13. Enunciate Clairaut's Theorem, and prove it as a consequence simply of the law of gravitation, assuming the surface of the earth to be a slightly oblate spheroid of revolution.

14. A prism of Iceland Spar, cut so that the plane bisecting its dihedral angle is perpendicular to the crystallographic axis, is used for forming a spectrum in the usual manner, the slit being in the focus of a collimating lens; if the prism be set for minimum deviation of a particular line, find the distance between the two images of any other line.

WEDNESDAY, *February* 2, 1876.

1. If the sides of a triangle or the sides produced cut a line at angles θ, ϕ, ψ in points at distances p, q, r from a point in the line, prove that

$$\Sigma (q - r) \sin \theta \sin (\phi - \psi) = \frac{4\Delta^2}{abc},$$

where Δ is the area, and a, b, c are the sides of the triangle.

2. Prove that

$$\int_0^n \left\{ 1 + (t - 1)\frac{x}{1} + (t - 1)^2 \frac{x(x-1)}{1 \cdot 2} \cdots \right.$$
$$\left. + (t - 1)^n \frac{x(x-1) \ldots (x - n + 1)}{1 \cdot 2 \ldots n} \right\} dx,$$

if arranged according to powers of t, has the same series of coefficients taken from beginning or end.

3. Given a particular integral $y = u$ of the differential equation

$$Y \equiv \frac{d^2y}{dx^2} + P \frac{dy}{dx} + Qy = 0,$$

express the complete integral of $Y = R$, when P, Q, R are functions of x.

4. Prove that, a being positive,

$$\int_0^\infty e^{-2ax} \cos x^2 \, dx = \int_a^\infty \sin (a'^2 - a^2) \, da';$$
$$\int_0^\infty e^{-2ax} \sin x^2 \, dx = \int_a^\infty \cos (a'^2 - a^2) \, da'.$$

5. Discuss the question whether $y = 0$ is a particular integral or a singular solution of the differential equation

$$2\left(x \frac{dy}{dx} + y \right)^3 = y \frac{dy}{dx}.$$

6. A needle with a torsion suspension is acted on by a slowly varying deflecting force, known only indirectly by the motions of the

needle, and the motion is checked by a slight resistance varying as the velocity; find the small corrections to reduce the observed azimuth to the statical azimuth, and shew how they may be completely determined by experiment and observation.

7. A radiant point in glass sends rays across a plane stratum of air interrupting the glass, and the rays afterwards emerge again into air so as to form a parallel system perpendicular to the stratum, find the form of the surface of emergence; and shew what it becomes when the thickness of the stratum is infinitesimal.

8. A pipe provided with an embouchure is of a form returning into itself, like the ring of an anchor; find the series of notes which it is capable of giving out.

9. A series of Leyden Jars of given capacities is charged by cascade from the prime conductor of a machine; find the capacity of the single jar which is equivalent to the series.

If one jar bursts under the strain, so that its coatings are put into connexion, what effect is produced on the other jars? Would the strain on any other jar, whether immediately or as the machine is worked on, be increased or diminished by the explosion?

10. If two streams of light from the same source interfere, prove that the breadth of the fringes will be inversely as the small angle between the courses of the two streams at the place where their interference is considered.

Verify this general principle by direct calculation in the ordinary way in the case of some simple instance of interference of the kind.

Why do Newton's Rings belong to a totally different class of interference phenomena from that here contemplated?

11. Find the expression for the velocity of propagation of a series of simple periodic waves in water of uniform depth, the motion being small and in two dimensions.

If two such series, of equal amplitude and nearly equal wavelength, travel in the same direction, so as to form alternate lulls and roughnesses, prove that in deep water these are propagated with half the velocity of the waves; and that as the ratio of the depth to the wave-length decreases from ∞ to 0, the ratio of the two velocities of propagation increases from $\frac{1}{2}$ to 1*.

12. A block of Iceland spar is formed of a pair of twin crystals, the twin plane being inclined to the axis, and a ray, ordinary or extraordinary, in one crystal is incident on the twin plane; give a geometrical construction for determining the courses of the reflected and refracted rays. How many of each will there be?

As regards the course of ordinary rays, Iceland spar behaves like glass, and the index is alike in all directions; therefore the two crystals behave, as regards ordinary rays, like two glasses of the same index in optical contact, and there can be no reflection of an ordinary ray at the twin plane. Is this true?

[* This is usually quoted as the source of the now well-known theory of wave-groups.]

THURSDAY, *January* 31, 1878.

1.   Given two concentric conics, find their conjugate diameters that have a common direction.

2.   Find the locus of the poles with respect to one conic of the tangents to another conic.

3.   A quantity the fluctuating part of which consists of two simple harmonic functions of the time, of which one has twice the period of the other, is registered continuously, but on a scale too much compressed as regards the time to permit of more than measuring the values of the maxima and minima; supposing that there are four real maxima or minima in the longer period, obtain an expression for the mean value in the particular case in which the function has the same values at equal intervals before and after a certain instant, giving the result free from ambiguity.

4.   Shew that there are three lines of curvature, of which two may be imaginary, passing through an umbilicus.

Investigate the lines of curvature in the immediate neighbourhood of an umbilicus in the particular case in which those lines intersect at angles of 120°.

5.   In a Kater's convertible pendulum, the times of oscillation about the two knife-edges are observed, and are found to be nearly but not exactly equal; determine the length of the seconds pendulum with the aid of the distance of the centre of gravity below one axis of suspension; and shew that this distance is not required to be known with extreme accuracy.

6.   Three unknown quantities are to be determined from three equations by numerical approximation, by the method of successive substitutions, the equations being taken in cyclic order, and the same equation being always used to determine the same unknown quantity. The initial errors being small, shew that the successive corrections to the same unknown quantity form a recurring series. Shew also that we cannot always ensure convergency with three, as we can with two unknown quantities*.

7.   A symmetrical rod is suspended horizontally by two vertical strings, of lengths l, l', attached at the distance x at each side of the centre of gravity; the rod being slightly disturbed so that its centre of gravity has no motion or displacement in the direction of the length, determine the motion; and examine its character in the case in which the two equations $l = l'$, $x = k$ are very nearly satisfied, k being the radius of gyration.

8.   The three angles of a plane triangle are repeatedly observed, the number of observations not in general being the same for each; find the most probable values of the angles.

9.   Find the most rapid diminution of temperature in ascending in the atmosphere that is consistent with stable equilibrium.

[* Cf. *supra*, p. 292.]

10.   One side of the quadrilateral of a Wheatstone's bridge arrangement is formed of a condenser of very great capacity, which is gradually charged from the battery.   Given the several resistances, that of the wires leading to the coatings of the condenser being insensible, compare the potential of the condenser at the moment when there is no current in the bridge with that of the battery supposed unconnected.

11.   Determine the principal small oscillation of a globule of liquid in free space consequent on the capillary force, namely the oscillation in which the globule becomes alternately prolate and oblate.

12.   A deep rectangular vessel nearly filled with water is continued at one end as a shallow canal of indefinite length; supposing the water of the vessel thrown into the condition of a stationary undulation, find approximately the rate at which the undulations would subside by communication to the water of the canal.

13.   An object-glass in construction is found on testing to have a small residual spherical aberration, which is measured, but to be right in other respects; find in what way and by what amount the form of the last surface must be altered to correct the aberration, without altering the curvature at the centre.

14.   A polarized spectrum is viewed through a Nicol's prism, and a tourmaline plate which polarizes only imperfectly is interposed; shew that, unlike the case of a transparent crystalline plate, there are certain limits of position within which if the Nicol be set, both of the systems of dark bands which alternately appear on rotating the tourmaline become perfectly black.

15.   A homogeneous solid extending infinitely on one side of a plane, and initially at the temperature 0, radiates into a medium at the temperature $f(t)$; find the temperature at any point and any time; and supposing $f(t)$ constant, express the temperature at the bounding plane by means of a tabulated integral.

TUESDAY, *February* 3, 1880.

1.   A vertical sheet of paper moves uniformly in a horizontal direction, and a tracing point moves vertically so as to describe the ordinate $a \sin nt + b \sin n't$, where n' is very nearly equal to $2n$; supposing the advance of the paper in the time n^{-1} insensible, but not that in the time $(2n - n')^{-1}$, find the boundary of the marked portion of the paper.

Shew what it becomes when $b = a$, and how it passes from the general form into that assumed in this case.

2.   If an umbilicus be defined as a point at which the principal radii of a surface are equal, the condition of equality, combined

with the equation of the surface, would seem to determine a curve; why then in the general case are there umbilici only at particular points?

3. Investigate a formula for the transformation of the slowly converging series

$$f(x) - f(x+a) + f(x+a+b) - f(x+2a+b) + f(x+2a+2b) - \ldots$$

so as to adapt it to numerical calculation. How would you practically apply the method to a series such as

$$\frac{1}{1^2} - \frac{1}{3^2} + \frac{1}{4^2} - \frac{1}{6^2} + \ldots ?$$

From the formula deduce another making the summation of

$$f(x) + f(x+h) + f(x+2h) \ldots$$

depend on $\int f(x)\, dx$ and its differences, and prove the same directly. Point out any advantages or disadvantages of this formula as compared with the more usual one which makes the summation depend on $\int f(x)\, dx$ and the differential coefficients of $\int f(x)$.

4. Prove that the expansion of $f(x)$ according to ascending powers of the real or imaginary variable x is convergent so long as the modulus of x is less than the smallest of those belonging to the values of x for which $f(x)$ or $f'(x)$ becomes infinite or discontinuous.

Illustrate this rule by reference to geometry, attractions, or hydro-dynamics.

5. In a spectroscope of several prisms, shew how to calculate the amount of curvature of fixed lines seen with a straight slit, simplifying as far as may be the calculation by approximation or otherwise.

6. An achromatic object-glass just finished is found on testing to be as good as possible as regards achromatism, but to shew a small residual spherical aberration. Shew that this may be corrected, without disturbing the performance of the object-glass in other respects, by repolishing one of the surfaces (suppose the fourth) in such a manner as to deviate a little from the spherical form, while retaining the figure one of revolution.

Give a rough numerical result for the distance between the altered surface at the edge and the sphere of curvature touching it at the middle point, supposing the aperture 15 inches, the focal length 15 feet, the index of the flint 1·62, and the longitudinal aberration the twentieth of an inch.

7. A quantity is repeatedly observed by two observers A and B, A having taken altogether m and B n observations, and the mean of the whole $m + n$ observations is taken as representing the most probable result: supposing that in reality the observers are of unequal skill, and that the probable error of an observation of A's is to one of B's as p to q, find under what circumstances the mean of the whole $m + n$ observations will be more or less trustworthy than the mean of the m observations taken by A alone.

8. A stretched string is constantly acted on at a given point by a small transverse disturbing force represented by $c \sin nt$; find the corresponding motion of the string.

From the result pass on to the solution of the problem to determine the motion of the string supposing it at rest to start with, and to be acted on at a given point from the time $t = 0$ onwards by a small disturbing force which is a given arbitrary function of the time.

9. Shew that from any possible irrotational motion of a perfect liquid in two dimensions, another may be deduced by interchanging ϕ and x, ψ and y, where

$$\phi = \int(u\,dx + v\,dy), \quad \psi = \int(u\,dy - v\,dx).$$

What motion is thus deduced from a flow from or towards a centre, alike in all directions?

10. Light coming from a luminous point is received on a plate of glass which is bounded by a straight edge, and the light is received on a screen at some distance. It is found that about the geometrical projection of the edge a band is seen on the screen which is much darker than the general ground on either side; account for this.

If the glass be thin, and the light which would otherwise fall on the screen be received on a narrow slit placed at the geometrical projection of the edge, and the light passing through the slit be viewed through a prism placed at some distance, a series of bands alternately black and coloured are seen; account for this.

11. A rhomb of glass of the general form of a Fresnel's rhomb is used in conjunction with a Nicol's prism for the scrutiny of elliptically polarized light: shew how to determine by experiment the difference of phase produced by the rhomb in the components polarized in and perpendicularly to the principal plane of the rhomb, investigating the formula employed.

12. A homogeneous spherical shell is heated to a uniform temperature, and then plunged into a liquid by which the temperature of the surface is kept to zero; determine the subsequent temperature at any point of the shell.

13. A galvanic current is led through a liquid contained in a rectangular metallic cell, which itself forms one electrode, the other being a wire only just introduced beneath the surface; determine the potential at any point of the liquid.

How may the forms of the equipotential surfaces be explored experimentally?

WEDNESDAY, *February* 1, 1882.

1. Chords of a circle touch another circle, and at the extremities of each chord are drawn tangents to the first circle; find the locus of the intersection of the pairs of tangents.

2. If an algebraic equation has a pair of equal roots, they can be found; shew how the process may be applied to indicate the existence and approximately determine the value of a pair of roots which are very nearly equal, and how the approximation may be continued.

3. Illustrate by reference to geometry the significance of the steps in the integration of the partial differential equation of the first order and degree

$$P \frac{dz}{dx} + Q \frac{dz}{dy} = R.$$

4. If three systems of surfaces cut each other orthogonally, prove that each surface is cut by the surfaces of the other two systems along its lines of curvature.

5. $f(x)$ is a function of x which fluctuates irregularly about an average value zero, and $F(x)$ is another function of x which is equal to $f'(x)$ between the limits $-a$ and a, where the range $2a$ is wide enough to comprise a great many fluctuations, and which outside those limits dwindles away, and soon vanishes. The functions $f(x)$, $F(x)$ being expressed by Fourier's theorem between the limits $-\infty$ and ∞ in the form

$$\int_0^\infty \phi(a) \cos ax \, da + \int_0^\infty \psi(a) \sin ax \, da,$$

point out the character of the difference between the functions $\phi(a)$, $\psi(a)$ for the one function and the other on which it depends that the functions $f(x)$, $F(x)$ agree within the limits $-a$ and a, though differing so widely at a distance from those limits.

6. A detached pendulum is swung with a cylindrical axis, instead of knife-edges, resting on agate planes; find the time of vibration, supposing the arc of vibration small, but not indefinitely small.

Draw any inference that occurs to you as to the effect of a wearing of the edge of the knife-edges commonly employed.

7. In the method of determining the velocity of light which depends on the use of a revolving toothed wheel and conjugate telescopes, investigate the relation between the brightness of the artificial star observed and the apertures and mutual distance of the telescopes.

8. Form the equation for the propagation of sound in a vertical direction in the atmosphere, supposing for simplicity the undisturbed temperature uniform and gravity constant; and thence shew that the velocity of propagation is, within practical limits, independent of the nature of the disturbance; and investigate the relation between the velocity of the particles in a given phase and the density at the different heights attained in the course of propagation.

9. Investigate the motion of a series of oscillatory waves of small disturbance in deep water, taking account of capillarity; and shew that the velocity of propagation, regarded as a function of the wave-length, has a minimum value.

10.   Assuming the equations of motion of a liquid when viscosity is taken into account, namely

$$\frac{1}{\rho}\frac{dp}{dx} = X - \frac{du}{dt} - u\frac{du}{dx} - v\frac{du}{dy} - w\frac{du}{dz} + \frac{\mu}{\rho}\left(\frac{d^2u}{dx^2} + \frac{d^2u}{dy^2} + \frac{d^2u}{dz^2}\right), \&c.,$$

investigate the conditions of the motion of a tall hollow cylinder containing liquid, which performs slow oscillations, by torsion, about its axis, which is vertical.   [N.B.   The motion is meant to be taken as in two dimensions throughout.]

11.   A closed surface contains within it gravitating matter, the potential of which is given all over the surface; shew that the mean internal potential, regarded as depending on the distribution, admits of a maximum, and thence that the matter may be condensed into open surfaces.

12.   A telescope is focussed for a luminous point, and the object-glass is covered with a screen containing a small aperture of the form of an oblique parallelogram; investigate the appearance, and shew that the field is crossed by two series of dark bands, leaving a central luminous patch which is similar and similarly situated to the aperture turned through 90°.

13.   Two plates of quartz, of equal thickness, are cut at an inclination of 45° to the axis, and being combined with their principal planes crossed at right angles are mounted in front of a Nicol's prism, the principal plane of which is inclined at 45° to the principal planes of the plates; investigate the appearance when light polarized at any azimuth is viewed through the system.

INDEX TO VOLUME V.

[The numbers refer to pages.]

CAMBRIDGE: PRINTED BY JOHN CLAY, M.A. AT THE UNIVERSITY PRESS.

Printed in the United States
By Bookmasters